T0138476

THE GOOD LIFE IN THE SCIENTIFIC REVOLUTION

The Good Life in the Scientific Revolution

DESCARTES, PASCAL, LEIBNIZ,
AND THE CULTIVATION OF VIRTUE

Matthew L. Jones

The University of Chicago Press Chicago & London

Matthew L. Jones is associate professor of history at Columbia University.

The University of Chicago Press, Chicago 60637
The University of Chicago Press, Ltd., London
© 2006 by The University of Chicago
All rights reserved. Published 2006
Printed in the United States of America

15 14 13 12 11 10 09 08 07 06   5 4 3 2 1

ISBN-13 (cloth): 978-0-226-40954-2
ISBN-13 (paper): 978-0-226-40955-9
ISBN-10 (cloth): 0-226-40954-6
ISBN-10 (paper): 0-226-40955-4

Library of Congress Cataloging-in-Publication Data

Jones, Matthew L. (Matthew Laurence), 1972–
    The good life in the scientific revolution : Descartes, Pascal, Leibniz, and the
cultivation of virtue / Matthew L. Jones.
        p.   cm.
    Includes bibliographical references and index.
    ISBN 0-226-40954-6 (cloth : alk. paper)—ISBN 0-226-40955-4 (pbk. : alk. paper)
    1. Science—History—17th century.   2. Mathematics—Philosophy—History—17th
century.   3. Descartes, René, 1596–1650.   4. Pascal, Blaise, 1623–1662.   5. Leibniz,
Gottfried Wilhelm, Freiherr von, 1646–1716.   6. Science–Moral and ethical aspects.
    I. Title.
Q125.2.J66   2006
509.032—dc22

                                                                                            2006040491

⊚ The paper used in this publication meets the minimum requirements of the
American National Standard for Information Sciences—Permanence of Paper
for Printed Library Materials, ANSI Z39.48-1992.

For Liz

Contents

Illustrations

Acknowledgments

Concerning those authors who speak of their works as "my book,"
"my commentary," "my story," and so forth, Mr. Pascal said that
they seem like well-established burghers, always with a chez moi *at*
their lips. This excellent man added that they would do far better to
say: "our book," "our commentary," "our history," and so forth,
since usually more of what is good in such works comes from others
than from themselves.

M. de Vigneul-Marville, *Mélanges de littérature et d'histoire* (1700)

This book began as a preface to a ridiculously grandiose project. Wise friends led me to pursue the preface on its own terms, guided me throughout, and made me recognize the merits of bringing it to a close, no matter how much I protested that some crucial piece of evidence still lurked on my shelves. A model advisor, generous reader, and good friend, Mario Biagioli has unstintingly offered me his insight and guidance in reading and improving many versions of my work. Ever since he warmly supported and directed my initial forays into the history of science, Peter Galison has taught me much about combining innovative methods with careful inquiry into technical practice. Many of my guiding ideas about early-modern science came in the course of a memorable year working with Simon Schaffer. Tom Conley steered me through the cartography of early-modern French literature. By introducing the long philosophical tradition of spiritual exercises to me, Arnold Davidson saved me from my parochialism; his intervention reoriented the project philosophically and historically.

Perceptive and copious comments from several referees helped me improve the style, argument, and evidence of the manuscript throughout. My thanks to Lorraine Daston and the anonymous referees, whose thoughtful care in responding to my work exemplifies the best of academic citizenship. Having agreed to read my dissertation at short notice, Dominico Bertoloni-Meli wrote an insightful commentary to which I often have returned. Many teachers, friends, and colleagues commented on chapters of the manuscript or on presentations of them, notably Peder Anker, Nick King, Bob Brain, Anne Davenport, Eric Ash, Heidi Voskuhl, Jamie Cohen-Cole, J. B. Shank, Cyrus Mody, Lisbet Rausing, Abby Zanger, Everett Mendelson, Mary Terrall, Anthony Koliha, Michael Binard, and Peter Miller. David Kaiser read many versions of this manuscript with a sharp eye for improving everything from equations to historiographical discussions. I reaped many rewards from the generous criticism of audiences at UCLA, the University of Minnesota, Stanford University, the University of Oslo, the University of Chicago, Cornell University, Harvard University, the New York Academy of Sciences, Bard Graduate Center, the Maison des sciences de l'homme in Paris, and several sessions at the History of Science Society.

More than reading histories of philosophy, teaching Contemporary Civilization to Columbia sophomores has helped me to understand the ancient tradition of considering philosophy as a way of life. The Department of History at Columbia has been an excellent place for a novice historian of science to learn something about teaching and writing history *simpliciter*; for their friendship and insight, I thank Ellen Baker, Elizabeth Blackmar, Caroline Bynum, Martha Howell, Joel Kaye, Adam Kosto, Gregory Mann, Samuel Moyn, William Harris, and Isser Woloch. Ever since I arrived at Columbia, Carol Gluck has been a wonderful interlocutor on all things historiographic. Beyond helping with several chapters, Pierre Force and Kathy Eden welcomed me into Columbia's broader community of early-modern scholars. Astute criticism from Christia Mercer prompted me to rethink every word concerning Leibniz. All errors that remain are mine.

Generous grants and fellowships from the National Science Foundation, the Dibner Institute for the History of Science and Technology, and Columbia University provided the financial support necessary for research and writing. A yearlong fellowship at the Institute for Scholars at Reid Hall in Paris provided a nearly ideal setting to rethink and expand the manuscript; Danielle Haase-Dubosc, Mihaela Bacou, and Charles Walton created a lively atmosphere conducive to scholarly work. The office staffs at the Department of History of Science at Harvard and the Department of History at Columbia facilitated countless tasks over many years. Megan Williams provided superior research assistance. The photographs in the book appear courtesy of the Rare Book and Manuscript Library at Columbia University, the Beinecke Library at Yale, and the Burndy Library of the Dibner Institute; thanks especially to Anne Battis and Jennifer Lee. Dr. Herbert Breger, of the Leibniz-Archiv, kindly allowed me to cite from a volume of Leibniz's mathematical works still in preparation. An early version of chapter 1 appeared in *Critical Inquiry*; a few pages of chapter 3 appeared in a preliminary form in *Studies in History and Philosophy of Science*. Christie Henry and especially Catherine Rice at the University of Chicago Press have supported this project with great enthusiasm. My copy editor Pamela Bruton improved the text throughout; her sharp queries led me to make some essential clarifications.

Two friends, whose compassion and wisdom I've treasured for many years, deserve special thanks. The little I claim to know about philosophy I probably misunderstood from Ishani Maitra. Michael Gordin has suffered through more poor writing and specious argumentation in half-baked drafts than anyone should have to endure. Without Ishani and Michael, graduate school would never have been as fruitful, fun, or meaningful, and this project surely would never have come to be.

My parents and sister always insisted that I pursue my interests wherever they lead. In braiding together history, philosophy, and mathematics, this book testifies to their boundless love and support as I have done so.

I dedicate this book to my partner and wife, Elizabeth H. Lee. Her perspicuity and discernment have improved every line of this book; her love, every moment of my life. 我愛你

Abbreviations

A Gottfried Wilhelm Leibniz. 1923–. *Sämtliche Schriften und Briefe.* Edited by Deutsche Akademie der Wissenschaften, etc. Darmstadt, Berlin, Munich, etc. Cited as, for example, A3,1:355 (3rd series, vol. 1: p. 355).

AT René Descartes. 1996. *Œuvres de Descartes.* Edited by Charles Ernest Adam and Paul Tannery. 2nd ed., corrected. 11 vols. Reprint, Paris: J. Vrin.

C Gottfried Wilhelm Leibniz. [1903] 1966. *Opuscules et fragments inédits de Leibniz, extraits des manuscrits de la Bibliothèque royale de Hanovre.* Edited by Louis Couturat. Paris. Reprint, Hildesheim: G. Olms.

Cc Gottfried Wilhelm Leibniz. [1914–18] 1986. *Catalogue critique des manuscrits de Leibniz.* Fasc. 2, *Mars 1672–novembre 1676* [only volume that appeared]. Poitiers: Société française d'imprimerie et de librairie. Reprint, Hildesheim: G. Olms.

CF Blaise Pascal. 1992. *Les provinciales.* Edited by Louis Cognet and Gérard Ferreyrolles. Paris: Bordas.

CSM René Descartes. 1984–91. *The Philosophical Writings of Descartes.* Translated by John Cottingham, Robert Stoothoff, Dugald Murdoch, and (vol. 3 only) Anthony Kenny. 3 vols. Cambridge: Cambridge University Press.

Ger Gottfried Wilhelm Leibniz. [1906] 1995. *Leibnizens nachgelassene Schriften physikalischen, mechanischen und technischen Inhalts.* Edited by Ernst Gerland. Abhandlungen zur Geschichte der mathematischen Wissenschaften mit Einschluss ihrer Anwendung XXI. Hft. Physikalischer Teil. Leipzig: B. G. Teubner. Reprint, Hildesheim: G. Olms.

GM Gottfried Wilhelm Leibniz. [1849–63] 1971. *Leibnizens mathematische Schriften.* Edited by C. I. Gerhardt. 7 vols. in 4. Berlin: A. Asher; Halle: H. W. Schmidt. Reprint, Hildesheim: G. Olms.

GP Gottfried Wilhelm Leibniz. [1875–90] 1960. *Die philosophischen Schriften.* Edited by C. I. Gerhardt. 7 vols. Berlin: Weidmann. Reprint, Hildesheim: G. Olms.

J Antoine Arnauld and Pierre Nicole. 1992. *La logique ou L'art de penser.* Edited by Charles Jourdain. Paris: Gallimard.

K Gottfried Wilhelm Leibniz. 1993. *De quadratura arithmetica circuli ellipseos et hyperbolae cujus corollarium est trigonometria sine tabulis.* Edited by Eberhard Knobloch. Abhandlungen der Akademie der Wissenschaften in Göttingen, Mathematische-physikalische Klasse, vol. 43. Göttingen: Vanderhoeck & Ruprecht.

L Gottfried Wilhelm Leibniz. 1969. *Philosophical Papers and Letters.* 2nd ed. Translated by Leroy E. Loemker. Dordrecht: D. Reidel.

LC Gottfried Wilhelm Leibniz. 2001. *The Labyrinth of the Continuum: Writings on the Continuum Problem, 1672–1686.* Edited and translated by Richard T. W. Arthur. New Haven: Yale University Press.

LG Blaise Pascal. 1998. *Œuvres complètes.* Edited by Michel Le Guern. 2 vols. Paris: Gallimard.

LH Leibniz Handscriften, Niedersächsische Landesbibliothek, Hanover, Germany. Partially cataloged in Eduard Bodemann. [1889] 1966. *Die Leibniz-Handschriften der Königlichen öffentlichen Bibliothek zu Hannover.* Hanover. Reprint, Hildesheim: G. Olms.

M Blaise Pascal. 1964–. *Œuvres complètes.* Edited by Jean Mesnard. 4 vols. to date. Paris: Desclée de Brouwer.

NE Gottfried Wilhelm Leibniz. 1765. *Nouveaux essais sur l'entendement humain, 1703–1705.* (In A6,6.)

Pk Gottfried Wilhelm Leibniz. 1992. *De summa rerum: Metaphysical Papers, 1675–1676.* Translated by G. H. R. Parkinson. New Haven: Yale University Press. (Partial translation of A6,3.)

S Blaise Pascal. 2000. *Pensées.* Edited by Gérard Ferreyrolles and Philippe Sellier. Paris: Livre de Poche. Abbreviation is followed by fragment number.

A Note on Conventions

Unless noted otherwise, translations are mine. Original orthography of quotations has been retained. Quotations from the standard English translation of Descartes' works are cited as "CSM volume:page number," along with the relevant citation to the standard edition of the original-language texts (AT). For my own translations, besides the reference to AT, I often provide a citation to CSM also, as in "cf. CSM volume:page number."

In the chapters on Leibniz, I have provided detailed titles and dates of documents, as many of them have only recently been published and are little known. Because the works of Descartes and Pascal are much better known, I have generally refrained from providing a similar level of detail. Document dates that are not certain and unexpressed signature and folio designations are given in square brackets. Question marks indicate greater uncertainty.

I have not uniformly used the past tense, as it can be unwieldy and verbose in extended textual analysis. In such analyses, I have often referred to past actions and deeds in the past tense and philosophical claims in the present tense: "Leibniz maintained that two plus two is four and that humans are fallible."

Introduction

This then will be the true teaching of method: not so much the
seeking after truth but, rather, of living.

Leibniz, "On the Art of Discovery"

What fields of learning should a noble and cultivated person pursue? In 1666 an anonymous author tackled the question. Many sciences were unacceptable means for cultivating oneself. Theology with its many mysteries should be left to "our superiors." The study of the natural world was far too uncertain and contested—even the greatest minds notoriously differed about nature. Although certain, the practice of mathematics came at too high a cost, for it "pulls you away from actions and pleasures and occupies you entirely." The author advocated a triumvirate of sciences—morality, politics, and belles-lettres—that could teach the true philosophy necessary for living well. History revealed the value of such guidance: once Rome attained refinement, every person of consideration was "attached to some philosophical sect—with the aim, not of understanding the principles and the nature of things, but of fortifying the mind through the study of wisdom."[1] No mere academic activity, philosophy should guide honorable people in pursuing the good life.

In portraying mathematics and the study of nature as unsuitable for a noble man, the author intimated that some contemporaries judged them appropriate for cultivating the mind and soul. Numerous early-modern philosophers, including René Descartes, Blaise Pascal, and Gottfried Leibniz, the subjects of this book, deemed knowledge and practices now considered scientific to be powerful tools for living a good and virtuous life. They developed and articulated their practices for improving the self in tandem with their mathematics and natural philosophy. Technical disciplines could perfect mind and soul; minds so refined could best extend knowledge of nature and mathematics.

Moralists, theologians, and philosophers in seventeenth-century western Europe worried that scientific disciplines could too easily divert one from engaging with the problems and demands of the world. The French writer Pierre Nicole diagnosed the underlying problem. Taken up as ends in themselves, mathematics and natural philosophy could be as vain and useless as critics complained. Not made simply to measure lines and the movement of matter, human beings are instead "obliged to be just, fair, judicious in all their speech, in all their actions, and in all the affairs that involve them; and it is toward these ends that they ought particularly to exercise and to form themselves" (J10). So often had the sciences been pursued as ends in themselves that their genuine potential remained masked. "[One] cannot say that they ought to pass for an amusement entirely vain and unworthy of wise people if they could not also serve as preparations for other knowledge that is truly useful. Thus, those who are attached to them for themselves, as if they were something great and revealed, do not know their true use. Such ignorance in those people is a much greater fault than if they knew nothing of these sciences."[2] The best sciences were not the most autonomous. We should not draw "on reason as an instrument for acquiring the sciences," Nicole maintained. We "ought to draw on the sciences as an instrument for perfecting reason" (J9). The sciences could develop the epistemic capacities and emotional strengths necessary throughout life; they could hone the skills and competencies essential for mental and spiritual nobility.

For numerous central figures of the seventeenth century, including Descartes, Pascal, and Leibniz, the natural and mathematical sciences offered just such exercise: mathematical and natural-philosophical practices could help one live the good life. Historians of science have of late described how new forms of etiquette, civility, and nobility were central resources in producing the new sciences. I will argue that a concern with developing new forms of nobility and civility guided some key scientific innovations in the seventeenth century. Their conviction that technical mathematics and experimentation could contribute to self-cultivation led Descartes, Pascal, and Leibniz to value certain facets of their inquiries over others, to accept some forms of practice as legitimate, and to perceive difficulties where others did not. Their efforts to improve techniques for living and thinking well helped shape the kind of mathematics and experimentation they did, what they thought made good mathematics and method, and what they saw as the lessons illustrated by their innovations.

As Nicole suggested, the sciences were potent but dangerous: they could train the mind to think nobler thoughts, or they could occupy it entirely with mechanical procedures; they could aid the intellect throughout life, or they could turn it forever away from the business of living well. Concerned with strengthening the mind, many early-modern writers distrusted procedures and techniques that appeared too automatic or mechanical, such as algebraic calculation, syllogistic logic, and the unreflective use of instruments. Rather than rejecting these techniques, Descartes, Pascal, and Leibniz sought to harness them to improve judgment and scientific acuity. In attempting to extend knowledge through sciences capable of perfecting the mind, they sought to integrate technical procedures into scientific and philosophical practice. Technical procedures needed the guidance of philosophy to bring out their potential; philosophy, likewise, required new techniques to achieve and regulate its goals.

The early-modern drive to create philosophical and scientific practices suitable for cultivating the self spurred tremendous effort to discern human limits and to fashion tools capable of compensating for those limits. In their mathematical and philosophical practices, Descartes, Pascal, and Leibniz worked to discern the abilities of human beings and attempted to produce techniques appropriate to those abilities. Descartes' geometry, Pascal's arithmetical triangle, and Leibniz's quadrature of the circle offered new tools to facilitate human reasoning; they linked these new tools to normative conceptions of mathematical and philosophical knowledge. All three mathematicians maintained that the limits of human ability had profound implications for the practices and objects that ought to be accepted as legitimate in mathematics and philosophy. For all three, mathematics and natural philosophy offered lessons about human nature in a postlapsarian world. In illustrating what philosophy and the sciences could achieve, mathematical and natural philosophy helped uncover what mental and spiritual nobility could be in this life.

Many of the most important new sciences of the seventeenth century involved new ways of transforming the traditional goals and techniques of natural philosophy with mathematics and craft knowledge.[3] This book tracks one major tradition of integrating mathematics with philosophy. In this tradition, hitching technical procedures to a project of self-cultivation was supposed to sever those techniques from the dangers inherent in their artisanal roots while retaining the power of the procedures. Hitching philosophy to mathematical techniques conversely kept philosophy from instilling a hubris grounded in a false picture of the practical nature of human reasoning.[4]

Although far from the only philosophers to integrate natural philosophy and mathematics with a striving for self-cultivation, Descartes, Pascal, and Leibniz offer examples of indubitable scientific and philosophical significance.[5] Considerations about how scientific practices could help effect living well did not *cause* Descartes, Pascal, and Leibniz to produce their innovations in natural philosophy and mathematics. The technical dimensions of their work cannot be reduced to such considerations. Likewise, these dimensions cannot be reduced to anything else supposedly "external" or "internal" to science, whether class, social interests, or more recent notions of pure mathematics, scientific method, or logic. Such forms of reductionism trivialize the mental and practical worlds historians must work to reconstruct.

LATE-RENAISSANCE HUMANISM AND PHILOSOPHY AS A WAY OF LIFE

The Roman philosopher Seneca bitterly denounced philosophy that was disconnected from its proper role: "I think no one has done a worse service to the human race than those who learned philosophy as a mercenary profession, that is, people who live other than according to the rules of life that they prescribe."[6] Such criticism of mercenary philosophy had wide appeal in early-modern Europe. Many Renaissance humanists denounced scholastic philosophy as useless in its isolation from the demands of living well.[7] Caught up in what humanists considered sterile subtleties, scholastic philosophy failed to consider the proper subject matter: ethical action. Worse yet, philosophy failed to lead the individual to love virtue and not just to understand it.[8] Humanists aimed to perfect forms of reasoning, speaking, and writing capable of effecting such moral education and reform.[9] However great the hyperbole of the humanists, their critique of forms of learning that did not inculcate morality and direct everyday life stimulated accounts of creating virtuous people for centuries and altered education across Europe.[10]

Among the greatest fruits of Renaissance philology were the recovery and reconstruction of the major philosophical traditions of antiquity other than Aristotelianism: Platonism, in its many variants, Stoicism, Epicureanism, and various forms of Skepticism. Many facets of these traditions appealed to early-modern readers.[11] These ancient philosophies share, in the words of John Cottingham, "a recurring conception of philosophy as searching for an all-embracing 'synoptic' schema of thought—one which could locate a programme for human fulfillment within a broad understanding of the nature of the universe and man's place within it."[12] Grounded in differing accounts of nature and the human role within it, each philosophy offered means to learn how one ought to live.

Seneca characterized philosophy properly understood: "Philosophy is not an occupation of a popular nature, nor is it pursued for the sake of self-advertisement. . . . It moulds and builds the personality, orders one's life, regulates one's conduct, shows one what one should do and what one should leave undone, sits at the helm and keeps one on the correct course as one is tossed about in perilous seas. Without it no one can lead a life free of fear or worry. Every hour of the day countless situations arise that call for advice, and for that advice we have to look to philosophy."[13] A practice appropriate for a small elite, philosophy was to help direct and structure one's thoughts, emotions, and actions—by weighing the validity of doctrines and by exercising the mind and affects—so that one could steel oneself to the vagaries of life and become capable of acting appropriately at all times.

In numerous important studies, Pierre Hadot has documented how widespread such a vision of the purpose and practice of philosophy was among the ancients. Reducible neither to formal philosophical systems nor to cognitive exercises, ancient philosophies involved practices necessary for cultivating an ethical self. Hadot points out that "the discourse of philosophy can take the form, not only of an exercise intended to develop the intelligence of the disciple, but of an exercise intended to transform his life."[14] Recognizing philosophy as an ethical practice—a set of spiritual exercises—underscores that philosophy was historically more than systematic knowledge of right and wrong, the good and the bad, the true and the false, and the criteria for making such distinctions. Producing and evaluating systematic knowledge were crucial activities for preparing a self that could follow the dictates of reason throughout life, in the face of any travail.

Their interest in encouraging virtue disposed humanist scholars of the Renaissance to seize upon these features of ancient philosophy and to put them into practice anew. No less than Renaissance humanists, seventeenth-century European elites sought consolation, succor, and direction in ancient philosophy and scholarship about the ancient world.[15] Philosophies that promised consolation, comprehension, and self-organization proved popular amid the deadly religious strife and social uncertainties of the late sixteenth and seventeenth centuries. In his still-foundational study of seventeenth-century French moralists, Anthony Levi argues that the "insistent search for practical guidance" helped to determine "both the interpretation of classical authors and the degree of esteem in which each of them was held."[16] In his consideration of the image of Socrates in seventeenth-century France, Emmanuel Bury contends that the century had "a vision of ancient philosophy that stressed consciousness of the existential character that ancient philosophy took on, so close, in many ways, to the spiritual exercises of the Christian religion: it is very much a question of a choice of a life, and philosophy has no meaning unless it is, in the final analysis, moral."[17] In an important letter discussing philosophy as an art of living, Pierre Gassendi, for example, wrote in 1642 of the power of philosophy "to make a healthy body, through the practice of virtues and especially of temperance, [and] a healthy soul, one tranquil before terrors and free from vain desires, through knowledge and the appropriate valuing of things, insofar as possible given human fragility."[18]

Early-seventeenth-century western European culture was awash in possibilities, new and old, for simultaneously reforming one's self and one's knowledge. These sundry "spiritual exercises" offered different modes and ideals for cultivating the self: Ignatius of Loyola's *Spiritual Exercises*, a central reference in the period, and its various reformulations; François de Sales's *Introduction to the Devout Life*; Michel de Montaigne's *Essays*; Pierre Charron's *On Wisdom*; Eustachius a Sancto Paulo's *Spiritual Exercises*; Pierre Gassendi's Christianized Epicureanism; Justus Lipsius's Stoicism; Cornelius Jansen's Augustinianism and its popularizations; popularized Scholasticism; and, finally, the frequently reprinted classics such as Seneca's *De vita beata* and Epictetus's *Manual*, to name but a few.[19] Guidebooks soon appeared to help one choose among the many options, new and old.[20]

Not merely cynical codifications of the abilities and beliefs of some social order as the good life, these spiritual exercises were attempts to create elites characterized by particular mental, spiritual, and moral virtues.[21] More than sets of doctrines, these spiritual exercises offered practices and objects of knowledge held to be intellectually and affectively appropriate for living well. Partisans of different exercises disputed the appropriateness and utility of different forms of self-cultivation and knowledge production.[22] Many stressed the study and emulation of historical exemplars, while others stressed literary ones; some recommended philosophical contemplation, elevated conversation, instructive travel, or even philological inquiry. Following Plato and his commentator Proclus, many asserted the power of mathematics to sharpen the intellect and turn it toward higher things.[23] Some religious, as well as libertine, writers denounced the focus on the intellect and cast doubt on its therapeutic potential; one poet wrote:

> Let us study more to enjoy than to know.
> I renounce good sense; I hate intelligence.
> The more a mind raises itself in knowledge,
> Better does it grasp the subject of its affliction.[24]

Still others extolled the intellect, while doubting the value of the more speculative sciences and of inquiries into nature.[25]

A few of these spiritual exercises, such as those offered by Justus Lipsius or Pierre Gassendi, focused on moralities grounded in knowledge of nature. For the ancient Epicureans and Stoics, such as Marcus Aurelius, the careful study of the natural world was important as a spiritual exercise. Natural philosophy uncovered, in principle, the naturalistic basis for ethics. Exemplified by the Stoic recognition of Providence or the Epicurean acknowledgment of the randomness of a materialist universe, knowledge of the natural order offered consolation for the pains of everyday existence and guidance in determining one's proper role in life. Working through natural philosophy helped in recognizing human limits and abilities and in discerning a coherent moral end. Such philosophical labor revealed the wellsprings of unhappiness and the means to avoid it. All these elements figured in the early-modern revivals of the ancient sects and in their new philosophical competitors that form the central subject of this book.[26]

As the examples of sundry spiritual exercises suggest, seventeenth-century Europe witnessed a blossoming of spiritual, philosophical, and scientific writing aimed at an elite reading public wider than learned readers of Latin and technically proficient philosophers and mathematicians. Descartes, Pascal, and Leibniz sought to perfect means for speaking and writing aptly to this public, a fickle one prone to fashion. All three philosophers sought to *publicize* as much as popularize: they sought not only to spread a set of doctrines but also to create a public that was physically, emotionally, and intellectually competent to understand and judge such doctrines.[27] Their relatively elite audiences were widely believed to possess good taste and good judgment, uncontaminated by the "artificial" methods and teachings of scholastic education. All three authors sought to draw on these putative cognitive competencies; they sought to heighten and to perfect them.

TRUTH, FALSITY, AND SELF-CULTIVATION

This study aims to contribute to the history of truth and falsity. Rather than a history of doctrines deemed true or false today, such a history investigates the notions of truth and falsity during the seventeenth century, the practices necessary to become someone capable of recognizing such truth and falsity, and the mechanisms and standards for being recognized as such a person. As a study in historical epistemology, it focuses upon the historical nature of definitions and demonstrations, of proof and evidence, and of means of argumentation.[28] As a study in the making of knowers, it focuses on the effects that scientific, mathematical, and philosophical practices putatively had on those pursuing them, and on the training and exercise believed necessary to become a competent knower.[29] The chapters that follow reconstruct the often-idiosyncratic standards for truth, proof, and evidence in Descartes, Pascal, and Leibniz, standards far removed from our own. All three stressed that prevailing philosophical accounts of truth, deduction, and evidence failed to capture their mathematical and natural-philosophical practices. All three drew heavily on their practices in developing new accounts of human knowledge in their methodological and epistemological writings. Without implying that their methodological writings correctly describe their practices, a study of their work can illustrate the productive interactions among their motivations, models, methodological theorizing, and practices.[30]

Early-modern thinkers maintained that usable forms of truth-telling had to be tempered by the real epistemic and social qualities human beings possessed after the Fall of Adam and Eve.[31] Rather than overcoming any supposed medieval insistence on human inability or removing otherworldly veils of ignorance, early-modern natural philosophers and mathematicians worked to acknowledge, to characterize, and to compensate for human epistemic and affective limits.[32] Like the Jesuits with whom they often quarreled, the early-modern natural philosophers examined here offered their work in part as clear indications of how human beings are limited and how they are not. Rightly or wrongly, all three major figures in this study criticized the laziness they perceived in Aristotelian empirical epistemology, which held, roughly, that ordinary experience of the world would produce knowledge of the fundamental principles of nature. Such a view offered far too rosy a picture of human epistemic capacity and posited too close a fit between those capacities and the structure of the world. Misunderstanding human ability after the Fall, those who adhered to this account of human nature failed in the eyes of their critics to recognize the labor and exercise necessary to gain knowledge and to become someone capable of gaining that knowledge.[33] Such failure precluded discovering the knowledge and practices potentially available to fallen human beings. Concerns about human nature helped constitute what we retrospectively consider major scientific achievements in the early-modern period. These concerns spurred the development of alternative practices of knowledge acquisition. The new mathematics and natural philosophies, and the methodological lessons gleaned from them, were defended based on visions of human capacity. Likewise—and circularly indeed—arguments for such visions drew on mathematics and experimental natural philosophy as resources.

In tandem with their efforts to work out new accounts of truth and legitimate reasoning, Descartes, Pascal, and Leibniz strived to perfect practices necessary for becoming someone capable of competently and safely reasoning. Becoming a competent reasoner and moral agent involved more than merely choosing to accept some set of philosophical concepts. Only considerable practice and exercise could produce the sort of person capable of using those concepts throughout life. An era of changing concepts of the self, the seventeenth century saw profound changes in practices of cultivating, identifying, and policing selves.[34] To many philosophers in the seventeenth century, pursuing detailed work in the sciences went hand in hand with developing one's moral and knowing person.

For some years now, historians of science have documented how the innovative forms of social organization of knowledge in the early-modern period were closely tied to new varieties of truth claims and means for establishing and reaching consensus about those claims. Courtly, gentlemanly, and legal modes of interaction offered important models for organizing collective activity and for fashioning the comportment, behavior, and linguistic styles of naturalists and philosophers. These forms of etiquette were important resources for developing new forms of knowledge and means for establishing them.[35] Polite conversational culture, to take an example pursued in this book, proved useful in envisioning new ways of collectively pursuing questions of natural and mathematical knowledge while avoiding strife and dispute.

Gentlemanly, noble, and legal cultures offered resources for creating new social and epistemic forms for gaining knowledge; these cultures delimited appropriate and inappropriate modes of discourse and standards of behavior. New technical practices in mathematics and philosophy in turn offered important tools for cultivating truer forms of spiritual and mental nobility.[36] These practices enabled mathematics and natural philosophy to transform, discipline, and train the intellect, the senses, and the affects, and they put these trained faculties at the heart of organizing one's life. The three figures of this study drew on the cognitive grasp perfected through philosophy and mathematics as a resource for the sorts of cognitive grasp necessary for living well. These cognitive and affective transformations were to help tackle the pressing problems of knowledge, belief, and order pervasive in early-modern Europe.

THE ARGUMENT

The book begins with René Descartes' *Geometry* of 1637, in many ways his most enigmatic text. Descartes saw a reformed mathematics as a central means for cultivation and excluded mathematical procedures and objects that he thought contrary to such cultivation. Geometric practice offered experience in recognizing true ideas—clear and distinct ideas. With practice, one could move from improper to proper reactions to the natural and social world, from uncomprehending astonishment at the world to temperate appreciation in grasping its structuring, unifying qualities. Descartes' earliest mathematical achievements showed, he maintained, that complex logical and mathematical relations, usually envisioned only over time through laborious deductions, might instead be conceived all at once, clearly and distinctly. Only such cognitions composed knowledge properly speaking. Descartes' demand for evident and certain knowledge came in part from the humanist disciplines of affect: rhetoric and poetry. Descartes seized on the rhetorical, humanist aspects of his Jesuit education, which taught forms of speaking and writing thought capable of moving people to virtue and godliness. He drew on the rhetorical goals and aesthetic standards of this humanist education in creating new philosophical and mathematical ones fulfilled by his new geometry and natural philosophy.

Throughout his life, Blaise Pascal created works in mathematics, natural philosophy, and religion within a series of rich conversational settings. In the early 1650s, he produced a number of short treatises on the properties of various kinds of special numbers; he was able to connect them all to an arithmetical triangle, now better known as Pascal's triangle. The rigor and clarity of those works he attributed in no small part to a conversation circle focused on mathematics. Working with the triangle and its associated numbers produced many surprising results—most surprisingly, a means for finding the areas of parabolic curves. This mathematical practice exemplified, for Pascal, the amazing discoveries that the human mind could make and the best means to exercise the mind. Yet this practice promised neither systematic discovery of all truths nor evident mathematics such as Descartes demanded.

Reflecting upon mathematical practice could quickly reveal the limits and powers of mathematics—limits and powers scarcely acknowledged by philosophers entranced by the certainty and form of mathematics but ignorant of the real grounds for its certainty and of the limits of formal reasoning. Drawing on this mathematical practice, as well as his earlier experimental work with the vacuum, Pascal offered an account of definition that was proper to limited human cognitive power. The evidence demanded by Descartes and his followers, including some of Pascal's conversational partners, simply rested on a misapprehension of what fallen human beings could know. Their counsels, whether in geometry, natural philosophy, or morality, could only be incorrect; their exercises, ill fitted for fallen human beings. By considering the infinitely small and infinitely great, in one of the most famous fragments of the *Pensées,* Pascal aimed to upset and to depress the polite audiences Descartes, the Jesuits, and others sought to console and comfort. Rather than revealing the greatness of human ability, the best mathematics and natural philosophy of his day ought to encourage the recognition of human incapacity; they ought to generate emotions of worthlessness and despair. Studying nature pointed to the insufficiency of human philosophy to describe the monstrosity of humanity and to prescribe any remedies offering genuine repose and happiness. The best human knowledge of nature ought to lead rational readers, as Pascal saw them, to seek alternative forms of knowing and feeling appropriate to the monstrous human condition, forms rooted in empirical human qualities and revealed scripture.

Whereas the diversions of natural philosophy and mathematics seduced, then pained, Pascal, the young Leibniz wished to put them to productive use. Sent on a diplomatic mission to Paris in the 1670s, he delighted in the panoply of intellectual and practical riches around him and soon took up the serious study of mathematics. In short order, Leibniz devised an innovative method of squaring the circle in which he could rigorously demonstrate that the area of the circle equaled an infinite series. Leibniz drew a number of morals and philosophical ideas from the techniques essential to his demonstration. Although clearly indicating the gulf between human and divine knowledge, the proof showed how new notational techniques, tempered to real human ability, could help human beings reach their true capacities for knowing. Leibniz's proof, he thought, underlined just how perspectival human knowledge was. God might see all views, but human beings were limited to sets of perspectival views. Overcoming such limitation without falling into presumption meant seeking to improve knowledge through palpable, written techniques for collecting, considering, and unifying different viewpoints.

At the ducal court and library in Hanover, Germany, after 1676, Leibniz sought out concrete means to aid human beings to grasp multiple things all at once and thereby discern the hidden unities among them. Drawing on his mathematical practice, Leibniz worked to extend such techniques to experimental natural philosophy, hermeneutics, the encyclopedia of knowledge, the state, and the moral reform of subjects. With these practices, the pessimism of thinkers such as Pascal about the human condition could rationally be dismissed. As means to help others live happier and more productive lives, Leibniz's innovations were a primary form of his knowledgeable charity: perfecting himself in helping to perfect others.

Recapturing the historical seriousness of early-modern philosophy and natural philosophy means recovering the texture of the motivations of Descartes, Pascal, and Leibniz. The truths of logic may hold in every historical moment and in every possible world, but the reasons for caring about them do not remain constant. To grasp a historically specific will to truth is to uncover something central to the emotional and rational world of a historical group. Understanding the rigorous philosophical systems and mathematical practices of Descartes, Pascal, and Leibniz properly, on their own terms, requires reconstructing why they valued rigorous philosophical and scientific inquiry and why such inquiry so appealed during their tumultuous time. They offered cures based on their diagnoses of the weaknesses and strengths of human capacities for knowledge and self-control. For all three, mathematics and natural philosophy offered both empirical evidence for humanity's capacities and, if correctly tuned, remedies for its faults.

PART I Descartes

CHAPTER ONE

Geometry as Spiritual Exercise

In some early notes, René Descartes commented on a surprising difference between the health of the body and the health of the soul: "Vices I call maladies of the soul, which are not so easily diagnosed as maladies of the body. While we often have experienced good health of the body, we have never experienced good health of the mind" (AT X:215). The language of spiritual malady harkened back to an ancient tradition, exemplified in Cicero's *Tusculan Disputations:* "Diseases of the soul are both more dangerous and more numerous than those of the body." Philosophy offered succor. "Assuredly there is an art of healing the soul—I mean philosophy, whose aid must be sought not outside ourselves, as in bodily diseases. We must use our own utmost endeavor, with all our resources and strength, to have the power to be ourselves our own physicians."[1] Philosophical exercise could overcome the sickness, the delusions, the sadness inculcated by institutions, tradition, and everyday affairs.[2] Attaining a healthy soul demanded work on oneself, exercise—*askesis.*

Descartes' famous quest to devise a superior philosophy should be understood within such a therapeutic model of philosophy. Descartes offered his geometry in part as just such a spiritual exercise, able to help one to attain a tranquillity grounded in an intellectual mastery in making all the choices of life. Along with other exercises, the practice of mathematics could help one to steel oneself against outside direction, confusion, and distraction. Influential works of Descartes' time echoed the diagnostic and curative model of philosophy seen in Cicero. In his widely read *On Wisdom* of 1601–4, Pierre Charron "calls man to himself, to examine, sound out, and study himself, so that he might know himself and feel his faults and miserable condition and thus render himself capable of salutary and necessary remedies—the advice and teachings of wisdom."[3] At the Jesuit school of La Flèche, Descartes went through a curriculum predicated in part on a model of the ennobling and curative virtues of philosophy.[4]

Descartes denied the therapeutic efficacy of the philosophies and models for spiritual health available in his day. However much he desired a therapy for the soul, the young Descartes could not see how to choose any particular program without means for gauging the health of the soul and the mind. Around 1619, Descartes lacked any certain means for choosing among the many treatments available in the early seventeenth century or for judging his own attempts.

The first glimmers of a new way forward appear elsewhere in his notebook where Descartes discussed a number of new mathematical discoveries involving instruments and machines.[5] Before long he applied these mathematical discoveries to the vexing question of the health of the soul. New forms of exercise, including geometric practice, could provide a way to measure and produce spiritual health. Descartes underscored the concreteness of these exercises a few years later when he advised the study of "the simplest and least exalted arts, and especially those in which order prevails—such as those of artisans who weave and make carpets, or the feminine arts of embroidery, in which threads are interwoven in an infinitely varied pattern." The same held for arithmetic and games with numbers: "It is astonishing how the practice of all these things exercises the mind," so long as we do not borrow their discovery from others but invent them ourselves (AT X:404).[6] For Descartes, such study habituated one to experiencing clear and distinct order: "We must therefore practice these easier tasks first, and above all methodically, so that by following accessible and familiar paths we may grow accustomed, just as if we were playing a game, to penetrating always to the deeper truth of things" (AT X:405; CSM I:36). Boldly Descartes announced, "Human discernment [*sagacitas*] consists almost entirely in the proper observance of such order" (AT X:404; CSM I:35). Upon such discernment of order rested the ability to make the will capable of accepting and following the guidance of the intellect. Much as Cicero had maintained, individual endeavor was necessary to foster the dependence on oneself central to spiritual health.

No activity could develop this discernment better than mathematics: "[T]hese rules are so useful in the pursuit of deeper wisdom that I have no hesitation in saying that this part of our method was designed not just for the sake of mathematical problems; our intention was, rather, that the mathematical problems should be studied almost exclusively for the sake of the excellent practice which they give us in the method" (AT X:442; CSM I:59). With his new geometry, Descartes did not offer algorithms that mechanically yielded certain knowledge about mathematical objects and a world governed by mathematical laws. His geometry offered exemplary practice in seeing and thinking clearly, in experiencing with a healthy soul. Such practice was essential both to improve mathematics and to heal the self. Geometrical exercise could help cultivate the discernment needed for choosing among philosophical doctrines and practices, for finding a purpose for one's life, and for prescribing the means necessary to achieve that purpose: in sum, for pursuing the good life.

This chapter studies one set of exercises Descartes deemed propaedeutic to a better life and better knowledge: his famous, if too little known, geometry. Commentators have often stressed how Descartes offered a mathematical model for knowledge, but they have not often inquired into the rather odd mathematics he set forth as this model. His geometry, neither Euclidean nor algebraic, has its own standards, its own rigor, and its own limitations.[7] These characteristics ought to modify our view of Descartes' much ballyhooed creation of the "modern subject" of knowledge. While the technical details of his geometry might seem interesting and comprehensible only to historians of mathematics, the essential features of his mathematics can be made readily comprehensible to anyone with high-school algebra. These fundamental characteristics reveal much about his concerns about what prevents and produces the right sort of subjectivity. More than simply providing a philosophical account of the modern, Cartesian subject in his *Meditations,* Descartes offered techniques for becoming a new kind of thinking, knowing, and acting subject. To pursue his philosophy was nothing less than to cultivate and to organize one's self. He offered his revolutionary but peculiar mathematics as a fundamental practice in this philosophy pursued as a way of life.

What does all this potentially anecdotal detail have to do with the substance of Descartes' mathematics? Descartes certainly may have held mathematics to be good for some sort of pedagogical exercise, but how could that possibly make the content, the essence, of that mathematics any different? After all, since Euclid had exemplified mathematical rigor in antiquity, what could be more obvious than the idea that mathematics retains an unchanging core despite all its varied uses, its manifold representations, and the motivations for doing it?

Few scholars of early-modern science and philosophy would now doubt the need to study rigorously the contingent constitutions of systems of thought or sets of epistemic practices and the embedding of those constitutions within their cultural and social roles. In nearly every study of such systems, however, a cordoned-off core of logic and rigorously argued philosophy remains relatively unexamined. So, one ought rightly to contend, however variable the cultural uses of philosophy and mathematics, there remains an invariable, autonomous essence in each to be studied in itself, an essence that can be abstracted from those uses. In the early-modern period, mathematics, logic, and natural philosophy were widely supposed to be rigorous, coherent, and certain, but there was considerable debate on what all those terms meant and implied.[8] After the criticisms of humanists, skeptics, and scholastics of the sixteenth century, the very definition and centrality of proof was contested. For his part, Descartes rejected standard mathematical proof. He was hardly alone. In his time, no consensus existed on the objects of mathematics, its proof techniques, its proper institutional settings, its place in the hierarchy of disciplines, or its relationship to "mathematical practitioners."[9] Recognizing these historical qualities of mathematics hardly implies the sophomoric relativism of "anything goes." It means that the very things that make a technical history of an object technical, systematic, or rigorous are themselves historical. The contingent history of logic and proof must be made evident and then explained. Each competitor in the variety of competing programs demands nonanachronistic technical inquiry. None is reducible to modern logic or proof, at least not without ignoring the historical understandings of rigor and proof and the historical motivations for pursuing mathematics and seeking certainty.

With his account of mathematics, Descartes called for changes in the objects of mathematics: so he limited proof processes, expanded the kinds of allowable curves, and added algebraic representation. Equally he called for altering the practitioners, the subjects of mathematics. Recent historical attention to the self-fashioning of mathematicians has paid insufficient attention to the variety of mathematical practices (and their metamathematical embedding) that were to help effect such fashioning. To assess correctly the contingent nature of mathematics demands examination of the changing social roles and statuses of its practitioners.[10] Likewise, to evaluate its practitioners and their social roles requires careful inquiry into technical content and practices, into the range of objects, tools, and allowed proof techniques of their particular mathematics.

What then is a spiritual exercise? As noted in the introduction, I take spiritual exercises to be practices held to be useful for cultivating the self. Specifying a spiritual exercise means something like determining (1) a set of practices, (2) a conception of the self, where the self need not be reduced exclusively to a mind or an intellect, and (3) the people the exercises are for, that is, the social field of the application of the exercises, either explicit or implicit.[11] For Descartes, his geometry offered a set of cognitive exercises that were extraordinarily useful, if not quite necessary, for achieving a self-regulated life grounded in apprehending natural and moral truths.

This chapter studies Descartes' *Geometry* of 1637, on its own terms, using his repetitive statements of its purpose, its contents, its foundation. In his writings from the 1620s until his death, he asserted regularly that mathematics is an exercise, perhaps the best that we have to practice thinking well. Taking this claim seriously helps to clarify problematic aspects of his geometry and philosophy of mathematics. Looking at both his mathematics and his understanding of its purpose, we will escape a long tradition of equating the subject Descartes aimed to create through his exercise with the so-called Cartesian and modern subjects.

A number of insightful scholars have rightly stressed the need to examine Descartes' works as forms of exemplary practice and not only as philosophical doctrine to be rationally reconstructed. They have largely avoided his mathematics.[12] His curious mathematics offers a key for understanding how Descartes' intended to have his philosophy *practiced*. I focus on the laborious nature of mathematics: while not the corporeal toil of a contemporary artisan or farmer, it is exercise, hard exercise—a point obvious enough to mathematicians but too often absent from histories and philosophies of the subject. Work using geometry as exemplar could produce the focused intelligence that Descartes thought was the heart of cultivation and was necessary for discerning and pursuing the good life.

ENVISIONING THE ANCIENTS

In *Discourse on Method* (1637), Descartes explained that the mathematics he experienced in the Jesuit school of La Flèche led him to conclude that mathematics was good only for clever tricks and mean trades.[13] Why then had Plato's Academy refused to admit anyone ignorant of mathematics, a science that seemed so "puerile and hollow"? The ancients, Descartes decided, must have had a "mathematics altogether different from the mathematics of our time" (AT X:376; cf. CSM I:18). Descartes caught glimpses of this true mathematics in the ancient mathematicians. Why only glimpses? In a remarkable piece of historical reconstruction, Descartes asserted that the ancients feared "that their method, being so easy and simple, would become cheapened if it were divulged, and so, to make us wonder, they put in its place sterile truths deductively demonstrated with some ingenuity, as the effects of their art, rather than teaching us this art itself, which might have dispelled our admiration" (AT X:376–77; cf. CSM I:19). Misled by the ancients' "ruse" of formal proof, Descartes' amazed contemporaries memorized the sterile truths of the ancients and failed to seek the relationships behind those truths and the methods used to discover them.

A friend of Descartes', the writer Jean-Louis Guez de Balzac, made a similar point about Cicero. Cicero's codified rhetoric for swaying the mob had been mistaken as his true rhetoric and dialectic and then made into strict rules.[14] The contemporaries of Guez de Balzac and Descartes had confused instantiations of technique with the essence of mathematics and rhetoric. While these ancient techniques certainly could deceive, move, and direct, they could neither dispel wonder nor produce novel mathematicians or genuine orators. In misapprehending the grounds of these techniques, their contemporaries had made themselves into people needing external discipline and rules, a mob to be swayed—when, presumably, they ought to have been swaying the mob. They lost the true mathematics and rhetoric and, with them, genuine knowledge and the ability to invent and discover on their own. Slaves to rules and outward appearances, they were not masters of their own disciplines, or of themselves.

Descartes and Guez de Balzac connected this slavish dependence on technique to the institutionalization of mathematics and rhetoric. Like many of their contemporaries, they envisioned the ancients as success-ful, stable, and productive precisely because they were *honnêtes hommes*—cultivated gentlemen—outside stultifying institutions.[15] Blaise Pascal cap-tured this common seventeenth-century view well: "We imagine Plato and Aristotle only in the long robes of pedants. They were *honnêtes gens*, much like the others, laughing with their friends. And when they amused them-selves by writing their *Laws* and *Politics*, they did so without trying. [Writ-ing these books] was the least philosophical and serious part of their lives; the most philosophical was to live simply and tranquilly. If they wrote of politics, it was to regulate a hospital of madmen" (S457). Generations had mistaken institutionalized technique useful for regulating others as essential philosophical doctrine and considered ways of living. The outer garments of rhetoric, mathematics, and philosophy had been substituted for the real methods of these disciplines. Like many of their contempo-raries, Descartes, Guez de Balzac, and Pascal assimilated the ancients to the seventeenth-century ideal of *honnêteté*.[16] Early in the century, the notion referred primarily to a vision of judgment and taste among the nobility of the robe. *Honnêteté* offered a normative ideal of self-cultivation that com-bined a genteel nonspecialization, proper manners, and truth making out-side formal institutions with elements of taste and judgment more broadly conceived.[17]

Whereas his contemporaries were trapped by final results and codified procedures, Descartes argued that the "same light of mind that allowed [the ancients] to see that one must prefer virtue to pleasure and the honorable to the useful . . . also gave them true ideas in philosophy and the method."[18] Descartes echoed Cicero's famous critique of Aristotle in phrase and intent: "Whereas Aristotle is content to regard *utilitas* or advantage as the aim of deliberative oratory, it seems to me that our aim should be *honestas* and *utilitas*."[19] The genuine knowledge and true skill of the ancients allowed them to be productive mathematicians and to recognize the proper basis of the good life, a life of virtue and duty toward others.

Moderns could surpass the ancients. Descartes and Guez de Balzac maintained that the ancient gentlemen-philosophers had failed to achieve their own ideal, whether in mathematics or in rhetoric. That they took the trouble "of writing so many vast books about" geometry showed that "they did not have the true method for finding all" the solutions (AT VI:376). While the ancients had the seeds of true method, Descartes argued, "they did not know it perfectly" (AT X:376; cf. CSM I:18). Guez de Balzac and Descartes likewise claimed that the ancients had the seeds of true rhetoric but lacked true rhetoric itself.[20] Ancient mathematics was far from perfect. Descartes' evidence? "[T]heir extravagant transports of joy and the sacrifices they offered to celebrate discoveries of little weight demonstrate clearly how rude they were" (AT X:376; cf. CSM I:18). The ancients' lack of self-control proved that some mathematical discoveries came as surprises—discovered not through methodic comprehension but with miraculous genius or just plain luck. Even if the lucky ancients had the right informal social structures, they lacked the complementary exercises necessary for eliminating imitation and surprise. They lacked, in sum, the true method for autonomous mathematical discovery and self-mastery.

Moderns needed better social forms and better exercises to renew and exceed the virtues of the ancients. Descartes' friend Guez de Balzac spearheaded a movement to civilize the unruly texts and practices of the Renaissance, to take the fruits of humanism and strip them of the extravagance and pedantry exemplified in the—to his mind—uncontrolled works of Montaigne and classical scholars such as Girolamo Cardano. Only then could humanism be properly deinstitutionalized, its essence revealed, and the true potential of ancient learning nourished. This return to the *urbanité* of noninstitutionalized higher philosophy needed new forms of writing and printing, which Guez de Balzac worked to produce.[21]

Returning to the true mathematics that the devious ancients had obscured meant creating new exercises capable of producing knowledge and civility alike. In the debate surrounding his *Geometry*, Descartes became enveloped in controversy with the famous Toulousian mathematician and lawyer Pierre de Fermat. Descartes condemned Fermat's mathematics as wondrous and uncivil: "Mr. Fermat is a Gascon; I am not. It is true that he has found numerous *particular* beautiful things and that he is a man with a great mind. But, as for me, I have always worked to consider things with extreme generality, to the end of being able to discern rules that also have utility elsewhere."[22] "Gascon" was a well-targeted snub: it suggested provincialism, extravagance, amusement, and a quest for advancement— in sum, incivility and disorder.[23] Descartes attacked how Fermat arrived at results: "without industry and by chance, one can easily fall onto the path one must take to encounter" a solution.[24] Fermat was doubly particular: he found pretty trinkets and he did so through genius—chancy genius. For

Descartes, chance findings typified mathematics as then practiced: such practices were an inferior means toward a lower form of mathematical knowledge, one that, nonetheless, often produced results. The aleatory nature of such discovery produced wonder, dependence, and extravagance, not clarity, independence, and self-control.

Fermat's mathematics was inferior *as mathematics* because it failed as an activity appropriate for cultivating the mind more generally. Descartes claimed that his own general geometric method would offer true understanding because an orderly constructive process produced all of its results.[25] Mathematical problem solving was not to proceed by some new ad hoc technique dependent on individual instantiations of skill, expertise, and genius for every problem.[26] For Descartes, *real* mathematical knowledge—knowledge capable of properly exercising the mind—demanded that chance and genius be eliminated. Mathematical practices that made the assent to inferential steps—deductive proofs—the essence of mathematics had to be abandoned, although Descartes called for the temporary use and then abandonment of such inferential steps. Descartes worked to sever a true mathematics of cultivation, the true method, from specious pretenders, which offered little but calculation, passive procedures, and, too often, deception.

DESCARTES' GEOMETRY OF 1637

The sixteenth and seventeenth centuries witnessed countless mathematical "duels," in which mathematicians tried to best one another with their ingenuity in solving particular problems.[27] Here is a very simple example of a typical problem: Given a triangle *ABC* and a point *D* outside the triangle, one must construct a line through *D* that divides the triangle in two equal parts (fig. 1.1). All solutions must comprise a series of constructions of circles and lines produced by standard compasses and straightedges. Usual solutions to such problems included no information about how to arrive at the solution or how to go about solving a similar problem.[28] Besting someone in a mathematical duel typically meant making others marvel at your ingenuity in solving but did not include offering them heuristic instruction. Descartes criticized his rivals, such as Fermat, for finding "particular things" instead of producing rules useful in geometry and elsewhere.

Descartes aimed to give a set of general tools offering a certain method for problem solving. His *Geometry* begins audaciously. "All problems in Geometry can easily be reduced to such terms that there is no need from then on to know more than the lengths of certain straight lines to construct them" (AT VI:369). By assigning letters to line lengths, Descartes could translate relationships among geometric objects into algebraic formulas.

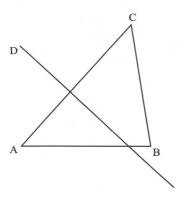

Figure 1.1. Example problem.

His geometry gained a powerful toolbox of algebra—largely why the book is so famous. In principle, although not in practice, every algebraic manipulation corresponded to a geometric construction. In Descartes' mathematics, geometric problems get geometrical constructions as solutions.[29]

Very schematically, Descartes' geometrical method involved the following steps:

1. Assuming the solution to the problem, naming the various parts involved, and putting them into algebraic equations;
2. Transforming the algebraic equations into an equation involving only one unknown or into a set of equations from which one must make additional assumptions to get only one unknown;
3. *Constructing* geometrically the solutions to the equation.

Algebra should serve only as a temporary means toward conceiving ever more clearly and distinctly the relations among geometric entities.[30]

To illustrate the power of his approach, Descartes addressed a key problem from antiquity—the Pappus problem.[31] In its simplest form, the problem is to find all the points that maintain distances from two lines such that the distances equal a constant. Solving the problem for a small number of lines was relatively easy. The traditional limitation to standard compass and straightedge constructions blocked the solution of the problem for a greater number of lines.

So Descartes needed more tools. He added a wider variety of curve-drawing instruments and defended their use (for an example, see fig. 1.2). His new instruments (or machines) generated a wider set of curves that could be used to solve geometric problems. With these new machines, Descartes believed that he could solve the famous Pappus problem for any number of lines, thereby far surpassing the ancients. More important, he could do so as part of his systematic method for solving geometric problems

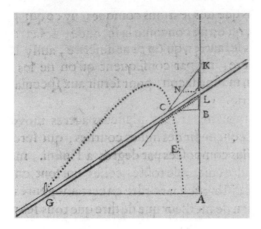

Figure 1.2. Geometric machine and associated curve. The machine rotates *GL* about *G*; *GL*, attached at *L*, pushes the contraption *KBC* along the line *AK*. The intersections at *C* of *GL* and *CNKB* make up the curve. From René Descartes, *Discours de la méthode pour bien conduire sa raison* . . . (Leiden, 1637), p. 320. Courtesy of Rare Book and Manuscript Library, Columbia University.

and not because of some ingenious insight or expertise.[32] Descartes concluded, perhaps foolishly and certainly prematurely, that he had provided the tools for classifying and systematically solving all geometric problems capable of being solved with certainty. No longer would slavish imitation of the big books of the ancients and those of their modern-day followers be needed.

Considerably more than mere tools for mathematical problem solving, his new geometrical tools offered essential exercise for sharpening the mind: "I stop before explaining all this in more detail because I would take away the pleasure of learning it yourself and the utility of cultivating your mind in exercising yourself on these problems, which is, in my opinion, the principal utility that one can take away from this science" (AT VI:374).[33] Solving the problems oneself would be more pleasurable and mathematically rewarding, and would have an even greater effect. Solving problems oneself aids in the development of the mind. In his autobiographical narrative in *Discourse on Method*—the preface to *Geometry* and other essays—Descartes explained that he turned to mathematics in the hope that "it would accustom my mind to nourish itself with truths and not be satisfied with false arguments" (AT VI:19). It worked. By following his method in mathematics "my mind accustomed itself little by little to conceive its objects more sharply and more distinctly." He set out to use the method on other sciences (AT VI:21).

THE INSUFFICIENCY OF OTHER CULTIVATING EXERCISES

For Descartes, cultivating oneself demanded developing the ability to permit the will to recognize and to accept freely the insights of reason and not to follow the passions or memorized patterns of actions. It meant recognizing the essential limits of human reason, that is, actively willing *not* to make judgments about things beyond the scope of human reason and learning to will only those actions the intellect judged to be proper.[34] From his *Rules for the Direction of the Natural Intelligence* of the 1620s to *Principles of Philosophy* and *Passions of the Soul* of the 1640s, Descartes criticized contemporary philosophical practices as deleterious to this proper self-cultivation. In his 1647 introduction to the French version of *Principles of Philosophy*, he argued that to live without philosophizing "is properly to have the eyes closed without ever trying to open them; . . . this study is more necessary to rule our manners and direct us in this life than is using our eyes to guide our steps" (AT IX/2:3–4).

Philosophy was far more than a logical system of interconnected statements or an academic discipline; philosophy should, in fact, direct one's everyday existence. With philosophy one could gain self-mastery and civility by discerning principles by oneself and then internalizing them through habituation. "[I]n studying these principles, one will accustom oneself, little by little, to judge better everything one encounters and thus become more wise" (AT IX/2:18). In traditional philosophy one simply memorized and followed external rules for reasoning and for acting. The true philosophy should never involve blindly following techniques and procedures. "[I]n this [his principles] will have an effect contrary to that of common Philosophy, for one easily notices in those people called Pedants that it renders them less capable of reasoning than if they had never learned it" (AT IX/2:18). Descartes roughly divided the human faculties into the intelligence, the memory, the imagination, and the senses. In his view, the attention can be focused on only one of these faculties or senses at a time.[35] For the will to receive the guidance of the intellect, the attention must be focused primarily on the intellect. Any epistemic procedure keeping the attention away from the intellect for too long must be rejected as antithetical to cultivation.

An exploding variety of new words, things, and approaches characterized and threatened Renaissance visions of knowledge. Descartes sought to overcome the fragmentation of attention caused by trying to understand individual, disjointed pieces of knowledge. Like Michel de Montaigne, he doubted that extant intellectual tools could ever unify these bits of knowledge. In an early notebook entry, Descartes complained that the art of memory was necessarily useless "because it requires the whole space [*chartam*] that ought to be occupied by better things and consists in an order that is not right." By helping to retain particulars in themselves, the art of memory precluded using the intellect to discern the underlying unities behind those particulars. The art diverted attention away from the discerning of order, a function of intellect, toward the recognition and cataloging of disjointed particulars, a function of the memory. In contrast, "the [right] order is that the images be formed from one another as interdependent" (AT X:230).[36] The pedagogic and reading practices associated with the art of memory, commonplace books, and encyclopedias promised to help discover such interconnection.[37] They all failed.

Focused upon individual disconnected experiences and historical facts, such techniques necessarily prevented the discovery of interconnections. Even worse, most narratives, theories, or syntheses that attempted to combine particulars generated less interconnection than monstrous mixtures that lacked any principle unifying the whole. Descartes described such narrative monsters as mere histories, not science.[38] Such histories always focused the attention on discrete elements in the imagination or memory and did not allow the intellect to grasp the fundamental unities behind the apparently discrete elements.

For Descartes, these epistemic failings led inevitably to moral ones. Deflecting the attention away from the intellect yields imitation, not introspection, the following of external rule, not self-mastery. Histories of unconnected facts mislead those "who rule their manners by the examples they take from them." Ruling manners by imitation and not introspection allows people "to fall into the extravagances of the paladins of our romances and to conceive of designs that surpass their strengths [*forces*]" (AT VI:7). Imitating Amadis de Gaul makes one into Don Quixote: unable to perceive the real world, incapable of knowing one's real strength, and forever questing after unconsidered goals with inappropriate weapons. Imitating Seneca or Cicero or Aristotle does the same. By imitating the models from "histories," his contemporaries made themselves unable freely to consider what "design" to follow in life and unable to rule themselves with their own plans, adjusted to their real strengths and abilities.

In a letter Descartes explained the range of disciplines he dismissed as mere history: "By history I understand all that has been previously found and is contained in books." True knowledge means, in contrast, the ability to resolve all questions "by one's own industry," to become *autarches*, that is, self-sufficient.[39] Only such self-sufficiency allows true inventiveness in reasoning, attentiveness to reasoning, and the moral life predicated upon the good use of that reason. Only such self-sufficiency allows us to know ourselves, so that we might plan for ourselves appropriately, to choose our goals and the means to fulfill them.

Mathematics did not escape Descartes' expansive condemnation of histories. In mathematics, as elsewhere, Descartes explained, imitating the works of the ancients precludes developing genuine competency: "Even though we know the demonstrations of others by heart, we shall never become mathematicians if we lack the aptitude, by virtue of our natural intelligence, to solve any given problem" (AT X:367; cf. CSM I:13). Standard mathematical proof is simply another form of imitation, as deleterious to reasoning and self-mastery as other histories. Why?

For Descartes, formal logical consequence, as in a syllogism or mathematical proof, rests on the possibility of surveying a formal deduction over time.[40] Considering a series of particular facts or observations in an enumeration demands the stepwise switching of attention as one reviews the series in memory or on paper. So too with the discrete steps of a formal deduction: "[When a deduction is complex and involved] we call it 'enumeration' or 'induction,' since the intellect cannot simultaneously grasp it as a whole, and its certainty in a sense depends on memory, which must retain the judgments we have made on the individual parts of the enumeration if we are to derive a single conclusion from them taken as a whole [*ut ex illis omnibus unum quid colligatur*]" (AT X:408; CSM I:37). If I know a series of relations $A:B$, $B:C$, $C:D$, $D:E$, then "I do not on that account see what the relation is between A and E, nor can I grasp it precisely from those already known, unless I recall them all" (AT X:387–88; cf. CSM I:25). The sequence in the proof offers good reasons to *consent* that the relation between the first and the fifth is such. In consenting in this way, however, we do not grasp the relation in anything like the way we grasp the more intermediate and immediately grasped relations.

Descartes admitted that formal deductions could be perfectly certain "in virtue of the form" (AT X:406).[41] Such formal *certainty* hardly makes the final result and its connection to the intermediate steps at all *evident*. With formal certainty one knows, to be sure, $A:E$, but not in the same way that one knows $A:B$ or $C:D$. The discrete steps are just like a set of particular observations about the natural world. Both syllogistic logic and mathematical demonstrations in their traditional forms, those products of ancient ruses, rest on memory. In slavishly imitating and assenting to proof, one allows reason to "amuse" oneself and thereby one loses the habit of reasoning. No less than history and traditional philosophy, traditional proof-driven mathematics hinders self-regulation in epistemic and moral matters.

LEGITIMATE KNOWLEDGE

Having rejected essentially all contemporary forms of knowledge production, what did the young Descartes have left? His early notebook hints at a "poetic" knowledge.[42] He contrasted the laborious processes of reasoning in philosophy with the organic unity of wisdom and knowledge that poets divined: "It seems amazing what profound thoughts are in the writings of the poets, more so than in those of the philosophers. The reason is that poets write through enthusiasm and the strength of the imagination, for there are sparks of knowledge within us, as in a flint: where philosophers extract them through reason, poets force them out through the imagination and they shine more brightly" (AT X:217; cf. CSM I:4). This poetic ideal promised knowledge of an intuitive character. Such knowledge stemmed from a grasp of the unity connecting objects; it was thereby appropriate for regulating oneself, as it focused the attention on unities perceived by the intellect, not on particulars stored in the memory.

In chapter 2 I discuss in some depth how Descartes turned this poetic claim of his early notebook into an epistemic standard of unity and interconnection. According to this standard, knowledge required a grasp of the unifying causal structure behind something to be known. Such knowledge of a unifying causal structure permitted one to eliminate the need for retaining particulars in memory, such as the steps of a formal proof. Once the causal structure was grasped, the original justification could easily be reproduced. Yet the knowledge of causal structure was neither secured by those formal steps nor did it include an enumeration of them. It comprised, rather, knowledge of an organizing principle underlying the *interdependence* of elements.[43] The correct "order" of knowledge, Descartes argued, was "that the images be formed from one another as interdependent" (AT X:230).[44] From the interdependence of elements contained within the cause, the series of connected images followed easily.

In this view, a true mathematics involves cognizing the interdependence of mathematical objects. Knowledge of a real mathematical cause should allow the trivial extraction of the procedures—the deductions or enumerations—helpful (or necessary) for attaining that knowledge. Mathematical knowledge would neither be grounded on a series of formal inferences nor contain such inferences spelled out, though it would always lead to understanding the causal structure sufficiently to make the spelling out of such inferences trivial.

A mathematics appropriate for self-cultivation offered experience in recognizing the interdependence and evidence of the steps of a formal proof and would make those steps ultimately superfluous. Under this constraint, the series of proportions $A:B$, $B:C$, $C:D$, and $D:E$ of a formal proof would have to be somehow grasped all at once.

What examples did Descartes have of mathematical knowledge unified in this manner? He had a new proportional compass (see fig. 1.3).[45] Descartes' compass begins with two straightedges YZ and YX (AT VI:391–92). BC is fixed on YX. Other straightedges perpendicular to YZ and YX respectively are attached but can move side to side along YZ and YX. As the compass is opened, BC pushes CD, which in turn pushes DE, which pushes EF, and so forth. As it opens, the compass produces a series of similar triangles YBC, YDE, YFG, and so on. This allows the infinite production of

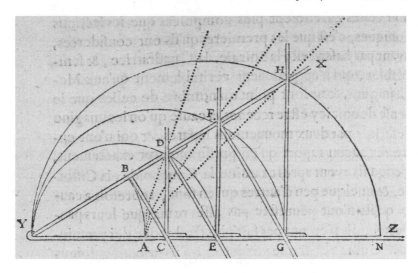

Figure 1.3. Descartes' compass. From René Descartes, *Discours de la méthode pour bien conduire sa raison . . .* (Leiden, 1637), p. 318. Courtesy of Rare Book and Manuscript Library, Columbia University.

mean proportionals, $YB:YC::YC:YD::YD:YE::YE:YF$. Numerous problems in geometry can be solved through finding such mean proportionals.

With the compass in mind, we can grasp the ordering principle—the causal structure—behind a sequence of continued relations. That is, we can grasp the relation between a first and last term ($YB:YF$) in a way similar to how we grasp an intermediate and more immediate relationship ($YB:YC$). We need not retain the individual proportions in memory to claim knowledge of any of the particular relations, since we can easily read them off the compass. The compass offered a crucial heuristic, a material propaedeutic, for Descartes' revised account of mathematics freed from memory and subject to a criterion of graspable unity.[46] A simple mathematical instrument offers a model and exemplar of the knowledge necessary for cultivating oneself.

EVIDENCE AND DEDUCTION

Descartes distinguished the evidence of a proof from its formal certainty. Formal demonstrations, like syllogisms or other logical forms of proof, could, in his eyes, produce a kind of certainty. They did not, however, make *evident* the connections one was proving. Descartes radically demanded that all real knowledge possess the same evidence as our knowledge of the simplest truths. In *Rules for the Direction of the Natural Intelligence* (1620s), Descartes formalized a new account of enumerative and deductive knowledge as subject to the criterion of evidence (discussed in greater detail in the next chapter). The new form of deduction extended evidence from single intuitions of particular things or simple truths to knowledge of simple and then complex unified systems.[47]

In *Rules*, Descartes reduced all true knowledge to intuitions: "two things are required for intuition: first, the proposition intuited must be understood clearly and distinctly; next, it must be understood all at once, and not bit by bit" (AT X:407–8; cf. CSM I:37). Descartes knew well that such an instantaneous intuitive grasp could hardly account for much complex knowledge. He demanded nevertheless that more complex knowledge retain the qualities of intuitions: "The evidence and certainty of intuition is required not only for apprehending single enunciations but equally for all routes" (AT X:369; cf. CSM I:14–15). Any sort of cognition more complex than immediate intuition would involve cognition over time using the memory. Such cognition would move the attention away from the intellect.

With his new model of deduction, Descartes proposed that one could raise demonstrations, including but by no means limited to mathematical and other traditionally deductive arguments, to certain and evident knowledge by revealing the occulted order underlying them: "So I will run through [all the particulars] several times in a continuous motion of the imagination, simultaneously intuiting one relation and passing on to the next, until I have learned to pass from the first to last so swiftly that no part is left to the memory, and I seem to intuit the whole thing at once. In this way our memory is relieved, the sluggishness of our intelligence redressed, and its capacity in some way enlarged" (AT X:388; cf. CSM I:25). Descartes' vision of intuiting "the whole thing at once" rested on there being an underlying structure or principle to be grasped at once, something that could guarantee the interconnection with the object and permit a continuous intuition of the object. His usual metaphor involved the chain: "If we have seen the connections between each link and its neighbor, this enables us to say that we have seen how the last link is connected to the first" (AT X:389; CSM I:26). For Descartes, attaining knowledge more complex than simple intuitions meant uncovering, then grasping, such connections much as if they were simple intuitions. Geometry offered practice in making such connections among geometric objects apparent.

His central example for his new deductions is the sequence of relations described above. To understand fully the final result, that *A:E*, we need not only to grasp the series of simple relations but the underlying order producing them. As we saw, the compass, in this example, leads one to grasp the causal structure, the ordering principle, the *how*, behind these relations. Producing curves through the motion of the compass allows the simultaneously grasped, clear, and distinct intuition of the entire chain of relations. With this little example, one might experience what it is like to have such an intuition, which, being basic, cannot be defined.

EXERCISES: EVIDENCE AND MATHEMATICS

Descartes' famous criteria of clarity and distinctness have long struck commentators as more aesthetic than epistemic, overly subjective, and therefore useless as criteria for knowledge. Leibniz, for one, famously complained that Descartes failed to provide adequate grounds for discerning the clear and distinct.[48] Something like this aesthetic quality attracted Descartes, for only such a criterion, drawn in part from poetry and rhetoric, could ensure the interconnection central to real knowledge—the interconnected knowledge divined by the poets. All human knowledge rested on the ability of individual subjects to discern the clear and distinct. In principle, the universality of the good sense made this ability available to anyone who should undertake to hone this good sense. Any human being who has cultivated this ability would discern the same truths as clear and distinct. All those who have cultivated the intellect would agree, upon reflection, that the same claims are true, because they would all perceive them as clear and distinct. Exercise honed the natural ability to recognize the clear and distinct and the truly interconnected; exercise made recourse to this ability habitual. By developing this habit of turning always to this ability of the intellect, one could recognize those truths that God had chosen to make available to human beings so that they could make all the choices in life.

In his *Rules for the Direction of the Natural Intelligence*, Descartes stressed the power of such exercise. The intelligence (*ingenium*) needed concrete practice in experiencing order, clarity, and distinctness in order to sharpen its ability to discern the true and to discover order. In a vein far removed from the image of the philosopher cogitating alone and without corporeal things, Descartes recommended first studying "the simplest and least exalted arts, and especially those in which order prevails" (AT X:404; cf. CSM I:35), such as the weaving and embroidery mentioned above. Such study habituated one to the experience of understanding order clearly and distinctly. Mathematics provided exercise in recognizing how things that were foundational for Descartes were indeed clear and distinct: the self, mind, extension, and God. It offered the practice that could allow one to choose among philosophical therapeutics. Geometrical practice was a potent exercise because it offered experience in knowing with certainty and evidence; such exercise could best refine the habitual ability to recognize the true and discern order amid apparent disorder. This august role for mathematics, however, set precise and troublesome boundaries for salutary geometrical and algebraic work.

ALGEBRA: DEVELOPING MATHEMATICAL HABITS
AND THREATENING TO SPOIL THEM

In his account of exercise, Descartes required a temporary use of artificial instruments and notations to make maximal use of natural abilities. Like manuals claiming to teach ostensibly natural manners, civility, or taste, such artificial means seem a bit paradoxical. Descartes' tools were supposed to hone a natural ability—the power of the *ingenium* (intelligence, good sense)—through artificial techniques. He attacked traditional formal reasoning because its forms promised natural knowledge through artificial means such as logic but never severed knowledge from those artificial means.[49]

Early-modern thinkers had some sophisticated theories for understanding the power of artificial means to bring out natural qualities, potentials, and abilities. Central in Descartes' *Passions of the Soul* (1649), the scholastic notion of habit offered a model for understanding how repetition of external acts could produce internal dispositions and skills for producing such acts.[50] A habit, as one dictionary defined it, is an "internal disposition acquired by the reiteration of many acts."[51] Aristotle and his followers distinguished "acting justly" from "performing a just act." If someone performs a just act, it is impossible to judge whether she has the virtue of justice or performed that act from that virtue. Such an act might be a mistake or mere imitation. Roughly, "acting justly" required performing a just act based on knowledge of its justness, choosing that act from that knowledge, and proceeding from a firm resolution to act justly.[52]

In comparing different forms of mathematics, we might say that Descartes distinguished, to draw on Aristotle's terminology, between acting geometrically and performing a geometrical act. Acting geometrically requires that one perform a geometrical act from knowledge of the underlying interconnections and that one choose to do so given the end of creating more intuitive knowledge. A formally valid calculation or geometric construction might either be merely a geometrical act or be a product of acting geometrically. Only by repeating geometrical acts can someone produce the underlying habits necessary for acting geometrically. Geometry was a powerful practice because it could provide a certain set of external acts useful for honing good epistemic habits.

This scholastic background helps to make sense of Descartes' account of the utility and danger of algebra and other arts for geometry. In *Rules*, Descartes explained that various arts temporarily aid reasoning by preparing one to intuit relations not immediately grasped. The greater part of human labor consisted in this preparation: "[A]bsolutely every cognition that one has not acquired through the simple and pure intuition of a unique thing is acquired by the comparison of two or multiple things among themselves. And certainly nearly all the industry of human reason consists in preparing for this operation; for when it [the operation] is open and simple, there is no need for any aid of an art, but the light of nature alone is necessary to intuit the truth, which is had by this" (AT X:440; cf. CSM I:57). Once the complex nature has been grasped intuitively, that is, all at once, the art no longer would be needed.

The early mathematical machines used by Descartes suggested that solutions to problems in mathematics would come from producing "means" that connected the objects one wanted to understand: "In every question there ought to be given a mean between two extremes through which they are conjoined explicitly or implicitly: as with the circle and parabola, by means of the conic section" (AT X:229).[53] As we saw, Descartes' compass exemplified how a string of proportionals are intimately connected, as are their algebraic representations.[54] Gaining knowledge of the connection necessarily demanded a movement from known things to unknown ones in order to fill the gaps sufficiently. This movement, in filling out a deduction, does not produce a "new kind of entity"; rather, "we are extending our entire knowledge of the question to the point where we perceive that the thing we are looking for participates in this way or that way in the nature of things given in the statement of the problem" (AT X:438; CSM I:56). He offered the example of someone who knows only basic colors but is able to fill out the remaining colors in an orderly fashion.

Whether consisting in the use of algebraic symbols or the proportional compass, art could help precisely because it helps to "evolve" (roll out) the relations "rolled" into "natures": "All the others are required for preparation, for no other reason than the common nature is not in both equally but rather is involved [enveloped] in them according to certain relations or proportions. The principal part of human industry is to reduce these proportions until the equality between the thing sought and the thing already known is seen clearly" (AT X:440; cf. CSM I:57–58). The art in question enabled one to see things clearly; but once seen, the art must no longer play any role, much as the scaffolding necessary for constructing a building is later removed.[55]

Descartes defended the use of algebra as a temporary means to help focus the attention on discovering the interconnection among a set of geometrical objects. Putting "down on paper whatever we have to retain" allows "the imagination to devote itself freely and completely to the ideas immediately before it." He explained how these symbols related to his account of intuition and deduction: "We shall do this with very concise symbols, so that after scrutinizing each item (following Rule 9), we may be able (following Rule 11) to run through all of them with the swiftest sweep of thought and intuit as many as possible at the same time" (AT X:455;

cf. CSM I:67). His algebraic practice in his *Geometry* (1637) embodied this goal of using algebra to enable the simultaneous grasp of a geometric interconnection. He used algebraic symbols to clarify the order at work but then, in principle, gave a geometrical construction that embodied that order. In his view, algebra helps to disaggregate a disordered collection of geometrical objects into a variety of distinctly conceived but clearly unified components. In so doing, one produces an apparently more complicated picture, replete with symbols and additional lines. Algebraic work produces a formula. This newly created algebraic formula guides the construction of a machine, which draws a curve. This curve/machine complex makes the interconnection among the geometrical objects evident. In this process, algebra enables one to get to this geometric order. An algebraic formula, however, should not substitute for knowledge of the geometric order it can help produce. Algebraic formulas should be effaced in the production of the final curve.[56]

His classification of curves illustrates how deeply this effacement went. Among third-degree curves, the one whose equation has no cross terms is simplest, algebraically speaking: for example, $y^3 + 2ay^2 + a^2y + 2a^3 = x$. For Descartes, by contrast, the "Cartesian parabola" $y^3 - 2ay^2 - a^2y + 2a^3 = axy$ was the simplest of third-degree curves.[57] The paradigmatic case of clear and distinct knowledge for Descartes, the paradigm of the order underlying a situation, was his proportional compass and the set of curves produced in its motion. The Cartesian parabola is the product of a line fixed at one point and a parabola along an axis.[58] It is produced by the most basic mechanical constructing apparatus; therefore, it is the simplest of its type of curve. Simplicity rests in the interconnecting geometrical intuition, not the algebraic form.

Descartes' reformed geometry offered habituation in experiencing the clear and distinct. It also offered habituation in filling in the intermediate relations of a set of discrete elements in order to transform a set of apparently independent elements into a clear and distinct unified order. This reformed geometry using algebra was thus a central practice for adding unseen and often invisible intermediate efficient causes to apparently disparate and definitely surprising phenomena.[59] Such a practice was central for natural philosophy, as a mechanical philosopher needed to provide the intermediate mechanisms connecting the basic mechanical building blocks of the world and the entire range of apparent phenomena. Descartes' geometry was supposed to cultivate the mind, then, by habituating one to discerning the clear and distinct, by habituating one to the temporary use of powerful tools, and by accustoming one to the process of producing mechanisms capable of showing the interconnections among phenomena and among geometrical objects.

For all its potency in helping human beings to discover interconnections among geometrical objects, algebra could easily turn mathematics into a mere craft or art, not a practice always creating evident knowledge. A crucial tool for producing intuitive mathematical habituation and for doing geometry, algebra threatened to eliminate self-sufficiency in thought and in action. Like other forms of formal reasoning, it threatened to allow one to perform mathematical acts more or less automatically, without attempting to understand what the symbols represent.

Drawing on Aristotle, the scholastic Scipion Dupleix used three criteria to distinguish the products of actions stemming from virtue and knowledge from those products stemming from "mere" art. The action must stem from knowledge of the good; the choice of it must depend on an end and "design" for action that is good and moral (*honnesté*); in choosing it, there must be a "firm resolution and perseverance of the will." In considering art, on the other hand, one considers neither "the intention nor the will of the workman in judging the perfection of the work, but only the sufficiency."[60] For Descartes, most mathematics of his day was a craft: one needed to consider only the final product, not the knowledge and motivations whence that product sprang. As a consequence, almost all mathematics of his day failed to illuminate, even if its practitioners solved geometrical problems and produced formally true statements.

Descartes condemned the technical procedures of the calculator and the dialectician alike. Neither offered anything of use for the project of self-mastery or the creation of knowledge: "I would not make so much of these rules, if they sufficed only to resolve the inane problems with which Calculators [*Logistae*] and Geometers amuse themselves to pass time, for in that case all I could credit myself with achieving would be to drabble in trifles with greater subtlety than they. . . . I am hardly thinking of vulgar Mathematics here, but I am talking of another discipline entirely, of which [these examples of figures and numbers] are rather the outer garment than the parts" (AT X:373–74; cf. CSM I:17). In only using the outer garment of mathematics—the rules of algebra and arithmetic—the calculator cannot cultivate the epistemic habits and abilities central to real mathematics and necessary to regulate manners and the mind. To work, as a calculator does, with empty numbers and imaginary figures without method means that "we get out of the habit of using our reason" (AT X:375).[61] Like the ancients who offered sacrifices when they chanced upon a discovery, the calculator, the essential imitator, is not his own master. His actions are always directed by external procedures. In performing calculations—mere mathematical acts—the calculator does not habitually turn to the intellect to grasp the unity of objects and to understand the objects represented by his symbols.

Vulgar, or ordinary, mathematics rests on blind and blinding procedures. "There is really nothing more futile than so busying ourselves with bare numbers and imaginary figures that we seem to rest content in the knowledge of such trifles." Such reckoning detracts from reasoning that is based on an intuitive knowledge of the properties of objects and their interrelations and from reasoning that is undertaken to obtain such knowledge. Nothing "is more futile than devoting our energies to those superficial proofs which are discovered more through chance than method." As is by now familiar to the reader, "the outcome of this is that, in a way, we get out of the habit of using our reason" (AT X:375; CSM I:18). With such superficial proofs, we do not use our intuitive grasp of objects that we can know clearly and distinctly and whose properties we can explicate systematically. When we use unreflexively the helpful procedures of art, we get out of the habit of turning to our knowledge. Users of such techniques prevent themselves from cultivating the intellectual habits necessary for autonomous, innovative mathematical production and for recognizing the true and the false more generally. Their mathematics is merely an art, not an activity grounded in honed intellectual habits, capable of methodical discovery of higher mathematical truths.

When the arithmetic of the calculator is made slightly more abstract by substituting variables for numbers, calculational practice can become a more suitable art for acquiring mathematical knowledge. Arithmeticians "usually represent individual magnitudes by means of several units or by some number, whereas in this context we are abstracting just as much from numbers as we did from geometrical figures." Doing so avoids "long and superfluous calculations" and allows one "to see that those parts of the problem which are the essential ones always remain distinct and are not obscured by useless numbers" (AT X:455–56; cf. CSM I:455).[62] Descartes gave a simple example. If one wants to find the hypotenuse to a triangle with sides 9 and 12, the arithmeticians will say the answer is $\sqrt{225}$, that is, 15. "We, on the other hand," use a and b for 9 and 12 and maintain the form $\sqrt{a^2 + b^2}$, "leaving the two parts of the expression distinct" (AT X:456; cf. CSM I:67–68). Descartes generalized the point: "We who seek to develop an evident and distinct knowledge of these things insist on these distinctions. Arithmeticians, on the other hand, are satisfied if the result turns up, even if they do not see *how* it depends on what has been given; yet that, quite simply, is what knowledge strictly speaking consists in" (AT X:458; cf. CSM I:69; my italics). This freeing of calculation from the attachment to specific numbers distinguished conceptual and social categories simultaneously. "We" who seek a better way steer clear of the tedious, obfuscating calculations of the arithmetician—the formal procedures of mathematical artisans. Their formal procedures focus their attention on narrowly directed mechanical processes: calculation and proof. Such techniques fail to distinguish the elements at play and thus prevent discovery. They fail to produce the evidence that makes mathematics good practice for thinking more generally. However amusing, ingenious, or useful their techniques are, they are not systematic means toward *scientia*, knowledge not only certain but fully comprehended and always reproducible—that is, evident. They are not means to gain the habits of recognizing, producing, and acting from such genuine knowledge.

In contrasting different ways of performing calculations, Descartes distinguished a mathematical practice ruled by technical procedures from one grounded in some form of higher knowledge that drew upon such technical procedures; he distinguished mathematics as a craft from mathematics as an activity with other ends. Throughout the early-modern period, such distinctions were central in the defense and creation of forms of artistic activity said to be liberal rather than mechanical or artisanal. Artisans, the justification went, certainly had manual habitual skills, but they lacked the real judgment necessary for producing works based on their intellect, good judgment, and taste. Numerous near contemporaries of Descartes defended mathematics and mechanics as liberal activities, on a par with philosophy.[63] Others derided them as mere arts. In mid-seventeenth-century France, a lively controversy erupted over the place of strict geometrical perspective in pictorial representation. The engraver Abraham Bosse, a disciple of Girard Desargues, was denounced as a mere artisan able only to follow geometrical procedures mechanically in producing his works. One polemicist attacked the pretension of "perspectival drawing in regarding the mechanics of the art rather than its spiritual content [*le spirituel*], and for being . . . only the instrument of the knowledge." Those who "apply their mind to" such mechanical technique "work rather as craftsmen [*gens métier*] than as those who study."[64] So too with Descartes' calculators: as mere craftsmen, they applied their minds solely to their technique rather than drawing upon that technique to expand and improve more fundamental knowledge—its spiritual content. Without such a goal, they could never move from the mechanical procedures of arithmetic to the genuine understanding that algebra could provide.

In distrusting formal procedures, Descartes seems to have left little room for the introduction of algebraic symbolism that so marks his geometry. Stephen Gaukroger has focused attention on an apparent contradiction in Descartes' thought between his conception of inference and the latent significance of his algebraic practice. Deduction, or inference, for Descartes, comprises a series of clear and distinct intuitions, first of elementary propositions (or objects), but then of the necessary interconnection between them.

The validity of inference is defined as the gaining of the clear and distinct intuition of the interconnection of the two propositions, and not, crucially, as the validity of a formal procedure. Descartes' algebra, his generalized science of quantity, offered, however, a potent paradigm for formal and symbolic reasoning, open to a real form of automation.[65] As we saw, Descartes rejected the symbolic and the formal as memory arts appropriate to imitation, not true self-cultivation. In line with a long tradition of understanding the power of repeated acts to produce internal dispositions, he permitted the introduction of certain arts to help develop the capacities of the mathematician and to help aid discovery and problem solving. His reintroduction of symbolic reasoning, both in practice and in theory, however, remains consistent, if in tension, with his critique.

At the center of Descartes' geometry is a simultaneous social and epistemic critique of contemporary mathematics and mathematicians. By no means do I claim that all of the features and tensions of his geometrical program can be so explained; indeed, geometry enticed Descartes precisely because of its rigidity and certainty once its boundaries had been defined. Understanding that the varieties of mathematics excluded by Descartes were those he viewed as antithetical to true cultivation, however, helps to show how numerous problematic aspects of Descartes' geometry are central characteristic features.

ALLOWABLE AND KNOWABLE CURVES

In his early notebook and in the eighth rule of his *Rules*, Descartes stressed that one must gain an adequate picture of human capacity—the real abilities and real failings—to avoid attempting to know what human beings cannot know. Only thereby could one gauge and temper the tools necessary to fulfill our potential (see AT X:215; cf. AT X:393–98). Descartes' account of curves, the heart of *Geometry*, offered a carefully selected middle path set methodologically and ontologically between abstraction and calculation: Descartes expanded the domain of knowable objects but included only those objects that he judged to be within the compass of human ability. The ancients had too low an estimation of human ability; they excluded too much. Descartes argued that they prohibited the use of curves other than circles and lines in constructions only because they, by chance, had happened upon a few bad examples and abandoned the whole lot.[66] In not methodically considering other candidates for mathematical objects, they failed to distinguish among the various curves and condemned all of them as "mechanical" and thus unknowable, rather than seeing that many were perfectly knowable and in fact necessary for systematic knowledge. Descartes argued for expanding the domain of the perfectly knowable while at the same time restricting this expansion severely. His exclusions and inclusions have long worried sensitive commentators, from Isaac Newton to Henk J. M. Bos.

The simple machines of compass and straightedge produce circles and lines. Descartes devised a compass that produces legitimate curves, described above. Though this compass can produce many different motions or curves, each is regulated by the first motion. Any of its curves can be known and understood by reference to any other. Along with producing a circle, this compass generates a number of additional curves that the ancients considered merely mechanical (AT VI:391–92): "in considering Geometry as a science that teaches generally how to know the dimensions of all bodies, one must no more exclude the most composed lines than the most simple ones, provided that one can imagine them described by a continuous movement or a number of movements which follow upon one another and of which the last are *entirely regulated* by those that precede them: for, by this means, one can always have an exact knowledge of their dimensions."[67] Mechanisms of interdependent motions were intimately connected to the constraints of Descartes' new account of mathematical deduction. Descartes apparently saw the connected chain of interdependent motions as more or less equivalent to a chain of clear and distinct mathematical reasons, each yielding something as clear and distinct as the previous.[68] Tracing a curve—the ability to create or imagine a machine of interdependent regulated motions tracing a given curve—and conceiving that curve came together. The material compass and its products offered the best exemplar of his deeply problematic notion of clarity and distinctness and his idiosyncratic view of deduction.

In contrast to the curves generated by machines like Descartes' compass, other curves cannot be so perceived or produced by interdependent, regulated, continuous motion. His prime example of an unknowable curve is the quadratrix, a curve useful for solving any number of problems, especially, for finding the quadrature of the circle (see fig. 1.4).[69] The quadratrix is "conceived as described by two separate movements, between which there is no relation that can be measured exactly" (AT VI:390). Quite characteristically, Descartes alternated between different standards for the acceptability of curves, from mechanical to proportional criteria. The key relationship between the two movements making up the quadratrix relates a segment of a circle to a segment of a straight line. Descartes retained the traditional notion that "the proportion between straight lines and curves is not known." "[N]or," he added, "do I believe that it can be known by human beings." As this proportion is not "able to be known," one could not conclude anything from considerations involving it (AT VI:412).[70] While we can imagine a machine producing the curve, one of the key motions of that machine cannot be known exactly in reference to the other. One can only

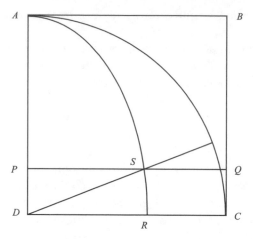

Figure 1.4. Quadratrix. The quadratrix is the locus *AR* formed from the intersections *S* of the radius from *D* uniformly rotating along the circular arc from *A* to *C* and the line *PQ* uniformly moving from *AB* to *DC*.

know the proportion *partially* through long arithmetical approximations. In contrast, "geometric" curves, "that is, those that admit of precise and exact measurement, have necessarily a certain relationship to all points of a straight line" (AT VI:392).[71]

The quadratrix remains a "mechanical" curve by numerous standards: it cannot be constructed by an interdependent, continuously moving machine; it requires long computation that produces only approximations; and it cannot be known all at once clearly and distinctly. Descartes maintained that, while the quadratrix might be useful for a vulgar calculator, it could never help to constitute interconnected, easily reproducible mathematical knowledge, knowledge both certain *and* evident. By using something too abstract, like a very broad definition of "curve," the mathematician would become intimately mired in the delusive calculations of the mere calculator; he would no longer focus his attention on the intellect, as necessary for true mathematical development.

The distinction between mechanical and geometric curves supplies a partial answer to the question of Descartes' inclusion and exclusion of curves. If Descartes were to be more abstract *ontologically* in his geometry and allow a greater set of curves, he would simultaneously become much less abstract *methodically* by dragging himself into the domain of blind and tedious calculation. In "progressively" accepting some geometrical objects and "regressively" rejecting others, Descartes may appear to have made inchoate and ad hoc choices from the perspective of modern, so-called Cartesian, analytical geometry. Within his program of avoiding the aleatory and laborious, in favor of a systematic method of producing clear and distinct knowledge graspable all at once, his choices cohere. The cognitive exercises necessary for his new geometrical subject required constraints on his geometrical objects. The curves he rejected involve one in unknowable proportions, laborious calculations, and aleatory problem solving. These curves would preclude genuine geometrical knowledge, necessarily divert the attention, and, ultimately, as the examples of the ancients and the mere calculators testify, contribute to the lack of proper self-control and self-knowledge among Descartes' mathematical contemporaries. As we saw, Descartes hardly believed this decadence accidental: a poor choice of mathematical objects leads invariably to impoverished mathematical subjects.

BEYOND GEOMETRY: HABITUATION, NATURAL PHILOSOPHY, AND THE GOOD LIFE

In a letter of 1645, the exiled princess of Bohemia, Elisabeth, thanked Descartes for clarifying some questions concerning Seneca's book *On the Blessed Life,* which he had recommended to her. "I hope," she wrote, "that you will continue, concerning what Seneca said, or what he ought to have said, to teach me the means of fortifying the understanding in order to judge better concerning all the actions of life."[72] One of Descartes' most insightful correspondents, Elisabeth characteristically cut to the heart of his philosophy in describing it as a way of life.

Among the various exercises Descartes had set out for Elisabeth was a fairly difficult mathematical problem, known as the problem of Apollonius. Given three non-overlapping circles, find the radius of the circle intersecting each other circle at one point (see fig. 1.5).[73] In November 1643, he sent her a solution that used the "keys" of his algebra.[74] He recommended introducing more than one unknown quantity so that the problem involves only right triangles. By introducing these quantities, "I see more clearly all that I am doing, and by thereby distinguishing [*demeslant*] them, I find the shortest paths better and exempt myself from superfluous calculations."[75] While one might happen upon a solution "by chance," adding such quantities helps one to solve the problem more systematically by clarifying the

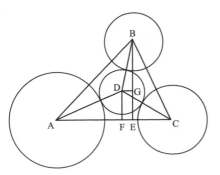

Figure 1.5. The problem of Apollonius. After AT IV:40.

relationships among the objects. Adding two quantities allowed him to produce right triangles, from which he quickly produced three equations in three unknowns. Once he reduced his two added quantities to the original quantity he sought, he declared the problem solved.[76] All that remained was to extract the roots, which he left aside as useless for the higher purposes of mathematical study: "For the rest serves not at all for cultivating or diverting the mind, but only for exercising the patience of some laboring calculator." Finding the roots themselves meant involving oneself in the mind-numbing processes of the *logistae*, the mere calculator.[77] Such labor was inappropriate for princesses and *honnêtes hommes*. In his solution for Elisabeth, Descartes offered an exercise appropriate more for clarifying and exercising the mind than for simply solving a problem by manipulating equations.[78]

In focusing on the importance of geometrical exercise for Descartes, I have not yet described the good life such exercise was supposed to help make possible. Descartes never claimed that geometry is necessary or sufficient for producing self-mastery, happiness, and the good life. Geometry is not necessary, because other forms of cognitive exercise can train the mind to recognize the clear and distinct, even if mathematics offers the best exercise. It is not sufficient, because simply having the ability to recognize the true and the false is not tantamount to pursuing the good life.[79] To complete the picture of the habituation provided by geometry or similar exercises, in this section I sketch the good life Descartes argued was possible once one had developed those habits. Rather than describing his ethical doctrines and practices in detail, which others have done well, I illustrate the role of cognitive exercises and the habits they engender within his conception of the good life.[80] In Descartes' later works, above all *Passions of the Soul*, printed in 1649, and his letters to Elisabeth, printed a few years after his death, he articulated the ethical implications of his philosophical practice, which previously he had only outlined. Even as the emphases in his writings shifted, the importance of exercises for cultivating oneself remained constant.

The 1647 French preface to *Principles of Philosophy* clarifies the important and limited roles of geometry and other exercises. Descartes there set out an order for "instructing oneself." First, one must adopt a provisional morality. Second, one must study a true logic—not "that of the schools," for the schools offer a logic that corrupts rather than improves the "good sense." A true logic, in contrast, "teaches us how best to direct our reason in order to discover the truths of which we are ignorant." Much as he had stressed as a younger man in his *Rules*, Descartes emphasized that mathematics is of paramount utility as an example of this true logic: because learning to direct one's reason "is very dependent on exercise [*usage*], it is good to practice the rules for a long time on easy and simple questions such as those of mathematics."[81] Once this habit of reasoning is developed through such habituating exercises, one should "apply oneself to the true philosophy," namely metaphysics, physics, and biology. Descartes called these the "highest and most perfect morality, which, in presupposing a complete knowledge of other sciences, is the highest degree of wisdom."[82] Such knowledge provides the tools essential for discerning the goals to pursue in life and the means for pursuing them.

Mathematics serves, then, as an exemplary exercise in cognitive habituation, a good prerequisite for more advanced stages of the philosophical life. While mathematics is an optional practice in pursuing wisdom, natural philosophy—the study of the physical and biological world—is an essential, more advanced stage. Only with natural philosophy can one come to know the causes producing the phenomena of the world. Descartes explained to Elisabeth that to "know exactly" how much something will make us more content, we must "consider the causes that produce it." Such causal knowledge is "one of the primary knowledges that can serve to facilitate the exercise [*usage*] of virtue." Only with such knowledge of the causes underlying things can we measure their value for increasing human perfection: "every pleasure ought to be measured by the extent of perfection that it produces, and it is thus that we measure those whose causes are clearly known to us."[83] True wisdom depends fundamentally on knowledge of the causal processes of the world; only with this knowledge can we select the best among the goods we might pursue. We must move from impressions that we have of their goodness based on their external appearances to the causal knowledge of how they are potentially or actually good for us.

The world is filled with things of potential benefit to human beings; pursuing a philosophical life involves evaluating such things with natural philosophy. The passion of wonder, Descartes argued, encourages the pursuit of new things and phenomena and thus can aid in finding things to help perfect us. Too often, however, we move from a proper form of wonder to mere astonishment. When astonished, someone fixates on admiring the appearance of something new rather than investigating the true worth of that thing: "one only perceives the first face that an object offers."[84] To live a life regulated by astonishment is to be ruled by these external appearances, not by one's own judgments about the value of the phenomena. Regulating oneself requires exercise in considering the nature and worth of all objects by investigating their mechanical constitution: "there is no other remedy for preventing oneself from admiring with excess than that of acquiring the knowledge of many things and of exercising oneself by considering all things that can seem the rarest and strangest."[85] Natural philosophy must become a habitual practice organizing everyday relations to the phenomenal world.

True wisdom for Descartes was not simply knowing the good or possessing a set of external techniques useful for improving human life. Such knowledge must be supplemented by habituating oneself always to turn to that knowledge in making all our choices. A true moral life rests on at least two habits: first, using the true logic to distinguish the true from the false, and the good from the bad; second, always turning to this refined judgment in making our choices. Descartes described the latter as "the habit that makes one think of and acquiesce to this knowledge any time an occasion requires it."[86] Only through "a long and frequent meditation" is the practice of judging "imprinted into our minds" and turned into a "habit."[87] The schools, Descartes noted, often had the right moral ideas, and even the proper insistence on habits. Yet they lacked the means to produce proper moral and intellectual habits.[88] The true logic, exemplified by his geometry, offers a paradigmatic practice for training the intellect infallibly to discern the true from the false, the good from the bad.[89] Training in the true logic is essential for evaluating everything one encounters in life.

Attaining the necessary habitual mastery over our lives requires exercise in knowing well and willing well: "From good use of the free will comes the greatest and most solid contentment of life."[90] The good use of the will and the intellect does not prescribe the ascetic existence of a rational mind detached from happiness and pleasure, either intellectual or physical. A properly philosophical life would be an enjoyable and happy one—a life, he claimed, of "contentment," that would involve neither the rejection of the body and its emotions nor the patient suffering of its ills until the peace of death.[91] By always using the trained intellect to choose actions, one would gain a profound peace of mind: "Those who know that they have never lacked what was necessary for doing their best, as much for knowing the good as for acquiring it, feel within themselves a repose of mind and an interior satisfaction that is a pleasure without comparison, [one] that is sweeter, more durable, and more solid than all those that come from elsewhere."[92] This true contentment and higher pleasure arise from knowing that one has worked to develop the habits necessary to live well and that one has worked to combine knowledge of the good with the willing of actions exclusively in accordance with that good. Human beings can do little more. In recognizing that we cannot possibly do more than this, Descartes claimed, we will have no legitimate intellectual grounds for being sorry. This knowledge will eliminate feelings of regret. In knowing that we have made our choices with the best possible resources available to us, we will experience a profound satisfaction qualitatively superior to any other.[93]

Preventing the passions from playing an improper role in decision making was central to Descartes' therapeutics. In contrast with Stoic accounts of *apathia*, however, the philosophical life entails no abandonment of the passions, just a proper relationship to them. For Descartes, embodied beings can never lose their passions, nor should they. They can alter some of the passions that arise whenever a given set of phenomena or thoughts presents itself; they can fine-tune their experience of them; they can reduce suffering from the negative passions.[94] All human experience, including our experience of thinking, involves them.[95] Before and after we pursue the philosophical life, we love those things we perceive to be good. In the philosophical life, we will love what we know to be the good, not something mistaken for the good. Such love never fades for embodied human beings—it pervades our experience of those things we believe to be good.[96]

For Descartes, the intellect can assess only a limited number of ideas as clear and distinct; the will, in contrast, has an infinite scope.[97] Making oneself philosophical and happy requires habitually willing only those things taken from the limited number of objects that the intellect can in fact assess and from those things within human compass to change. Making oneself philosophical and happy requires coming to recognize and to accept the limitations of human beings in knowing and in acting on the world. Pursuing the good life means habitually estimating human intellectual, volitional, and physical ability properly.[98] More than just a fundamental goal of philosophy, such estimation is a foremost virtue.

Descartes called the virtue of properly estimating oneself "generosity," the "key of all the other" virtues.[99] This virtue comprises, first, recognizing that nothing belongs properly to oneself other than the "free disposition of one's willing" and, second, feeling in oneself "a firm and constant resolution" to use the will only to pursue those goals judged to be the best.[100] Only careful meditation can turn these beliefs into habits—into virtues in the traditional Aristotelian sense.[101]

At first glance, this meaning of generosity appears to overlap little with our ordinary sense of the term. Descartes' usage seems distant from its premodern grounding in notions of the innate nobility of superior social orders. As Descartes described it, the habitual proper estimation of oneself provides the springboard for treating others well throughout life. All those who have understood human abilities and limits will do great things insofar, and only, as possible: "Those who are generous in this fashion are naturally led to do great things and never to attempt anything of which they feel they are incapable." Not specifying what these great things might be, he continued: "And since they hold nothing higher than to do good to other men and to distrust their own interest, they are always perfectly courteous, affable, and dutiful toward each person."[102] The dismissal of overzealous and mistaken self-interest admittedly yields here something more like a friendly sociability than grand acts of charity.[103] Redefining the noble virtue of generosity to mean treating all others civilly, regardless of their social station, was no small point in seventeenth-century Europe. Making oneself happy and content through a philosophical life should include making one an appropriate member of society; with the intellectual virtues come social ones, of an admittedly limited character.[104]

Just as we should use natural philosophy to measure the value of natural things for increasing human perfection, we should use philosophy to compare the value of narrow self-interest with that of social life. Descartes offered Elisabeth an important exercise of the imagination to help her overcome a mistaken conception of self-interest, and thus to fortify her understanding of the social virtues. He urged her to imagine our inability to live without others. Although individuals indeed have distinct interests, "one must always think that one does not know how to subsist alone, and that one is, in fact, one of the parts of the universe, and more particularly still one of the parts of this earth, one of the parts of the State, of the Society, and the family to which one is connected by one's home, by one's confession, by one's birth." These imaginative considerations lead both to the social virtues and to much more heroic actions in times of need: "in considering oneself as part of the public, one takes pleasure in doing good for everyone, and one even does not fear putting one's life in danger for the service of another when the occasion comes about; or even, one would want to lose one's soul, if it could be done, to save others. . . . [T]his consideration is the source and origin of all the most heroic actions that men do."[105]

Elisabeth confessed to Descartes that she had difficulty accepting his therapeutic claims—for example, that life affords more good than bad. Descartes admitted to her that the therapeutic power of belief in such a claim could come only after one ceased to care about the things that are outside our free will and our power to change, "in comparison to those [that are] dependent on us, which we can always make good, when we know how to use them well." To reach the state of self-development where external evils no longer make us suffer sadness in our soul, he admitted, is difficult indeed: "I admit that one must be a strong philosopher to reach this point."[106] With the fortified understanding necessary for the good life, a strong philosopher will have both knowledge and the habit of turning to and producing the knowledge necessary for life.

Descartes offered Elisabeth some remarkable imaginative exercises for fortifying the understanding. As one might expect from his *Meditations*, he told her to focus first on the existence of God and then on the nature of soul. She was then to imagine other things. In urging Elisabeth to consider nature, he insisted that it is helpful "to judge the works of God as they deserve" by accepting "this vast idea of the extent of the universe, which I tried to make one conceive in the third book of my *Principles.*"[107] In his *Principles of Philosophy,* at the beginning of book 3, Descartes considered some of the constraints necessary for framing good mechanical explanations of natural phenomena, given our inability to perceive the mechanisms involved directly. He insisted on several things "to [be] put everyday before the eyes." In framing our mechanical explanations, we should keep before us "that the power and goodness of God are infinite, as this makes us recognize that we must not fear failing by imagining his works to be too great, too beautiful, or too perfect." But we ought always to recall "that the capacity of our minds is very mediocre." We must not presume to know the limits of creation; and we absolutely must not presume to know God's purpose for the entire universe or any of its parts.[108] In making arguments about the mechanisms underlying things, then, we must never posit that those things have as their purpose the serving of human beings. Both to accept our limitations and to honor God properly, we must avoid anthropocentric and teleological thinking in framing these mechanisms.[109]

To imagine a finite world created for human beings offers a false consolation, one grounded in human presumption and stemming from a fundamental deception about real human abilities—a total failure of generosity in his sense. Imagining a universe made for humanity leads to a large number of scientific and moral mistakes. In believing in a creation made for us, we mistakenly take as the proper end of the good life "that this earth is our principal home and this life our best." In believing in a finite creation accessible to human knowledge, we misunderstand human limitation and human greatness: "instead of knowing the perfections that are truly in us, we attribute to other creatures imperfections they do not have, for raising ourselves above them; and entertaining an impertinent presumption, we want to be the steward of God and to take up with him the direction of the world."[110] Our cognitive mistakes about the constitution of the world lead inexorably to an improper set of beliefs about the good life, our own self-constitution, and our relationship to the divine.

By accepting God's infinite goodness and rejecting teleological reasoning, one can better frame causal mechanisms to explain, to understand, and to value the world. Descartes claimed that even the criteria used to create better mechanical explanations could console. Why did he not jump to a Pascalian expression of despair at being lost in a limitless universe whose purpose is forever opaque and unknown to us? Descartes maintained that properly imagining the greatness of creation would not diminish human beings. In a letter to another correspondent, Descartes explained something of his counterintuitive reasoning: "On the contrary, when we love

God, and when through him we join our will with all the things he has created, as we conceive them all the more great, more noble, more perfect, then we esteem ourselves all the more also, because we are part of a more accomplished whole."[111] To accept that human beings are not the end of all creation is not to diminish us but to recognize, through our epistemic humility, the potential scope of divine perfection, human perfection, and the perfection of all created things. Recognizing our limits plays a fundamental role in consoling us about our place in creation: it guides us in coming to perceive the perfections we in fact have and motivates us to strive to grasp the perfections of all other things. In the exercises Descartes offered Elisabeth, recognizing human limits and real capacities provides the fundamental springboard for the acquisition of both knowledge and the virtues necessary to live well, for attaining the true virtue of generosity.

We have apparently moved far from Descartes' geometric exercises, which are useful, but not necessary or sufficient, for living well. Two continuities between Descartes' geometry and his moral account stand out, one concerning practice, the other concerning the implications for human life. His exercises for the imagination, his natural philosophy, and his geometry all move from apparently isolated objects to underlying, interconnected mechanisms. Descartes moved from disparate geometric objects to their underlying geometrical connections. He moved from the surface phenomena of nature to their underlying unifying mechanisms. He moved from the mistaken belief in a narrow self-interest to the recognition of the real self-interest of being a member of an overlapping series of communities. His imaginative exercises, his natural philosophy, and his geometry all involve a striking insistence on discerning the real capacities of human beings. Proper mathematical activity depends on neither rejecting the curves human beings can know nor including those beyond human ken. As later commentators such as Malebranche, Pascal, and Leibniz remarked, good mathematics exemplifies the proper estimation of oneself—what Descartes called generosity. Just as in his geometry, in the exercises for fortifying the understanding, Descartes sought tools to help make one recognize the greatness and limitation of human beings, in order to provoke one to proper and humanly possible action.

MATHEMATICAL PEDAGOGY AND THE TRUE LOGIC

In his writings toward the end of his life, Descartes commented on the power of reformed mathematics for exercising the mind less often than he had in his *Rules* and other early writings. As seen from the French preface to *Principles of Philosophy*, Descartes nevertheless still opined that mathematics, while neither necessary nor sufficient for allowing one to achieve the good life, was among the best of all cognitive exercises for gaining self-mastery in making decisions throughout life. A number of influential seventeenth-century thinkers and teachers developed this central strand in his thinking. In pedagogical materials and philosophical doctrines, they focused on the centrality of mathematical exercise for helping lead to the good life. More emphatically than Descartes himself, they made a reformed mathematics central to their accounts of producing good citizens and good Christians. Their works had a long-lasting impact on the teaching of mathematics.

In his *Confessions,* Jean-Jacques Rousseau described his early mathematical education: "I did not like the taste of Euclid's [geometry], which seeks a chain of demonstrations rather than the connection of ideas." This demand on mathematics, redolent of Descartes, came from another source. "I preferred the Geometry of Father Lamy, who became, from then on, one of my favorite authors, whose works I reread with pleasure." He turned also to Lamy's *Conversations on the Sciences.* "It was a sort of introduction to the knowledge of books that consider" the sciences and religion. "I read it and reread it a hundred times. I resolved to make it my guide."[112]

In his books, the Oratorian Bernard Lamy reaffirmed Descartes' account of the power of a reformed geometry to help exercise the intellect; Lamy attempted to put this account into practice. In his *Conversations on the Sciences,* first published in 1675 and revised several times, Lamy considered the utility of various realms of learning for producing good Christians. Some forms of learning he defended, insofar as they offered intellectual, moral, and spiritual advance; others he castigated. Lamy sought out the disciplines capable of improving the mind. The rules of logic, even good ones, do not suffice to make "a mind sound [*juste*]." More important than knowing the set of logical rules was habituating oneself to logical thinking. "There is a difference indeed between knowing and doing. This soundness is acquired by exercise and practice." Which practice? "No study is more appropriate for these exercises than geometry and the other parts of mathematics," for "geometry offers models of clarity and order." Exercise with such mathematical models can perform the tough task of habituation well. Without "giving the rules of reasoning, which belong to logic," mathematics "imperceptibly accustoms the mind to reasoning well."[113] He compared the training of the mind to the training of a merchant: "Just as merchants are able to judge the quality of fabrics because they often see excellent versions of them, so he who applies himself often to consider clear truths knows best how to discern the truth from the false. If beyond this, he has exercised himself to draw from these truths all the consequences that one can deduce, he acquires by this exercise a penetrating mind, which allows him to easily work his way into things, so that he knows all that is hidden within them ."[114] As Descartes had suggested, mathematics offers the best exercise, first, in learning to discern the true and, second, in grasping the hidden connections among distinct phenomena.

The geometry text that so delighted Rousseau was designed to fulfill this pedagogical imperative of habituating the mind to thinking well. In his *Elements of Geometry* (1685), Lamy explicitly aimed to reveal connections among ideas rather than merely proffering chains of formally true demonstrations. "My principal purpose," Lamy claimed, was to make it so that his textbook "could serve as a logic; that is to say, that one can learn there the manner of conducting oneself in the quest for truth."[115] Historians of elementary mathematics have noted the wide influence and importance of Lamy's geometry, which saw frequent reprintings into the eighteenth century.[116]

For all its influence and importance in the history of pedagogy, Lamy's work was largely derivative in its goals, if not its execution. He drew heavily not just on Descartes but on his fellow Oratorian Nicolas Malebranche and on the Port-Royal thinkers Pierre Nicole and Antoine Arnauld. They all developed Descartes' account of the power of mathematics to train the intellect and sought to put it into institutionalized practice. Like him, they largely scoffed at the formal rules of logic and doubted the value of erudition without a cultivated intellect capable of passing judgments on it. All of them stressed the power of mathematics to train the mind to distinguish between the true and the false and to gain a habit of judgment that would be useful throughout life. Although they did not produce groundbreaking mathematics, they were extremely important for *pedagogical* change. As the example of Rousseau illustrates, Descartes' exercises long remained important in philosophy and in mathematical pedagogy.

In writing both an account of the power of mathematics to exercise the mind and a textbook of mathematics, Lamy followed Arnauld and Nicole's attempt to produce a new logic and geometry along Cartesian lines. In the famous *Logic, or The Art of Thinking* (1662), Nicole insisted that the sciences are "all useless if one considers them in themselves and for themselves." By drawing "on the sciences as an instrument for perfecting one's reason," one can use reason in all the decisions of life (J10, 9). In the preface to Arnauld's *New Elements of Geometry*, Nicole explained that young people easily fall into a love of sensible things. "Only grace and the exercises of piety ... can cure it truly; but among the human exercises that can serve most to diminish it, and to dispose the mind to receive the Christian truths with less opposition and disgust, it seems that there are hardly any more appropriate than the study of geometry."[117] Nicole cast geometry as a central exercise helpful for spiritual (meaning here Christian) elevation.

Geometry trains the mind to recognize the truly evident—to distinguish the true from the false. It is "very useful and very important for accustoming oneself to love the true, to have a taste for it, and to feel its beauty."[118] Profoundly interested in aesthetics, Nicole seized on its idiom to describe the power of geometry to perfect an appreciation for truth. Having developed this ability, a student will be prepared to struggle with the difficult questions of science, life, and faith. While "important truths for the conduct of life and for salvation" are often hard to comprehend, "the study of Geometry is indeed a remedy," for "in applying the mind to abstract and difficult truths, it makes easy all those things requiring less application."[119]

Unfortunately, however perfect geometry was for forming minds, Euclid's geometry was "so confused and mixed up" that, "far from being able to give the mind the idea and taste for the true order, it could only accustom it to disorder and confusion."[120] Hence, there was a need for Arnauld's new geometry, which was designed to instill order and lacked formal procedures and objects that confused rather than clarified. The new textbooks of Lamy and Arnauld were both explicitly supposed to make evident the interconnection of ideas rather than simply provide a series of formally true propositions.

Lamy's fellow Oratorian Malebranche, for his part, focused particularly on the limits of human knowledge that geometry made clear and the remedies it offered.[121] Aristotle and the scholastics never appreciated the limits of human knowledge, and, therefore, "they neglected to seek the means for augmenting the capacity of the mind." They failed signally in properly estimating themselves. In applying the intellect to matters in which it cannot judge, they misjudged its power and mixed the knowable with the unknowable. As a result, they could discover only "numerous plausible things," not "truth with evidence."[122]

Everyone, Malebranche averred, avows the piety that the human mind has limited powers. Most people know this limitation only "confusedly and confess it only in words alone." Their practice in their studies betrays the emptiness of their confession, "since they act as if they believe that their mind has no limits." After all, they do not really know those limits clearly. In sharp contrast, mathematicians alone "have well recognized the small extent of the mind; or at least they pursue their studies in a manner that marks that they know it perfectly." Whereas Lamy and Arnauld largely ignored or disliked much of the new mathematics of their time, Malebranche saw the new algebra and analysis in particular as remedies allowing the mind to overcome its inability to apply itself to multiple figures simultaneously. In contrast, ordinary logic is "more appropriate for diminishing the mind than improving it," as it divides the capacity of the mind. Such a mind "has less [capacity] for being attentive and for understanding all the extent of the subject it examines." The true logic, in evidence in mathematical practice, offers techniques allowing the attention to remain focused.[123]

Late-seventeenth-century figures saw Descartes as more than a mathematician, epistemologist, natural philosopher, and metaphysician; he was understood as a philosopher offering a way of life with a central mathematical practice. An array of influential pedagogical and philosophical thinkers seized upon his rich account of the centrality of mathematics as an exercise for the good life and developed it into a larger pedagogical program with a suitable elementary geometry. In varying degrees, they focused on three major emphases taken from Descartes. First, mathematics is a fine activity for disciplining the mind to grasp truths and to grasp multiple things all at once. Second, mathematics can make evident the limits and power of the intellect, a realization that was necessary both for doing good geometry and for living well. Third, mathematics should be transformed by eliminating objects and proof standards that are incapable of disciplining the mind.

The pedagogical reforms organized around these Cartesian notions of the power of mathematics as exercises for living well also worked their way into Jesuit teaching. Discussing one eighteenth-century Jesuit professor, L. W. B. Brockliss remarks, "Like the good Cartesian he was, he insisted that the study of geometry had the same value as the study of verbal logic. It was a practical science that trained the intellect to argue tautly and move from the simple to the obscure."[124] Both the eighteenth-century Jesuit schoolboy and the autodidact Rousseau shared in the legacy of Descartes' vision of the power of mathematics to prepare one for the good life.

*   *   *

In introducing his 1657 edition of Descartes' letters, Claude Clerselier took pains to remind readers that Descartes had sought to find the best way to live and had not simply pursued mathematics and physics as ends in themselves. Descartes' letters reveal that "he was not so greatly occupied with the consideration of things in the air, nor in the secret ways that Nature observes below," that he neglected the good life. Rather, "he often reflected upon himself" and "he took his greatest and foremost efforts to instruct himself and to regulate the actions of his life according to true reason."[125] Eschewing a chronological approach, Clerselier began his edition with Descartes' extraordinary letters from late in life to Queen Christina of Sweden and Princess Elisabeth. After the publication of his correspondence with Elisabeth, the editor asserted, "there will no longer be any of those who in their writings have accused him of vanity in his studies, as if he were attached entirely to the seeking out of vain things, and whose knowledge inflated his mind, rather than to those studies that instruct and perfect man."[126] Rightly stressing Descartes' philosophical quest for living well, this editorial work underestimated the tight connection between Descartes' recondite technical work and his philosophy for living. Natural philosophy and mathematics were no vain curiosities.

Descartes' works stimulated sundry different visions of Cartesianism. Lamy and Arnauld extended Descartes' scheme for self-cultivation through methodic geometric practice but paid little heed to Descartes' own geometric practices. The *Geometry*, in a series of ever more massively annotated Latin translations, quickly took on an autonomous life, still marked by its exclusion from most editions of *Discourse on Method* and its essays.[127] Descartes' natural philosophy became in time an institutionalized replacement for Aristotelianism. Other versions of the philosophy became fixtures of salons and polite culture.[128]

A number of historians have rightly maintained that Descartes wanted to replace traditional philosophizing with practice and knowledge useful for the *honnête homme* and *femme*, the civilized gentleman and gentlewoman. But they have understated the centrality of geometry at the level of practice. Descartes said that only the third essay following *Discourse on Method*, that is, *Geometry*, *demonstrated* his method at work.[129] Traditional mathematics and natural philosophy, with their pretty stories and turgid deductions, merely increased the number of wonders in the world; they made imitation necessary. In contrast, one who masters the three essays following *Discourse on Method—Meteors, Dioptrics,* and *Geometry*—or who masters *Principles of Philosophy* gains ascendancy over a large class of wondrous objects that previously had captured thought.[130] *Geometry* offers habituation in filling in the intermediate relations missing from a disparate picture. A grasp of the coherence of objects filling the world replaces the loss of attention and of self-control that everyday phenomena produce in the unlearned. One regains the ability of the will to consider freely the guidance of the intellect. Imagining the mechanisms of God's creation as ever more perfect only underscores the perfection of our being, even as we acknowledge our limitations. Philosophy returns to its proper role of improving our lives, ruling our manners, and guiding our steps. With philosophical exercise, human beings can bring forth their occulted ability to regulate themselves in choosing only those actions that their intellects identify as best. In their dominion over their will alone can human beings "legitimately esteem" themselves.[131] Only in properly valuing themselves can human beings set forth their own designs for living the good life.

CHAPTER TWO

A Rhetorical History of Truth

Midway through his *Treatise on Enthusiasm* of 1654, Meric Casaubon trumpeted the dangers of mystical theology, a set of practices promising divine illumination that the "heathen Philosophers" had recommended as "the highest and most perfect way" to true knowledge. To recommend this path "to ordinary people, and to women especially, is to perswade them to madnesse; and to expose them to the illusions of the devil." To teach that anyone might gain unmediated access to divine truths was to invite insanity and to perpetuate religious and political chaos. Conspiratorial powers were spreading this anarchy: "The use of this Theologie, doth most properly belong unto *Jesuits*," who instruct people and monarchs alike in "this mysticall art."

Casaubon turned next to a dangerous new fad: "Neither can I have any better opinion (in point of *Sciences*) of that Method, which of late years hath been proposed by some, and by many . . . gladly entertained." In the guise of directing others, René Descartes had tried to enhance his reputation and to instill his beliefs as divine oracles. He had opened the floodgates to dangerous claims to knowledge: "I believe he saw much in the Mathematicks: . . . though I would not have any man rely on his demonstrations, concerning either the being of a God, or the Immortalitie of the Soul. But his abilities I question not: his *Method*, having so much affinity with this *Mysticall Theologie*, against which I think too much cannot be said, I could not passe it without some censure." While Casaubon readily recognized the power of natural knowledge in aiding spiritual health, he feared the damage a single innovator such as Descartes might wreck: "I honour and admire a good Physician [natural philosopher] much more, who can (as God's instrument) by the knowledge of nature, bring a man to his right wits again: and I tremble . . . when I think that one Mad man is enough to infect an entire Province."[1] Descartes' method threatened more than academic philosophy: by empowering the untutored mob, it threatened to undermine religion and deserved the same suppression meted out in the previous century to the Spanish *Alumbrados*.

Casaubon was more correct than he knew in connecting Descartes' method and Jesuit mystical theory. I contend in this chapter that Descartes' famous epistemological standards were in part his creative adaptation of the rhetorical tradition of the Jesuits in which he was trained as a school-boy. In his first publication, a curious broadside of 1616, Descartes listed the questions involved in his thesis for his law degree. In florid Latin, he sang the praises of poetry and rhetoric and told of his long desire to find true knowledge. Poetry and rhetoric set high standards and offered lauda-tory ideals. "I thirsted for the broader rivers of eloquence most ardently. But as they make one crave more knowledge rather than quench one's thirst, they could not satisfy me in the least."[2] The "broader rivers of elo-quence" offered epistemic and moral goals that those rivers—rhetoric and poetry—proved sadly incapable of fulfilling.

Complementing the now much revitalized tradition of illustrating the connections between Descartes and the scholastics he explicitly repudi-ated, this chapter stresses another side of his education at the great Jesuit school of La Flèche: rhetoric and the exercises associated with developing the ability to speak and write well. Rhetorical exercise, scholastic philo-sophical training, and moral instruction in a Christianized Stoicism: these formed the heart of Descartes' schooling at La Flèche. The moral disci-pline the Jesuits wished to instill in the nascent elites of Europe had a Christian, scholastic, and rhetorical core; philosophy and rhetoric together could constitute the new Catholic citizen-orators Europe needed to govern itself politically and spiritually.

In *Discourse on Method* (1637), Descartes remarked: "I took pleasure above all in mathematics, because of the certainty and evidence of its rea-sons" (AT VI:7).[3] In his earlier *Rules for the Direction of the Natural Intelligence* (1620s), he had demanded those qualities of all real knowledge: "All true knowledge is a certain and evident cognition" (AT X:362). Few commen-tators have seriously investigated what demanding "evidence" implied in mid-seventeenth-century Europe.[4] In Descartes' early work, something "evident" was something known clearly, distinctly, and simultaneously. In this chapter I will build on Stephen Gaukroger's suggestion that the classical rhetorical tradition provided a crucial resource for Descartes in developing his account of clear and distinct ideas.[5] Descartes' creative de-velopment of this ancient tradition, much articulated and developed by his Jesuit teachers, anchored his radical transformation of intuition and de-duction. His geometry, as well as his natural philosophy and metaphysics, fulfilled the standards he creatively appropriated from late-Renaissance rhetoric and poetry. With his account of truth—of definition and deduction—Descartes worked to satisfy his boyhood desire for therapeutic means of curing the soul through higher knowledge.

Much important scholarship within the history of science in recent years has examined the development of the philosophical, technological, and social means for producing intersubjective knowledge of the natural (and spiritual) world. For Robert Boyle, as well as Pascal and Leibniz, the dogmatic certainty of individuals about their knowledge, judgments, and opinions was one of the key pathologies afflicting contemporaneous natural knowledge. For Boyle, securing knowledge required abandoning the pretence to certainty and creating collective means of coming to modest empirical and theoretical claims about natural objects.[6] For other natural philosophers such as Pascal, Isaac Newton, and Robert Hooke, new experimental procedures, mathematical demonstrations, and means of communication were required. Whether a product of enthusiasm or syllogisms, dogmatic certainty in knowledge threatened such attempts to create legitimate natural knowledge and a secure polity for coming to consensus about the natural world.

In this chapter I will examine a slightly earlier moment in the history of truth, the most famous subjectivist solution to problems of knowledge: Descartes' criteria of clarity and distinctness. ("Subjectivist" here means grounded in a single person, not the pejorative sense of being subject to the arbitrary discretion of an individual.) For Descartes, neither creating consent among individuals nor following standardized rules of inference and deduction could secure knowledge. Only properly cultivated individuals, people who have engaged in the set of exercises necessary to recognize the clear and distinct knowledge that was in principle available to all, could claim to possess knowledge. As shown in the previous chapter, even though Descartes' geometry exemplified such exercises, they were not limited to mathematics. Intersubjective agreement would stem from the identity of clear and distinct ideas in all human beings. Consensus could be produced simply by agreeing to include only cultivated individuals, those people who have properly honed their natural ability to recognize clear and distinct ideas through exercise. A developed epistemic personality would legitimate subjective experience and reasoning and prevent the dangerous excesses of enthusiasm and dogmatism. The problem of creating legitimate assent could properly be reduced to the philosophical and pedagogical problem of cultivating this legitimate subjective judgment. Figures as diverse as Boyle, Pascal, Leibniz, and Causabon came to view Descartes' insistence on individual judgment, and particularly on the certainty of his own judgment, as a primary pathology in the learning of their time.

This history of truth focuses on its apparent opposite: rhetoric. Scholars of classical and Renaissance rhetoric have taken great pains to separate rhetoric from sophistry, to show that rhetoric need not be antithetical to speaking truly.[7] This chapter shows one way in which rhetoric and its exercises came to provide standards and practices for defining truth itself.

POETICAL KNOWLEDGE

In a dream on 10 November 1619, Descartes saw a table on which were an encyclopedia and a collection of poems, the collection he had used as a schoolboy. His biographer glossed the images: "[Descartes] considered that the *Dictionary* meant nothing other than all the sciences amassed together, and the collection of Poems, called *Corpus poetarum*, showed separately and in a most distinct manner Philosophy and Wisdom joined together."[8] In his private notebook of 1618–19, poetry was an ideal that combined knowledge and wisdom. The knowledge of poets possesses a beauty, depth, and power surpassing that of philosophers: "It seems amazing what profound thoughts are in the writings of the poets, more so than in those of the philosophers. The reason is that poets write through enthusiasm and the strength of the imagination, for there are sparks of knowledge within us, as in a flint: where philosophers extract them through reason, poets force them out through the imagination and they shine more brightly" (AT X:217).

Descartes called for a knowledge capable of grasping the entire encyclopedia all at once: "All the sciences are linked together; one cannot have a complete science without the others following spontaneously; one grasps the entire encyclopedia together."[9] The poetic ideal combined wisdom with knowledge truly fused together, interconnected and integrated. An encyclopedia or a set of commonplaces, mere "amassed knowledge," involved simply the concatenation of disparate singular bits of knowledge with no guarantee of coherence. Rather than remaining satisfied with concatenating such elements, Descartes insisted that a higher, simultaneous cognition of the entire chain is possible. He claimed that this sort of sublime knowledge was instantiated in some mathematical examples: "The sciences are now masked; once the masks are removed, they will appear very beautiful. To him that sees completely the entire chain of knowledges, it will not seem more difficult to retain them in the soul than to retain a series of numbers" (AT X:215). Unmasking the true sciences will reveal the beauty and power inherent in their interconnection.

In the *Compendium of Music* (1618), his first completed work, Descartes related interconnected knowledge to emotional power. Assimilating a song meant moving from discrete notes to grasping its fundamental unity: "thereupon we again conjoin the two first with the latter two so that we conceive these four simultaneously as one. And thus our imagination proceeds all the way to the end, where at last it conceives the song as one thing fused out of many members" (AT X:94).[10] Fragments necessarily unveiled in time must be synthesized into something grasped all at once. Rationally appreciating a beautiful song exemplified the cognition of interconnection much more generally. As Dennis Sepper has shown, for the young Descartes, this synthetic imaginative work was essential for music, poetry, and knowledge.[11]

For the young Descartes, music had the end of "moving" the emotions (AT X:89). His treatise belongs among the Renaissance accounts of music as a rational means for controlling and directing the passions and, in some cases, leading people to divine furor.[12] Much music theory provided a rational aesthetics for directing the will of others toward enlightenment and a "well-tempered social world."[13] Interconnection was essential for higher forms of knowledge, forms capable of moving people to wisdom.

In praising interconnected poetic knowledge and the power of music and other cultural forms to stir the emotions, and in wishing for a higher knowledge, Descartes reiterated many of the emphases of modish Renaissance Platonism. Scholastics, Stoics, and rhetoricians in France continued to draw on Marsilio Ficino and his epigones well into the seventeenth century and to underline the Platonist elements within Aquinas and Augustine.[14] The sixteenth-century group of poets and rhetoricians called the Pléiade offer one example of such hopes for higher poetic knowledge. They sharply distinguished an esoteric poetic wisdom unifying all branches of knowledge from knowledge attained through dialectical means, as Frances Yates described some time ago: "The poet is a being divinely inspired who grasps the encyclopedia as one whole." In poetic knowledge, disparate bits of knowledge are reunified, so that the knower "begins to perceive a kind of coherence and unity in the whole encyclopedia of the separate arts and sciences."[15] For the Pléiade, the movement into this integrative sphere required figurative and symbolic languages. By granting access to the harmonies of the world and of human beings, poetry plays an essential role. The poet seeks unmediated access to the truths of nature but temporarily requires the arts of poetry, rhetoric, and dialectic while making his ascent.[16] One member of the Pléiade, Pontus de Tyard, argued: "Who can doubt but that the sciences are the most proper steps by which to rise to the highest summit, and that, without them, the human understanding can but hardly divest itself of its heavy clothing and raise itself with dexterity to the enterprise to which it is called?"[17]

Illumination demands hard work: the transition from mere discursive, rational knowledge to intuitive knowing is, as Yates comments, "reached not by a wild, undirected, or amoral mysticism, but after a long and severe intellectual and moral discipline."[18] Such discipline transforms its subject into someone capable of higher knowledge. As seen in chapter 1, Descartes demanded a similar discipline: an aspirant must use a variety of arts of order, such as his geometry, to attain interconnected knowledge and to gain certain experience of true knowledge. Like the members of the Pléiade, Descartes demanded *askesis* in order to become a properly disciplined knower.[19]

We need not postulate any direct influence of the Pléiade on Descartes.[20] He studied at the Jesuits' premier secondary educational institution in France and there encountered a subtle combination of the rhetorical tradition with a Platonist poetic one.[21] His rhetoric professors aligned Renaissance eloquence with an account of the furor necessary for powerful rhetoric and poetry.[22] The greatest northern Jesuit rhetorician, Nicholas Caussin, taught at La Flèche while Descartes likely attended the school. In 1619, having relocated to Paris, Caussin published *On the Parallels of Sacred and Human Eloquence*, a text on rhetoric based in part on his teaching. He stressed the power of images to move, to effect real emotional transformation in the listener: proper rhetoric could move the reader into "a realm of invisible realities."[23] In his early works, I suggest, Descartes did much with such teaching.

The evidence from Descartes' early years permits some conclusions important for understanding how he developed his mature philosophy. He had a long-standing, if vague, poetic ideal that knowledge should be interconnected and psychologically moving. Certain mathematical objects, namely, series of numbers possessing some unifying, productive principle capable of being grasped all at once, exemplified this ideal. He viewed algebra, logic, musical theory, and rhetoric as artificial techniques that could help yield this interconnected poetic knowledge.[24] In both his early and his mature writings, he evidently drew on the central Renaissance rhetorical, poetical, and musical tradition of developing techniques to move listeners. His famous criteria of clarity and distinctness were his development of Catholic Reformation rhetoric and poetics; the key subject matter that could fulfill, exemplify, and refine his poetic criteria was mathematics.

FROM POETICAL DREAMS TO EVIDENT INTUITIONS

These early yearnings for an interconnected knowledge soon gained a philosophically richer expression. In his *Rules for the Direction of the Natural Intelligence*, written at different times during the 1620s, Descartes offered an account of interconnected knowledge as instantiated by his new mathematics, in which all knowledge would be *evident*. "The evidence and certainty of intuition are required not only for apprehending single enunciations but equally for all routes" (AT X:369; cf. CSM I:14–15). All forms of knowledge, including the most complex and most indirect, those stemming from logical derivations as well as those based on sensory impressions and imaginative mechanisms, must possess the quality of evidence. This more mature form of his youthful, vague hope for interconnection within knowledge was among Descartes' most radical and productive demands.

In *Rules*, Descartes admitted only two forms of evident knowledge, *intuitus* and *deductio*. Deciphering the elusive quality that is evidence must begin by unpacking the term *intuitus*, a term difficult to translate well.[25] (*Deductio* is treated below.) In his fine account, Thomas Vinci justly describes "intuitions as nonpropositional awarenesses of simple natures (properties or attributes)."[26] In Rule 11, Descartes explained that "two things are required for intuition: first, the proposition intuited must be understood clearly and distinctly; next, it must be understood all at once, and not bit by bit" (AT X:407–8; cf. CSM I:37). Interpreters should neither over- nor underestimate the importance of the "clearly and distinctly." Given the concern with explicating the later Descartes, commentators often collapse the requirements for intuition to clarity and distinctness. Equally important in *Rules* is the simultaneity condition.[27]

Before offering these criteria concerning the things intuited, Descartes offered a description of the process of intuiting: "By intuition I understand . . . a conceptual act of a pure and attentive mind, one so easy and distinct that no doubt whatsoever remains about that which we are understanding, that is, what is the same, an indubitable conception of a pure and attentive mind" (AT X:368).[28] This account stresses a particular state of mind before and during understanding. A number of ramifications follow. First, an intuition does not necessarily involve the production of a logical proposition, a bearer of truth or falsity, or, even more loosely, a statement—Descartes used the general term "enunciation." An intuition allows one to produce a statement; the intuition might have been gained in part through understanding such an enunciation; the intuition is not however reducible to such a statement. Second, recognizing and therefore having an intuition requires attentiveness and purity, which are properties of the mind. Not some simply and easily expressed statement, a logical proposition, or such, an intuition is a simple and easy conceptual act of a mind in an attentive state. Ease and simplicity here do not primarily concern the formal description of something but rather concern *how* that thing is cognized. For Descartes, something expressed in a simple logical or algebraic form might not be simply cognizable; likewise, some things that have complicated algebraic, logical, or formal expressions might be cognizable with ease and simplicity. For example, in his account of geometric simplicity, as noted in chapter 1, he maintained that apparently more complex forms of algebraic expression reflect simpler geometric entities. Such formulas describe simpler curve-producing machines capable of being more easily cognized all at once.

The ease and simplicity of conceptual acts depend on the qualities of the things being cognized. Attentiveness and purity in thinking demand rejecting all forms of argument and any perceptions that distract the mind away from easy and simple acts of intuiting. At the ellipsis in the quotation above, he stressed that intuitions must include neither "the fluctuating testimony of the senses" nor "the deceptive judgment of an imagination that composes things badly" (AT X:368; cf. CSM I:14). With his account of intuition, Descartes appears to reject sensation and imagination completely. He did not. Descartes only rejected sensation and imagination if their products were held by minds that were not clear or attentive.[29] In discussing the imagination, Descartes made the grounds of this rejection clearer. He did not dismiss the products of the imagination entirely. He rejected the products of an imagination that put things together poorly, that is, the products of any imagination that created disjoint admixtures or concatenations lacking some unifying interconnection. Likewise, in considering sensory perceptions, he refused to accept any that were disjoint and fluctuating. At numerous points in his *Rules,* he stressed not trying to intuit complex things immediately. Perceptions of disjoint and complex things, whether produced by the imagination or the senses, require the attention to move about the various disconnected aspects of the perceived thing. Such disjoint and complex things are not perspicuously interconnected mixtures capable of being cognized with an undivided attention: "whoever tries to look at many objects together with one and the same intuition sees none of them distinctly. Likewise, whoever is inclined to attend to many things at once by means of a single act of thought possesses" a confused mind (AT X:400–401; cf. CSM I:33).

The poorly connected products of the imagination and of the fluctuating senses disrupt the attention. The attention cannot fix itself, because such products include too many distinct parts without any proper principle of interconnection that would permit cognition of the whole. As it is unable to grasp these products in an easy conceptual act, the attention must instead wander about examining various aspects of the imaginative mixture or the stream of sensations rather than focusing on any particular simple thing or discerning the unifying principle connecting and structuring the whole. To claim the status of knowledge for the perception of ramshackle or fluctuating wholes would be to accept knowledge that involves a deeply divided attention and conceptual acts that are not simple, as we saw with the curve called the quadratrix. It would be to abandon evidence.

Dividing up—analyzing—the complex offerings of the imagination or the senses, however, could very well yield a series of evident intuitions.

What prevents an evident intuition is not the source of the information but the disruption caused by taking too much all at once from that source. The mind needs first to be attentive and clear to be able to recognize those products of its sources capable of being known with clarity, distinctness, and simultaneity. Once the mind has reached this point, the source, whether the imagination, pure intellect, or the senses, no longer matters; only the clarity, distinctness, and simultaneity of the intuiting act do.

Human beings are not uncertain because they cannot have true knowledge. They simply have not disciplined themselves to have the tightly focused attention necessary to discern those things human beings can in fact perceive clearly, distinctly, and simultaneously. Such things, capable of being known intuitively, are cast among the many ramshackle things humans attempt to cognize using the intellect, imagination, and the senses. Humanity needs practice to discern them and to discover the connections among them.

All of this appears to lead to a vicious circle, where possession of clear, distinct, and simultaneous knowledge allows attentiveness, and attentiveness allows clear, distinct, and simultaneous knowledge. How can one be certain that one is cognizing something with clarity, distinctness, and simultaneity, that is, cognizing something in an easy conceptual act with an undivided attention? Descartes insisted on practical exercise to escape the impasse. Mathematics and other arts of order offer experience in cognizing those things that can be known clearly, distinctly, and simultaneously; thus, these arts offer certain experience in having the focused attention characteristic of the easy and simple cognitive acts of intuiting such objects. Descartes compared such exercise with the development of the acumen of craftsmen, who avoid the confusion and lack of attention plaguing contemporary knowledge: "Those craftsmen who are employed in minute operations and who have become accustomed attentively to focusing their vision on singular points . . . acquire, by means of practice, the capacity to distinguish perfectly between things, however fine and subtle. So also do those become perspicacious who never let their thought become distracted by various and sundry objects at the same time, but who rather always devote their undivided attention to the consideration of the simplest and easiest matters" (AT X:401; cf. CSM I:33).

Descartes could not provide any set of standardized rules for gauging whether something was an intuition. At best he could describe some of its experiential attributes, as in *Rules,* and, far more importantly, offer a series of exercises that allow one systematically to experience evident intuition, to experience what having such true knowledge is like, to feel and recognize how different evident knowledge is from all that human beings habitually mistake as knowledge. The capacity to perceive something as clear and distinct is natural to everyone, though the ability and willpower to do so varies from person to person. Given the postlapsarian and historically decadent state of humanity, this ability demands cultivation through cognitive exercises. Only with exercise could anyone become a productive and competent knowing subject.

THE RHETORICAL AND THE EVIDENT

The philosopher and historian Stephen Gaukroger has advanced an important thesis on the origins of Descartes' criteria of clarity and distinctness.[30] Rather than focusing on the philosophical candidates of Stoicism and Scholasticism, Gaukroger points to classical rhetoric, particularly the Roman rhetorician and teacher Quintilian (first century CE). Gaukroger's argument rests upon seeing Descartes' account of clear and distinct ideas, in its early formulations at least, as a form of psychological grasp of images. Rightly stressing that the predominant field for studying psychological grasp in the Renaissance was rhetorical theory, Gaukroger notes: "Just as Aristotle and Quintilian are concerned with the vividness and particularity of the images employed by the orator, dramatist, or lawyer, so Descartes is concerned with the clarity and distinctness of the mental images he refers to as 'ideas.'"[31] In strikingly parallel ways, the classical, particularly Roman, rhetoricians and Descartes grounded self-conviction in the clarity and vividness of images.

Clarity and distinctness, however, appear together only once in Descartes' *Rules*.[32] On this basis, one critic has argued against Gaukroger that "one can hardly say there is a 'doctrine of clear and distinct ideas'" in *Rules*.[33] In *Rules*, Descartes articulated less an account of clarity and distinctness than one of *evident* intuitions. This absence is deceptive, for clarity, distinctness, and simultaneity were the qualities that Descartes used to articulate evidence at different points in *Rules*. Evidence was the prior concept in Descartes' early work. Later, especially by the 1640s, in *Meditations* and *Principles of Philosophy*, clarity and distinctness became more central. Evidence never completely disappeared, even though Descartes had repudiated his early stance that clear and distinct ideas were images.

Focusing on evidence rather than simply on clarity and distinction strengthens Gaukroger's carefully argued thesis. In both Quintilian and Descartes, clarity is a characterization of evidence. *Evidentia* is Quintilian's translation of the Greek rhetorical term *enargeia*, following Cicero: "There are certain experiences, which the Greeks call *phantasias*, and the Romans *visions*, whereby things absent are presented to our imaginations with such extreme vividness that they seem actually to be before our very eyes. . . . From such impressions arises that *enargeia* which Cicero calls *illustratio* and *evidentia*, which seems not so much to narrate as to exhibit the actual scene, while our emotions will be no less actively stirred than if we were present at the actual occurrence."[34]

In using his imagination to produce a vivid picture first for himself, a speaker can reach higher levels of eloquence. Such vividness—evidence— in the imagination is the precondition for producing descriptive language with the power to reproduce the scene in the imagination of listeners. Such speech can produce images that sway the emotions of listeners more powerfully than ordinary descriptions or narrations. According to Quintilian, the greatest poets drew on such visions: "Is it not from visions such as these that Virgil was inspired to write: 'Suddenly her fingers let the shuttle fall / And all the thread was spilled,' or 'in his smooth breast the gaping wound,' or the description of the horse at the funeral of Pallas, 'his trappings laid aside'?"[35] By creating an intensely vivid image within themselves before composing, poets provoke themselves to their greatest heights of eloquence and emotional power, much like the poets provoked to wisdom in Descartes' extolment of poetic knowledge. Indeed, Quintillian noted, "the man who is really sensitive to such impressions . . . will have the greatest power over the emotions of his listener." Everyone "may acquire" this power "of vivid imagination, whereby things, words and actions are presented in the most realistic manner."[36] Quintilian argued that convincing speech could be produced only in those people able to stimulate their own imaginations and their own passions through the internal production of images.[37] Descartes similarly insisted on the practice necessary to acquire a sensitivity for clarity, distinctness, and simultaneity.

Rhetorical evidence shares far more than just a name and a focus on images with Descartes' evidence. It shares with Descartes' intuition an insistence on clarity and the simultaneity of vision necessary to produce self-conviction. In Descartes, interconnection within the intuition of something is necessary to achieve simultaneity in knowing it; in Quintilian, interconnection among profuse details produces the transcending of those narrative details necessary for self-conviction and the conviction of others. In Descartes, language is certainly useful for acquiring intuitions, but intuitions are not propositions in language. Language, including algebra, is a temporary tool for getting to intuitions. Likewise, in Quintilian, vividness can be produced by, and produces, affective and effective language, but it cannot be reduced to that language. Language is a tool for gaining visions and transforming listeners. Both Descartes and Quintilian insist on the practice necessary to produce a sensitivity for the evident. Such mere parallels of ideas rarely make convincing history, however.

Enargeia had a much more central role in Renaissance rhetoric than it had had in classical rhetoric.[38] The parallel demands of the Reformation and the Catholic Reformation encouraged a flourishing of preaching and the study of its rhetorical and psychological foundations. Protestants and Catholics alike developed Augustinian accounts of the faculties that stressed the emotive nature of faith and belief.[39] More important than instructing people about the divine was moving them to love God, his actions, his creation, and his church. Long considered dubious in the courtroom, skill in manipulating the emotions and moving the listener was essential for the pulpit.[40]

The praise of the evident by Cicero and Quintilian gained new importance and legitimacy in the rhetorical teaching of the Protestant Reformations and the Catholic Reformation. Maximizing emotion required vividly representing excellent objects by amplifying their *magnitudo* (greatness) and *praesentia* (presence). Sharpening belief and intensifying emotion demanded, not an explanation of abstract concepts, but a description of concrete details. By clearly describing concrete images, an orator could bring listeners to a state where they felt themselves in the presence of the supersensible; detailed images moved them from the sensible and quotidian to the unseen and awesome, encouraging love of God and hatred of sin and godless life.[41] "All the teachers of rhetoric advise," wrote one prominent Jesuit rhetorician, that one must first move oneself by imagining vivid images in order to move others with such images.[42] Ignatius of Loyola made such imaginative practices central to the lives of every Jesuit. Making the divine present in the Reformations mattered far more than the ancient forensic need of making mere criminal acts present.[43] Making the divine present in listeners was essential for inspiring proper beliefs, for securing the ecclesiastical order, for organizing a godly people, and for preventing heresy and idiosyncratic inspiration. While rhetoric was a major preoccupation of thinkers and preachers in all the major confessions, few people pursued this rhetorical track in theory and in practice as avidly as the Jesuits, Descartes' teachers. Descartes' writings throughout his life bear witness to his deep conviction in the importance and power of affective images, for knowledge and virtue alike.

PHILOSOPHY, RHETORIC, AND HUMANIST PEDAGOGY, JESUIT STYLE

As scholars from at least Etienne Gilson onward have reminded us, Descartes developed his philosophy in dialogue with the scholastic philosophy he encountered while in Jesuit school and afterward.[44] Often lost in such accounts is the demand that philosophy and theology be pastorally relevant in the Jesuit tradition. From the start, Jesuits were instructed to "join speculation with devotion and spiritual understanding." They were ordered to avoid both mere logical games and mere grammar.[45] They produced an important fusion of humanist ideals with Scholasticism. Even treatises on logic, such as that of Pedro Fonseca—the so-called Portuguese Aristotle—offered a surprising, and little studied, fusion of humanism and scholastic philosophy.[46] Jesuit schools put this fusion into practice. Philosophy was emphasized as necessary for the new Catholic citizen-orator that the Jesuits sought to create. The great student of Jesuit pedagogy François de Dainville characterized its overriding emphasis: "our Jesuits ... were not far from thinking, like [the great humanist Lorenzo] Valla that the greatest philosophers, the greatest orators, the greatest lawyers, even the greatest writers were 'those ones who have applied themselves to speak well.'"[47] Philosophical training, like rhetoric, was to cultivate active elites, both secular and religious.

The school that Descartes attended, La Flèche, spearheaded the renewed Jesuit educational mission in France.[48] La Flèche was founded after Henri IV permitted the Jesuits to return to France, and its curriculum was designed to create a new elite fit for governing early-modern France politically and spiritually. The Jesuits hoped to run similar schools across France, and they soon did.[49]

Quintilian offered the foremost ancient model for creating citizen-orators and was a central source for the great Jesuit order of studies, the *Ratio studiorum*, finalized in 1599. Quintilian taught the necessity for constant exercise of mind, mouth, and body, and the Jesuits duly followed his recommendations.[50] Dainville stressed that "the Jesuit method insisted finally on practice and exercise more than precepts." They had numerous rhetorical exercises for their students: make sentences with two, three, or four phrases; change poems into prose and vice versa; express the same theme, now with abundance, now with concision; write an exordium, a peroration, a panegyric; compose commonplaces; practice *pronunciatio*.[51] Jesuit pedagogy strongly emphasized the use of dramatic presentation and colloquies to produce more convincing orators. Whereas Quintilian suggested that orators try to feel their roles as if they were actors, the Jesuits achieved this in their students by having them actually act in plays. Descartes participated in these plays and practiced the oratorical exercises.[52]

The rhetorical text Descartes almost certainly used while at La Flèche, Cyprian Soarez's *On the Rhetorical Art*, exemplified the rhetorical strands just described. Soarez explained the role of the orator: "To speak is to talk elegantly, with gravity and copiousness. The job of rhetoric is to speak appropriately to persuade; its end is to persuade by the use of words."[53] Soarez praised *amplitudo* and the use of images. His ideal orator possessed the power to produce vividness: "He has a miraculous power in being able to make images of absent things fill our souls, so that we seem to see these things with our eyes and have them as if present. Whoever can well conceive such things will thereby become capable of tremendous emotional power."[54] Images could make the distant and absent things immediate and present. These images were praised for their ability to generate the greatest amount of affect, and thus were most appropriate for moving and properly controlling the passions. In citing and glossing Quintilian, Soarez stressed the need for skills in conceiving such images in order to move others with vivid descriptions.[55]

The anonymous classical text *Rhetorica ad herennium* offered further instruction about producing vividness: "It is Ocular Demonstration when an event is so described in words that the business seems to be enacted and the subject to pass vividly before our eyes. This we can effect by including what has preceded, followed, and accompanied the event itself, or by keeping steadily to its consequences or the attendant circumstances."[56] The anonymous author suggested delineating immediately antecedent causes and precedent effects, among other details. The rhetorical production of vividness, the rising up of a narrative into a moving picture, though certainly lacking necessary and sufficient conditions, demanded something like a unified profusion of attendant details and a coherent local causal picture. The rhetorics developed in the Reformations amplified the association of deduction and demonstration with plenitudes of detail, including apparent causes and effects. *Déduire* in early-modern French often meant to "describe in detail"; much the same held with the Latin *demonstratio*.[57] The Jesuit teacher Caussin, for example, argued that a demonstration "is when a thing is expressed by words so that it seems to be carried and to appear before the eyes."[58]

At Jesuit school Descartes learned about both logical and rhetorical forms of "definition," which were appropriate for different audiences and types of speaking or writing. Jesuit scholastics and rhetoricians had a nuanced account of the decorum involved in using different sorts of definitions and descriptions of things. Dry definitions of essences were appropriate for philosophy. They were not at all appropriate to poetry or rhetoric or for encouraging piety or for speaking to most people. Jesuit logical texts, from which Descartes' teachers were instructed to draw, explained that dialectical definitions were verbal descriptions of essences, such as the famous example that man is a rational animal. These texts also recognized lower forms of definitions that involve nonessential descriptions: man is an animal who talks or man is a bipedal animal without plumage, and so forth.[59] In his rhetorical textbook, Soarez explained these various sorts of definition: "For definition unrolls that which, as it were, is rolled up about what is sought. But this the orator does not do as plainly and concisely as is usually done in those very learned disputations, but rather with more explanation and more copiousness, better accommodated to common judgment and the people's intelligence."[60] Likewise, the great philosopher Pedro Fonseca, in the introduction to his dialectic, which was reprinted and used at La Flèche at this time, noted that explanations are either brief, as with philosophers, or longer and more luxurious, as with orators and poets: "Certain sketches and metaphorical explanations . . . put the thing easily before the eyes," as in Ovid and Horace such descriptions "sketch out the primary features," not the "nature of a thing."[61] Metaphorical transfer enhances lay understanding and feeling: "Orators and poets often define with a certain charm by means of a movement [*translationem*] of words from a similar thing."[62]

In Caussin's text, *On the Parallels of Sacred and Human Eloquence*, intended in part as a summary of his teaching at La Flèche during Descartes' time there, he discussed such nondialectical definition and enumeration in great detail. In a series of chapters, he considered definitions based on causes and on effects, before considering the enumeration of parts. He deemed bringing forth all the parts of something to be essential. "From this distinction of parts" arises "dignity, and elegance . . . and a flowing clarity surge forth." Such an enumeration "you may rightly call the economy of the entire speech." The orator had not only to enumerate the parts but also to make vivid their organization. Interconnection had to be illustrated. Without instituting an organized concordance (*concinnus*) among the parts, the orator produces mere darkness, much as a "confused body" is monstrous and vile.[63] The Jesuit rhetoric taught to Descartes transformed a relatively minor aspect of ancient eloquence into a most prized tool. The Jesuit dialectical manuals equally stressed the utility of enumeration.[64] A definition through copious description, a unified, organized, interconnected enumeration of parts, would make both speaker and listener imagine and marvel.

The Jesuit logicians had little doubt about the epistemic status of these kinds of definitions: definitions of essences alone constituted *scientia*—causal knowledge of essence strictly speaking—however inspiring, appealing, delightful, and appropriate metaphorical description might be. In a challenge to disciplinary hierarchy, their rhetorical and poetic colleagues pointed to the possibility of gaining knowledge higher than that of dialectical definitions: truly inspired *intellectus*, not mere *scientia*, the knowledge of divine Plato, not of Aristotle.[65] Marc Fumaroli has illustrated the growth in the theory of the power of rhetorical description as a means of higher definition: "the imaginative enthusiasm of the rhetor-poet that, through a game of metaphors [*translationes*], makes a definition explode into a copious, ornate description, speaking to the senses and to the heart."[66] Poets and orators could not necessarily arrive at the essence of the thing traditionally understood, but they could nevertheless make all the details come alive in a vivid image that perhaps was more truthful than any dialectical definition. Many Jesuit teachers combined *enargeia* with a Platonist enthusiasm to give an account of description capable of generating instantaneous, poetic grasp, severed from language, much as the Pléiade and countless other Platonists had long dreamt.[67] Rhetorical description might be able to replace traditional definition, long under attack by humanists, and produce knowledge both more certain and clearly morally improving. Descartes' early poetical dreams illustrate his hope for such unified poetic knowledge, so unlike the forced knowledge of philosophy. His formal account of intuition and mathematical practice demonstrate his belief that he had found it.

At the heart of Descartes' rhetorical education was the development of persuasive and image-producing rhetorical abilities, precisely the aspects of Quintilian's rhetoric that most closely parallel Descartes' epistemic standards. As in his poetical knowledge, rhetorical definition entailed making clear the interconnection among complex entities, to make them appear all at once to the listener. His Jesuit rhetorical preceptors taught a vision of rhetoric that challenged the disciplinary and epistemic superiority of dialectical philosophy. Descartes made good on that disciplinary challenge.

In his early discussions of clarity and evidence, Descartes stopped before attempting to define clarity; he insisted that one needed to experience certain examples of evidence, and therefore of clarity and distinctness. The exercises with geometry discussed in chapter 1 offered such certain experiences of evidence, experiences of the clear, distinct, and simultaneous. Some years later, when Descartes attempted to write a textbook to supplant scholastic philosophy and natural philosophy term by term, definition by definition, he explained "clarity." He drew on the rhetorical language he had long ago learned. Consider his definition in *Principles of Philosophy* (1644): "I call a perception clear when it is present and manifest to an attentive mind—just as when we say that we see things clearly when they are present to the eye's gaze and move it with sufficient strength and accessibility." The Latin here is telling: "Claram voco illam, quae menti attendenti praesens & aperta est: sicut ea clarè à nobis videri dicimus, quae, oculo intuenti praesentia, satis fortiter & apertè illum movent."[68] Besides the key words of his own technical vocabulary (attentive, intuition), nearly every important word in Descartes' definition was a standard term used in rhetorical theory: *movere, praesens, oculus*. To *move* someone required making something *present*, to make speech into *vision*, to put it as if before the eyes, *velut sub oculos subiecto*. The underlying heuristic of this epistemology remained the ability to imagine something vividly before bringing it to life for one's listeners, that which the orator ideally did through *enargeia*. Descartes retained such descriptions long after he had stressed that many truths are intelligible without being imaginable.[69] Many things visually unimaginable, such as a many-sided polygon, are intelligible and can be understood clearly and distinctly even though we cannot conceive them as images.

Even after he abandoned images themselves within his epistemic criteria, however, he kept the rhetorical model of affectively powerful representations at the center of his practices of self-cultivation. In the *Passions of the Soul* (1649), he explained that sometimes "we imagine certain things so vividly that we think we see them before us or feel them in our body, although they are not there at all."[70] Changing one's habitual intellectual, emotional, and physical responses to everyday experiences is not merely a question of willing change: "Our passions cannot be directly excited or removed by the action of our will, but they can be indirectly by the representation of things customarily joined with the passions we wish to have." Epistemic and moral reform demands the production of representations capable of moving us. As his sundry exercises for Elisabeth underscore, we must use our reason and imagination to arouse and persuade, to cause the proper passions we ought to feel, to develop proper habits of reacting and acting when we so will.[71]

Such moral reform through images was central to Jesuit rhetoric and Jesuit spirituality. The focus of Descartes' Jesuit teachers on imagining detailed scenes to move oneself and one's listeners was not simply lifted from Quintilian; nor was it only part of Catholic Reformation practice more generally. An ordered process of such imagining is central to Ignatius of Loyola's *Spiritual Exercises*; every Jesuit practiced these exercises regularly. Loyola directed one to imagine a scene, such as the nativity or the first sin of the angels. One imagines first the "place," then inserts the actors into the "composition," and finally ruminates on its details—its texture, their conversation. Intellectual and affective change springs from these intense meditations on scenes and their implications.[72] Jesuits after Loyola worked to clarify and standardize these procedures.[73] Other Jesuits worked to transform the exercises into something appropriate for larger reading audiences. Jesuit rhetoricians such as Caussin and Pierre Le Moyne implemented the theory of emotional and moral transformation through vivid images in a series of vernacular books intended to guide the French elite.[74]

Regardless of whether Descartes ever encountered or performed the exercises himself, in theory and in practice he offered a radical new version of the Ignatian procedure of bringing oneself into proper epistemic, emotional, and spiritual states through imagining concrete, detailed situations and pondering their implications.[75] Descartes' epistemological standards of clarity and distinctness and his project of personal reform using them belong among Catholic Reformation traditions of transforming oneself through potent imaginative exercises. All these procedures aimed at morally effective knowledge and the proper disposition of emotional states. Descartes creatively transferred these practices to natural philosophy and mathematics more generally and used them to undermine the scholastic synthesis upon which they rested.

AN APTITUDE FOR PRODUCING *ENARGEIA*

From his Jesuit training in the rhetorical tradition Descartes quite plausibly drew heuristics and the language, not just for his epistemic criteria, but also for the human faculty needing cultivation. Descartes' *Rules* was aimed to help direct the *ingenium*—the natural intelligence, often translated "wit" in seventeenth-century English and *"esprit"* in seventeenth-century French.[76] *Ingenium* was classically the aptitude or skill for any particular endeavor, as well as mental acuity more generally. The ancient rhetoricians used *ingenium* to account for the ability to produce vivid—evident—images within oneself, necessary for producing the discourse capable of reproducing these images in others. In no domain, the Jesuit Caussin argued, was *ingenium* more needed than in eloquence.[77] The great humanist and educational theorist Juan Luis Vives illustrates the late-Renaissance usage well: "They have in the first place a fruitful *ingenium* who easily and suitably conceive images of things."[78] Vives specified that the parts of *ingenium* included "keenness in intuiting" (literally, "in visually grasping") and "a capacity for bringing together, collecting for judgment."[79] Keen intuition and descriptive agility came together in *ingenium*.

In a move characteristic of humanist critics of scholastic logic, Vives translated this rhetorical facility into the heart of discovery, of *inventio*, in finding commonplaces appropriate for speaking and in discovering the principles of nature. He portrayed Aristotle and Galen as productive philosophers because they possessed what otherwise would seem a mere "rhetorical" ability. An aptness for describing and narrating allowed them to produce their entire edifices, which had subsequently and deceptively been encased in formal dialectic.[80] Such rhetorical-dialectical uses of *ingenium* were widespread by the late sixteenth century and appear prominently in Jesuit rhetoricians and logicians.[81] By the early seventeenth century, *ingenium* was strongly associated with natural power or ability to discover arguments from the rhetorical tradition.[82]

According to the Jesuit Caussin, the instructor ought to perfect the *ingenia* of his students: "it is appropriate in eloquence to bring about and diligently to observe the movement and speed of the *ingenium*." Like Vives, he stressed the insight and power of historical figures with great *ingenia*: certain natures are "aroused by the divine fury . . . and like something rubbed with a loadstone, bathed in a certain hidden power, they are thus inspired by the divine mind." Having learned Aristotle's logic and rhetoric by age twelve, Saint Augustine had such a mind. His abilities stemmed from "a power of the *phantasia* that takes up impressed images of things most easily, just as if it were soft wax, and hangs onto them steadfastly."[83] The Jesuit rhetoricians held that the highest forms of truth-telling speech stemmed ultimately from a real connection to divine truths. Able to retain images as if their minds were wax, great figures like Augustine could trivially achieve the vividness necessary to discover and to communicate such divine truths.

In his book of rhetoric, Caussin detailed the qualities of *ingenium* necessary for the orator and offered techniques for perfecting those qualities.[84] Although Quintilian claimed that *ingenium* was "beyond all art," he stressed nevertheless the necessity for practice in imagining scenes to hone this natural, but obscured, ability: "In the schools it is desirable that a student should be moved by his theme, and should imagine it to be true."[85] Descartes' preceptors, likely including Caussin, put Quintilian's admonition to practice rhetorical exercises at the heart of their pedagogy.[86]

This rhetorical usage of *ingenium* appeared in Descartes' only extensive discussion of classical rhetoric, the 1629 defense of the *Lettres* of his friend Guez de Balzac. In a historical account of the fall of rhetoric into mere sophistry, Descartes praised preclassical *ingenium:* "In the primitive uncouth ages, before there were any quarrels in the world and when speech willingly followed the inclinations of a guileless mind, there was, in fact, in persons of greater *ingenium* a force of godlike eloquence that poured forth from zeal for truth and an abundance of feeling" (AT I:9).[87] Like his Jesuit rhetoric teachers, Descartes envisioned that the rhetoricians of a primitive age possessed morally powerful truths and had a language capable of communicating them without sophistry. Much like Quintilian, Soarez, and Caussin, Descartes focused on the need for self-persuasion to produce conviction: Guez de Balzac only ever spoke of things "of which he had previously persuaded himself" (AT I:10).[88] Self-conviction energized his speech, allowing him to communicate eloquently and from an ardent desire for truth. Guez de Balzac's style signaled the beginning of a new age of sincere rhetoric, one combining elegant artifice, insight into truth, and persuasive power perfectly.

Like Quintilian and Caussin, Descartes held the *ingenium* to be essentially beyond art. To bring out its natural abilities nevertheless required exercise and cultivation, ever since the true rhetoric and true knowledge of a more primitive age had disappeared. In his work on directing the *ingenium,* Descartes advised a difficult path between the artificial and the natural. Rather than a new art of eloquent figures, his techniques in mathematics involve the deft, but temporary, use of artificial notations to create natural and mathematical knowledge of interconnected wholes. Just as rhetoric could quickly turn into ornate emptiness and sophistry, algebra could quickly devolve into a meaningless game concerned only with "naked numbers" and "imaginary figures." Such devolution into mere formalism would make it as useless and deceiving for cultivating oneself as syllogistic or sophistic rhetoric. It would never partake of the qualities of poetic knowledge.

DESCRIPTION, ENUMERATION, AND CARTESIAN DEDUCTION

In 1643 a brutal attack on Descartes and his method was published in the Dutch city of Utrecht.[89] Amid a farrago of complaints, calumnies, and serious arguments in *The Wonderful Method of René Descartes*, the author, one Martinus Schoock, argued that Cartesians gussied up rhetoric in the guise of philosophy. In a chapter entitled "The Cartesians make a simple narration of facts pass for the demonstration," Schoock explained that when Descartes and his followers attempt to explain the phenomena of the world, "they invent some principles and affirm certain hypotheses without any proof, without being able to say whether they are true or not; next, they deduce and demonstrate, in appearance, . . . certain things to the end of persuading the imprudent and ignorant that all things can be deduced and demonstrated by them in the same way the earlier things had been deduced from these false hypotheses." Anyone who reads Descartes' *Meteors* will perceive this "ruse."[90] There, after all, Descartes promised to render his suppositions about the principles of nature "so simple and easy that you will perhaps not have difficulty in believing them, even though I have not demonstrated them at all" (AT VI:233).

Schoock's next chapter, entitled "In the eyes of the Cartesians, comparisons take the place of demonstrations," denounces this substitution of rhetoric for philosophy. In his *Dioptrics,* Descartes claimed that comparisons "help in conceiving in the manner that seems to me most convenient" to explain all the properties of light that we know from experience, as well as many that we do not experience (AT VI:83). Schoock demurred: "serving above all to elucidate certain claims," comparisons "are instruments for the orator, who seeks to use words proper for persuading, rather than for the philosopher, who pursues the essence of things that he seeks to prove with solid reasons." Descartes' philosophy rests on a comparison between machines and the world, on a "dogma" that the essence of nonhuman things consists entirely in their extension. "Presented with no demonstration, this dogma is proved more or less like this: just as clothes are made of silk and a clock is built from iron, by the disposition of parts alone, thus all natural things exist through the disposition of the parts of the corporeal substance alone." Like many commentators on Descartes since, Schoock maintained that his analogy between machines and natural beings was unproven, unjustified, and philosophically problematic: these ersatz demonstrations are "more an affair of beautiful presentation than solid demonstration."[91]

Schoock's philosophical points were acute; so were his claims that Descartes and followers substituted a form of rhetorical description of external appearances for traditional philosophical demonstration based on essences. In practice, Descartes replaced traditional definitions of essences with detailed descriptions of physical and geometric processes. In a fine analysis of demonstration and deduction in Descartes, Doren Recker characterizes his deductions as "explanatory narrations."[92] The exemplary deduction in *Discourse on Method* (1637) outlines the physical disposition of the parts making up the heart. To show the philosophical legitimacy of his practice, Descartes needed an account of deduction that could encompass his twin demands of explaining all physical processes mechanically and grounding all knowledge in evident intuitions.

As we saw above, scholastics and rhetoricians recognized two major types of definition. True definition, proper to philosophers, captured the essence of a thing in language. Rhetorical definition involved describing external attributes or metaphorically evoking the thing. The first involved substantial and essential features, such as purpose and final ends, or at least genus and species; the second, accidental attributes, such as appearances.[93] Merely describing the accidents, appearances, external features, and activities of something, as poets and orators do, could never constitute genuine knowledge of something. Descartes upset this hierarchy of definition. Like other mechanical philosophers, Descartes replaced such definitions and explanations involving forms, qualities, teleological motions, and efficient causes with detailed descriptions of material structures and matter in motion—basically efficient causality alone. Definitions would hitherto legitimately include only matter in motion, including its positions and local motions. He reduced the essence of things to their accidents, as they had been traditionally understood. To communicate the essence of something is then to *describe* its physical dispositions and local motions. By Descartes' time, the use of morphological description, a form of definition much lower in the epistemic hierarchy, had already become widespread in medicine and other endeavors, such as zoology and botany.[94]

For Descartes and other mechanists, defining something involves adequately describing its attributes of extension, many of which are not immediately obvious, always observable, or perhaps ever observable. Definition in terms of accidents—a rarified variant of the definitions of poets and orators, not philosophers—ought to be the only legitimate kind of natural-philosophical definition. Just as natural philosophy had to be stripped of its unintelligible substantial forms, it needed to be stripped of the forms of argumentation and definition that defended them. An intelligible natural philosophy had to gain the certainty and evidence lacking in traditional philosophical demonstrations; a reformed philosophy required mechanical explanations and new concepts of deduction and definition.

Descartes maintained that descriptions in terms of external characteristics and dispositions have a necessity like that of a mathematical demonstration:

> so that those who do not know the force of Mathematical demonstrations and are not accustomed to distinguishing the true reasons from the seemingly true do not risk rejecting this [explanation] without examination, I want to instruct them that this movement, which I have just explained, follows as necessarily from the disposition of organs alone that one is able to see by eye in the heart, and from the heat that one can feel with the fingers, and from the nature of blood that one can know through experience, as that of a clock follows from the force, the situation, and the shape of its counterweights and wheels. (AT VI:50)

In this passage Descartes seems to conflate the apparently different necessities of mathematical demonstrations, of mechanical devices, and of explanations of natural processes in the body in terms of physical structures and matter in motion.

In his earlier *Rules*, Descartes provided an account of deduction. Recall that Descartes demanded the qualities of evidence from all knowledge: "The evidence and certainty of intuition is required not only for apprehending single enunciations but equally for all routes" (AT X:369; CSM I:14–15).[95] Since deductions clearly involve cognition over time and usually require the memory to recall earlier steps, it is difficult to see how any deduction could meet Descartes' criteria for evident intuitions. Turning the attention from the intellect to the memory, in order to recall an earlier inferential step or premise, means that one no longer grasps something in an easy and simple cognitive act of understanding.[96]

Descartes attempted to provide an account of deduction that satisfied his tough standard for evidence with something called "sufficient enumerations." Such enumerations require a movement of thought: "many things are known with certainty, although they are not in themselves evident, provided only that they are deduced from true and known principles through a continuous and in no way interrupted movement of thinking which transparently intuits each part separately" (AT X:369; cf. CSM I:15). This movement of thought is "needed to make good any weakness of memory" (AT X:387; CSM I:25).[97] The connection linking the individual intuitions in the deduction must itself become an intuition of sorts: "So I shall run through them several times in a continuous motion of the imagination, simultaneously intuiting one relation and passing on to the next, until I have learned to pass from the first to last so swiftly that no part is left to the memory, and I seem to intuit the whole thing at once" (AT X:388; cf. CSM I:25). Deduction, then, involves a set of connected intuitions, a continuous set of acts that do not break the attention and that eventually lead to a nearly simultaneous understanding of the whole, its parts, and their interconnection. A deduction comprises a set of intuitions of individual elements and an intuition of the interconnection of those individual elements.

Deduction generally means something like a connected series of legitimate logical inferences. While considering logical inferences as a form of sufficient enumeration, Descartes viewed other enumerations as legitimate deductions, such as the detailed description of the physical parts making up the heart. Any grouping of particulars might become a deduction, as long as that grouping contains exclusively elements susceptible to being evidently intuited by themselves and the grouping as a whole is susceptible to being intuited in a continuous sweep.[98] Without some underlying ordering principle, no enumeration—whether of logical propositions, observations of the heart, or a series of algebraic steps—could possibly be intuited in such a continuous sweep. Descartes used the metaphor of the chain: "we cannot distinguish all the rings of a long chain in a single glimpse of the eyes [*uno oculorum intuitu*]; nevertheless, if we have seen the connection between each link and its neighbor, this suffices for us to say that that we have observed how the first is connected to the last" (AT X:389; CSM I:26). In his Rule 10, Descartes praised the study of "weaving and carpet-making." These activities provide suitable exercise in practicing deduction and other sorts of enumerations because "they present to us in the most distinct way innumerable instances of order" (AT X:404; CSM I:35).[99] How, in general, are we to discover such a principle of interconnection among other sets of intuited things? Most enumerations fail because they lack an appropriate level of detail permitting one to grasp an ordering principle in a continuous mental sweep. To produce an evident deduction, one needs to move from an incomplete enumeration of discrete particulars to a sufficient, but not necessarily a complete, enumeration. With an insufficient level of detail, not enough elements of the enumeration are available for us to grasp what connects them all. Reaching sufficiency requires seeking out unknowns to fill the gaps between known things.

As described in chapter 1, Descartes' geometry offered valuable practice first in isolating simple elements and then in producing the intermediate elements necessary for connecting those elements. In natural philosophy, one begins with obvious structures that one sees and feels. Filling the gaps might include detailed empirical investigation of the less apparent internal structures of the heart to discern how the immediately apparent structures

are connected. Investigating a sufficient number of the less apparent structures eventually permits one to grasp sufficiently how the structures work
together to pump blood. This investigative work results in a descriptive
causal picture—an explanatory narration—of natural processes, such as
Descartes set forth in his early texts *World* and *Man,* in the essays accompanying his *Discourse on Method,* and in his *Principles of Philosophy.*

Deduction for Descartes, then, is a well-organized description that details the structures of something in their interconnected organization; such
a description permits one to grasp the entire complex in a continuous
sweep without losing sight of the distinct elements or their interconnection.
Descartes' rather bizarre account of intuitions and deductions resounds
strikingly with the most powerful forms of rhetorical definition. Definition
through enumeration, as described by Caussin, for example, involved the
well-organized description of an entity that details all its parts in their hierarchical organization; such a description put before the eyes a vision of all
the parts and the composition and interrelations of the whole. Descartes'
form of deduction presents an ordered enumeration leading one to see the
economy of the entire thing and the qualities of all the parts that are always
conceived distinctly.[100] In the rhetorical tradition, a well-ordered speech
organized disparate members into a unified whole, as Quintilian argued:

> it is not enough merely to arrange the various parts: each several part has
> its own internal economy, according to which one thought will come first,
> another second, another third, while we must struggle not merely to place
> these thoughts in the proper order, but to link them together and give them
> such cohesion that there will be no trace of any suture: they must form a
> body, not a congeries of limbs. . . . Thus different facts . . . will be united with
> what precedes and follows by an intimate bond of union [*aliqua societate*],
> with the result that our speech will give the impression not merely of having
> been put together, but of natural continuity.[101]

The best "economy" of speech will be like the human body: organized and
unified but with all the parts set out distinctly.

Many of Descartes' key works contain just such an ordered enumeration;
and he was explicit that his descriptions should lead his reader to conceive
the unified "economy" underlying a mechanical explanation of physical
processes.[102] Among many examples from Descartes' natural philosophy,
consider *Man* (c. 1632), in which he enumerated the animal parts and
functions of imagined beings that look and act like humans. He described
a series of mechanisms capable of producing all of the externally visual
effects. "Conceive," "see," "think," "remark upon," Descartes instructed
his readers in the course of making them imagine his sundry mechanisms,

to the end of making the reader "understand all this distinctly."[103] With the help of descriptions, diagrams, and analogies with machines, grottoes, and so forth, the reader is supposed to work through a series of exercises of imagining in order to grasp the mechanisms and then to understand how they produce the various effects. Descartes seems to have followed the counsel of his old rhetoric textbook: "images of absent things fill our souls, so that we seem to see these things with our eyes and have them as if present."[104] Descartes ended by restating his purpose: to show that you (reader) can conceive of the human body as a machine. The dispositions and motions of the parts suffice to explain the full range of animal activities and reactions of humanlike beings:

> I want you to contemplate, after this, that all the functions that I have attributed to this machine, such as the digestion of food, the beating of the heart and its arteries, the nourishing and growth of its members, respiration, waking and sleeping; the reception of light; . . . the interior movement of the appetites and passions, and finally the exterior movements of all the limbs, which follow so exactly the actions of objects that offer themselves to the senses, as well as the passions and the impressions found in the memory, that they imitate as perfectly as possible those of a true man. I want you to contemplate that these functions all follow naturally, in this machine, from the disposition of its organs alone, no more or no less than those of a clock, or another automaton. (AT XI:201–2)

Recapitulating the organization of the text, this enumeration of the parts, functions, and effects of the body pushes the reader to conceive the text and thus the machine as a unified whole; it offers an enumeration that reveals the economy of the whole.[105] *Man* is *enargeia* in action: the text brings the entire scene of a mechanical man before the eyes and does so in such a way as to portray the body as a unity of parts conceived distinctly. This is *enargeia*, rhetorical to be sure, in the service of comprehending and understanding a complex entity, in the service of a deduction, in the service of philosophy.

In his later works, Descartes separated his epistemic criteria of clarity and evidence from the production of visual images and yet retained fundamental features of this early account of deduction. He insisted that works must be read actively, intensively, and repetitively in order to be understood and internalized. Just as with his *Man*, the reader must go through a process of imagining in order to understand the whole and its subordinated parts distinctly. *Meditations* offers a series of striking images and recondite reasoning intended to effect a moral and epistemic transformation of the attentive reader. *Meditations* is strictly an enumeration in Caussin's sense:

truly to benefit, Descartes explained, the reader must "grasp [*intueri*] the entire body of my *Meditations* while at the same time discerning all its singular members."[106] *Meditations,* like all Cartesian deductions, needs to be understood simultaneously in a sweep and distinctly in its parts.[107] In a roughly contemporary document, he claimed that "whatever is to be taken from the writings of men of excellent *ingenium* consists, not in this or that thought that can be pulled from it, but from what rises out from the *whole body* of the work." Only regular and sustained rereading will allow us to grasp this unity and thus "so to speak we convert it into our own spirit [*succum*]."[108] Like such great works of the greatest minds, Descartes' *Meditations* exemplifies the deductive account he had long before enunciated: these works are unities whose transformative power comes only in grasping them *as* unities while keeping their parts distinct.

Demanding from deduction something like the movement from narrative to the vision of *enargeia* could provide the poetical, interconnected knowledge Descartes had long before deemed essential for wisdom. Descartes' new geometry, a mathematics of interconnected machines, showed the reality of this new form of deduction based on these descriptive ideals. It showed how explanatory narrations could produce a vision of the entire thing and make clear the relationships of all the parts, the economy of the whole. It showed that human beings were capable of such vision, and it pointed toward the expansion of interconnected, human knowledge with appropriate linguistic tools. It provided a basis for a proper set of exercises useful for producing a good moral and epistemic subject, one capable of ascending from mere dialectical procedure to true wisdom.

In the rhetorical tradition, chains of logical reasoning were seen as persuasive by virtue of the plenitude of affect and clarity of vision that they created. According to Descartes, chains of reasoning produced true knowledge by virtue of the plenitude of evident vision that they provided. Descartes turned the traditional rhetorical view of dialectic as one among many kinds of arguments designed to persuade into a new set of criteria for reasoning itself. A precondition for being able to speak and write effectively—the unified vividness of that being depicted—became the quality characterizing certain and evident knowledge.

To say mathematics is certain and evident meant, roughly, that mathematics possesses the certainty characteristic of formal logic and that it possesses intrinsically the qualities of vividness an orator usually would have to add to formal logical or dialectical proof. A kind of superdialectic, mathematics, as Descartes had delimited it in his *Geometry* and in his *Rules,* was supposed to have been in itself vivid and thus evident. Descartes' radical move was to make formal certainty an aspect of mathematics' intrinsic evidence. Mathematics was not evident because it was certain; it was certain because it was evident. Thanks to that, it offered experience of something akin to the poetic wisdom that Descartes had desired from his youth. The labor of gaining that experience opened the path to a philosophical life of regulating oneself.

A SCHOLASTIC BACKGROUND FOR EVIDENCE AND INTUITION?

Many readers may find this rhetorical byway an insufficiently parsimonious explanation. Following the enormously fruitful work on the scholastic background to his metaphysics and natural philosophy, why not simply connect Descartes' epistemic standards directly to the scholastic philosophical tradition? The scholastic notion of *"intuitiva notitia,"* which one scholar describes as "the nondiscursive immediate certainty created by sensory perception," seems a rather good candidate.[109] Scholastics after Aquinas, including Jesuit sixteenth-century scholastics, regularly speak of "evident intuitive cognitions." Since recent secondary literature has tended to stress William of Ockham's notion of evident intuitions, I focus my analysis there, although nearly all of the doctrines discussed here were commonly held by scholastics of the late-medieval and early-modern period.

Ockham defined evident knowledge as "a cognition of some complex truth from the noncomplex knowledge of terms that are either immediately or mediately sufficiently caused to arise."[110] Something evident is so causally determined that one's own intellect, given the set of cognitive acts it is undergoing, has no choice but to assent to it. Anyone else undergoing such a set of cognitive acts or experiences would equally have no choice but to consent to the same thing. Evident intuitive knowledge, in turn, comprises primarily knowledge of contingent facts about the world, at least in this life. The authors of a recent study note that, for Ockham, "Intuitive knowledge is a knowledge of what is present to us as actually present."[111] In an evident intuition one assents to a proposition about a contingent situation being perceived. Confronted with a blooming tree, for example, one would necessarily assent to the proposition "the tree is blooming" upon observing the tree blooming.

The central characteristic of the scholastic account of evident intuitions is human *in*ability to dissent from evident propositions. For scholastics, the key role of the notion of evidence was to distinguish between two kinds of certain knowledge: certain philosophical knowledge (*scientia*) and faith. Evidence captured the difference between the certain knowledge of natural truths (including naturally knowable theological truths) and the certain knowledge of revealed theological truths.[112] Faith involved certainty without evidence. Faith was meritorious because it was not causally determined, unlike the evident knowledge of the natural world. Merit came, for the scholastics, in accepting the certainty of a privileged set of nonevident beliefs, not in the necessary, causally impelled assent to evident intuitions.

In his account of intuition, Ockham stressed the essential obscurity and confusion endemic to postlapsarian humanity (*in isto statu*): "Perhaps when something is known intuitively clearly and perfectly, it can be discerned from another specifically distinct thing known intuitively clearly and perfectly. . . . But our intellect knows nothing clearly, distinctly, and perfectly in this life, and hence it cannot discern that thing from everything else."[113] An intuition might be able to provide only a small number or perhaps no contingent truths about that thing in itself; the intellect nonetheless assents to propositions about the unclear and indistinct things being perceived. Moving from obscure and confused intuitions toward clear and distinct ones may not be possible for human beings in this world.

Why is this account of evidence not that of Descartes? First, and most importantly, an obscure and confused awareness never constituted a genuine intuition for Descartes. A primary tenet of his method is never to judge without first having a clear and distinct perception of the matter at issue.[114] For the scholastics, as just shown, evident intuitions often were obscure and confused; evidence, for them, is not a question of clarity or distinctness but of the presence of the object intuited. For example, human beings have no evident intuitive knowledge of the divine in this life, not because our knowledge of the divine is obscure and confused, but because the divine is not immediately present. Even if one comes to have distinct knowledge of the divine, as scholastics such as Scotus maintained at least some human beings could do, that knowledge is never an evident intuitive cognition but only an abstractive cognition gained through discursive reasoning.[115] Only in heaven might human beings come to have evident intuitions of the divine, because they would be in the presence of the divine.

Second, intuitions were not causally impelled for Descartes. Judging the truth or falsity of something should always involve a free act of the will, a choice of affirming or denying. Under different terminology, Descartes consistently held that the will freely acts to assent to the intellect, all the while maintaining that the will infallibly assents to clear ideas held by an attentive intellect.[116] Grasping this ability to withhold assent was essential to understanding the greatness of humanity. This free will, the "empire that we have over our willing," offered the only just reason to "esteem ourselves."[117] The freedom to dissent to even the most evident idea captured human proximity to God. Although human beings lack God's amplitude of knowledge, they possess his amplitude of willing. In both *Meditations* and *Passions of the Soul,* recognizing true human nature meant grasping the amplitude of the will and its fundamental independence from the intellect. For Descartes, knowing human nature meant rejecting the scholastic account of causally impelled intuitions.

Third, one did not have to be a knowing subject whose faculties have been honed through cognitive exercises in order to have intuitions, as the scholastics understood them. In his mature texts of the 1640s, Descartes amplified his early insistence on the exercises necessary for learning to recognize clarity and evidence, which was so important in *Rules*, written in the 1620s. In the responses following the *Meditations*, Descartes stressed how the clarity with which we grasp a good or a position depends on our readiness to cognize that clarity. Only a long series of acts of the will allows one to be certain about having clear and distinct cognition. He emphasized the practice necessary to reach the state where one can consider things clearly and distinctly. One must first will oneself not to judge anything obscure or confused. One must will the practice necessary to come to the state where the intellect can gauge clarity and distinctness; only then, perhaps, might the will itself properly be impelled by the correct perception of clarity and distinctness.[118] In *Rules*, Descartes praised the practice necessary to gain certain experience of the clear and distinct, for example, with mathematics or tapestries.[119] One must will oneself through bootstrapping exercises to a higher state of freedom, a state where one's options are not indifferent, or falsely different, but guided by the discernment of the clear and distinct. In Descartes' later considerations of the passions that accompany our impressions of things, he emphasized again the necessity of the self-discipline that trains our ability to judge freely.[120]

The scholastic account of having evident intuitions has no place for the cognitive training necessary to recognize clarity and distinctness that was at the heart of Descartes' account in his early and later writings. According to Ockham, people assent to evident intuitions of contingent facts and evident deductions all the time, and they assent regularly to confused and unclear objects. Evidence in the scholastic sense demands no real training or exercise of the intellect whatsoever; the evident impels necessarily and without practice. Not all human beings, to be sure, will have evident intuitive cognitions of the same things. Through divine revelation, a lucky few may gain temporary intuitive cognition of God, a state of knowledge otherwise left to those in paradise, where beatific vision is precisely an unmediated, intuitive knowledge of an ever-present God. While the attainment of abstractive knowledge of the divine certainly depended on cognitive and moral development for scholastics after Scotus, that development did not require refining any faculty for evaluating whether perceptions were in fact evident intuitions.

Recent good scholarship on Descartes rightly focuses less on medieval philosophers than on his immediate scholastic context, much as this chapter has focused on the immediate rhetorical context. The scholastic account of evidence appears prominently in the scholastic texts Descartes knew. The Jesuit *Ratio studiorum* ordered professors to introduce topics taken from the dialectic and logical manuals by Francisco Toledo and Pedro Fonseca. Toledo gave a detailed exposition of the qualities of knowledge along the lines just seen. Knowledge properly speaking is a certain and evident habit, produced from certain causes. Evident knowledge excludes the knowledge of faith.[121] In Fonseca's great commentary on Aristotle's *Metaphysics*, he wrote that "an act of knowing is evident . . . [if it is] such that the intellect could not judge otherwise."[122] Of all the scholastics, Descartes claimed to like Eustachius a Sancto Paulo the best. Eustachius used *evidentia* to distinguish knowledge from faith.[123] Evidence, for the late scholastics such as Ockham and Scotus, captured the inability of the *intellect* to judge otherwise. For Descartes, the realignment of one's life came about by training oneself to assent only to the evident; for the scholastics, assenting to the evident required no exercise, as it was automatic. The scholastics certainly had a fundamental place for intellectual striving and training; but this striving simply had nothing to do with honing an ability to have evident cognitions; that ability is simply given to human beings.

Rhetorical evidence was, to be sure, not the only resource useful for Descartes in developing his account of evident knowledge; he certainly drew on scholastic philosophy as well as rhetoric. The rhetorical and scholastic aspects of Jesuit education were supposed to be mutually supporting. The rhetorical picture of using words to produce an image as if before the eyes in principle rested on solid philosophical understanding of the cognitive, emotional, and imaginative faculties of human beings. Insofar as Descartes employed resources from scholastic philosophy, he did so from elsewhere than from doctrines of intuitive cognition. Focused largely on the terms "evidence" and "clarity," my account of evidence in Descartes has hardly discussed "distinctness," a quality of knowledge important for scholastics and, as discussed at length below, central for Leibniz. Moving from confused cognitions to distinct ones was something intellectual effort could achieve in the scholastic picture.[124] Important to scholastic discussions was the extent to which God, angels, and human beings could attain simultaneous, distinct knowledge of things, including the infinite.[125] It remains to be investigated more fully just how central scholastic accounts of distinctness were for Descartes.

ENTHUSIASTIC AND PSYCHOLOGISTIC CONCERNS

Some philosophical commentators have worried about a specter haunting the so-called rationalism of Descartes.[126] "Clarity and distinctness" seems not to be a standard with which we disinterestedly judge things. Worse yet, ideas possessing clarity and distinctness seem to seize us; they appear to have an apparently irrational grasp on us. Following in the steps of Leibniz, these critics see the standard as deeply psychologistic—subjective in the pejorative sense. Other commentators maintain that Descartes' standards must have some objective, external logical core.[127] It is no accident that advocates of both positions have supporting evidence from across Descartes' works. Beginning with a literal dream of poetic knowledge, Descartes' account of knowledge partook in the extrarational and the enthusiastic; if the argument of this chapter is correct, he developed his epistemic criteria in part from rhetorical theory and practice, which was dedicated to moving the passions through vivid images. As the previous chapter has underlined, his account of knowledge drew from the formal reasoning of algebra, albeit within his geometry. This double historical background illuminates why interpreters carefully and painstaking sifting through Descartes' works find so many apparently contradictory psychologistic and logical aspects of his account of knowledge. Both were central in its genesis.

Just as Descartes' account has roots in means for producing affect, its troubling subjective character has deep roots in Catholic Reformation debates about securing religious knowledge. Despite the dangers stemming from its subjective nature, affective spirituality invigorated the Catholic Reformation and provided some of its greatest internal political and theological difficulties, as the struggles within the burgeoning Society of Jesus well illustrate. By the early seventeenth century, the highly personal affective spirituality at the core of Ignatius of Loyola's *Spiritual Exercises* seemed dangerous to many Jesuit leaders as a standardized practice for all Jesuits and their lay followers. For all the claims of its rhetoricians and spiritual directors to be able to control and channel affect, allowing religion to include such individual idiosyncratic transformations threatened the order with dissent and internal disagreement. By the late sixteenth century, the order produced an official guidebook for directing the exercises. It emphasized safe reasoning over potent but dangerous illumination: "Without doubt the more excellent and higher way is when the divinely illuminated will precedes and transmits to the intellect. . . . But the second way, by reasoning and discursive thought, is more secure and safer."[128] The confessional mission of the order needed external apostolic activity far more than interior contemplation.[129] Yet the Jesuit authorities well recognized that affective spirituality and rhetorical power were essential tools in the internal mission of the Jesuits to turn the people of Europe to proper Christianity.

Such political, epistemic, and spiritual problems were endemic in a Church that stressed hierarchical authority and also promoted the possibility of direct interaction with the divine. Theological and spiritual authority could not rest exclusively with sanctioned theologians so long as the possibility of direct divine inspiration remained. The sixteenth and seventeenth centuries are seen as golden ages of mysticism in both Spain and France, particularly for women.[130] While the Catholic Church did not, and could not, monopolize access to the holy, it monopolized the right to validate such access. The Church desperately needed criteria for sorting and authenticating subjective experience, for only then might the potential anarchical ramifications of direct inspiration be quelled and mystical vision turned to bolster, not to undermine, church and governance. These tensions were undeniably productive theologically, spiritually, and philosophically, as any reader of Teresa of Avila can attest. Descartes' effort to secure a subjectivist epistemology deserves a privileged place as perhaps the most philosophically interesting and certainly the most fruitful of these manifold early-modern efforts to distinguish among and to secure subjectivist knowledges. The debates of Descartes scholars today stem in no small part from the creation of his doctrines amid such fundamental tensions within Catholic spirituality. Meric Casaubon was right to put Descartes' method immediately after Jesuit mysticism.

Following Descartes himself, many commentators have portrayed him as quintessentially breaking with the major trends of the Renaissance. Descartes, however, effected a remarkable epistemologization of Renaissance rhetorical practices and the psychology that accompanied them. His account of the human intellectual and emotional faculties and his suggestions for improving them offer one summation to the humanist encounter with Scholasticism. In Descartes, the new rhetoric and psychology that armed the Catholic Reformation came finally to attack the philosophical, epistemological, and ontological bases underlying the official theology of the Council of Trent. The need for a pastorally relevant rhetoric and psychology had, then, as one highly creative effect, the production of something regarded as a touchstone of modern epistemology and destructive of traditional philosophy.

Descartes' solution was to offer a series of practices for developing the discernment necessary to guide oneself through life, aided here and there by temporary arts, exemplified in a new constructive geometry of knowable curves. Only tough individualized practice, not the standardized path of formal dialectic or the passive acceptance of the appearances of the world, would produce the discernment and moral character central for an *honnête homme*. For all his disagreements with Descartes, Pascal seized upon this individualism to stress the capacity of cultivated men and women to question theological and natural philosophical authority— including especially the authority and certainty of one René Descartes.

PART II   Pascal

CHAPTER THREE

Mathematical Liaisons

Toward the end of his life, Pascal wrote to Pierre de Fermat: "To speak honestly to you about mathematics, I find it the greatest exercise of the mind." The letter continues in a more negative vein: "but at the same time I know it to be so useless that I can see little difference between a man who is only a mathematician and a skilled artisan. Thus, I call it the most beautiful craft in the world; but in the end, it is only a craft; and I have often said that it is good for making an effort, but not at all for the employment of our force: so much so that I would not take two steps for mathematics, and I assure myself that you are very much of my opinion."[1] Much like Descartes in his *Rules for the Direction of the Natural Intelligence*, Pascal stressed the intellectual and social opposition between mathematics as a craft and mathematics as an exercise useful for other ends. Whereas Descartes disparaged Fermat as a mere *"gasçon"* capable of doing some tricks, Pascal admired Fermat's mathematics and his person alike: "although you are the person in all of Europe that I hold to be the greatest mathematician, that quality would not have attracted me. But I find so much of the mind and of the *honnête* in your conversation that I would seek you out for those [qualities]."[2] Fermat had a *mind*—a much more generalized good judgment—in addition to his skill in mathematics. No mathematical artisan, he was an *honnête homme*.

Pascal had long admired those capable of combining mathematical acuity and *honnêteté*—a key seventeenth-century ideal of noble cultivation. Six years previously, he had encouraged Fermat to try to convince a paragon of *honnêteté* and amicable conversation—the chevalier de Méré—about the infinite divisibility of a mathematical line. "If you could do it," Pascal quipped, "he would be made perfect."[3] In his works from the mid-1650s onward, Pascal insisted that mathematical and logical competencies be supplemented with other forms of judgment: the "mathematical mind" (*esprit de géométrie*) needed to be complemented by the "intuitive mind" (*esprit de finesse*). In a fragment of *Pensées* entitled "*Honnête homme*," Pascal envisioned widely cultured men, not mere artisans of thought limited to mechanical production in some narrow field: "One must be able to say that he is neither mathematician nor preacher nor rhetorician but that he is an *honnête homme*" (S532). While poetry and mathematics were crafts pursued by specialized artisans, a universal *honnête homme* could in principle judge products of those two fields among others. Such a man might occasionally produce examples of poetry or mathematics without such activities dominating either his ways of thinking or his path in life. "Universal men are not called poets or mathematicians or the like. But they are all of that and judge all those things" (S486).[4] Their skills deployed in writing poems and proofs do not define and habituate them, for their judgment and creative abilities rest on more general qualities of mind. One can turn to them for their opinions on all things, not only in some specialty: "it is a poor mark when one does not turn to a man when it is a question of judging some verse" or indeed almost any other question (S486). Like poetry, mathematics was one activity among many appropriate for the *honnête homme*, who was potentially capable of both judging and meaningfully contributing to it.[5]

In this chapter I consider mathematics and other pursuits deemed suitable for the *honnête homme*. Pascal's account of the cognitive exercise provided by good mathematics developed in tandem with his distinctive mathematical practice. Pascal's *Treatise on the Arithmetical Triangle* (early to mid-1650s) illustrates what form he thought proper mathematical development ought to take: he explicitly sought to show how good mathematics exercises the mind. Rather than helping to develop a habit of discerning the clear and distinct, proper mathematics enhanced the ability to relate apparently disparate objects by multiplying enunciations about them. Although mathematics certainly involved the production of truth, practicing mathematics most significantly involved learning to reason well, given the abilities and limits of human beings. Considering the practice of mathematics would show what reason maximally could accomplish and what it could not. Pascal came eventually to stress the power of mathematics to lead *honnêtes hommes* to recognize and to accept the human condition.

Other exercises were as essential as mathematics for developing good judgment. Like Michel de Montaigne and many of his contemporaries, Pascal praised conversation as a crucial exercise for developing skills in judging, speaking, and writing. He developed his innovative mathematical practice and his account of mathematical knowledge while participating in a series of rich conversational spheres. In this chapter I set some of Pascal's mathematical writings in the conversational contexts that helped refine his work. Pascal's friends—mathematicians and *honnêtes hommes* alike—created conditions in which he came to develop his innovative views about mathematics and its power to cultivate the mind and to end dispute. I also explore the implications for choosing a way to live that Pascal drew from his mathematical practice and discoveries. Properly considering mathematics and philosophy revealed the limits of reason for orienting one's life. The conversation of his friends helped him develop and sharpen his thoughts and writing. Pascal attempted in turn to make his words develop and sharpen their opinions, above all, their opinions about the greatness and wretchedness of humanity.

MATHEMATICAL CONVERSATIONS

In a letter written to Thomas Hobbes in 1654, François du Verdus described the mathematical sociability he was experiencing in Paris: "We can even from time to time tell you something about the ancient Porisms restored by Mr. Fermat; the magic numbers of Mr. Pascal, and other things which we converse about Saturdays after dinner at Mr. Le Pailleur's on the rue St. André."[6] Le Pailleur's conversational group, or "academy," continued the informal group led by the great correspondent Marin Mersenne. An educated, respected mathematician as well as bon vivant, Jacques Le Pailleur epitomized the nonpedantic savant, the true *honnête homme*, someone capable of applying his taste and judgment to music, mathematics, and poetry, as well as creating in all those domains.[7] As du Verdus' letter makes clear, mathematics was a subject fit for learned conversation in settings such as the Saturday meetings at Le Pailleur's. Learned sociability figured centrally throughout Pascal's life, despite his retreats for reasons of health and piety. In Paris in the late 1640s and early 1650s, Pascal participated actively in various circles interested in natural philosophy and mathematics, among other topics: the conferences of Mersenne, later those of Le Pailleur and Bourdelot, the salons of Jansenists, and the *hôtel* of the duchess of Aiguillon.[8] While living in Rouen, Pascal visited the Le Pailleur and Bourdelot circles when he traveled to Paris.[9] Even after he had abandoned most worldly concerns and the pursuit of glory through his scientific work, he continued to discuss theology, natural philosophy, and mathematics in groups centered on Port-Royal.[10]

Throughout his life, Pascal extolled the clarity, efficacy, and accessibility of the spoken over the written word, whether in defending his arithmetical machine or his experiments with the vacuum.[11] His claims about the power of the spoken word echoed convictions about the power of conversation widely shared within his intellectual milieu. Members of the early-modern republic of letters often remarked on the superiority of conversation over solitary reading and intellection.[12] Peter Miller has recently shown that erudite humanists understood such learned conversation as mental and ethical self-cultivation.[13] In his massive anatomy of the republic of letters, the *Polyhistor* of 1688, Daniel Georg Morhof argued that "nothing contributes more to cultivating our minds than frequent conversation with learned men; it is, of all the disciplines, the best, and it insinuates itself better than in the tedious way of reading and personal meditations."[14] Pascal's friend the chevalier de Méré wrote, "Three things make a knowledgeable and skillful man: reading, conversation, and imagination." Whereas reading "enriches the memory," conversation "polishes the mind."[15] Montaigne, whom Pascal called the "incomparable author of the *Art of Conversation*," wrote that the "most fruitful and natural exercise of our minds is in my opinion conversation." A superior exercise for the mind, conversation could move, even transform, its participants: "The study of books is a languishing and feeble movement that completely fails to move us; whereas conversation teaches and exercises all at once."[16] Far more than a channel of information, good conversation pushed its participants to improve themselves emotionally and intellectually; it encouraged them to do more than collect sundry facts, clever ripostes, and laconic quips.

Noting how lucky he had been in his choice of conversational partners, Pascal emphasized the power of conversation to cultivate or to ruin: "One forms the mind and the *sentiment* through conversations; one spoils the mind and the *sentiment* through conversations" (S658). In a speech intended for a group of mathematicians, likely those of Le Pailleur's academy, Pascal credited much of the quality of his work to his interactions with the group: "I confess to be yours what I would not have made mine if I had not been formed among you; and recognize for mine alone that which I see unworthy of such eminent mathematicians." Pascal noted that they had embraced him even as a youth. The members of the academy pushed him to a level of rigor and elegance he otherwise would never have reached: "For nothing pleases you unless it is significant and perfectly demonstrated. Rare is the gift of audacity in discovery; rarer yet, in my opinion, is the gift of elegance in demonstration; rarer still is having both" (M II:1031–32).[17]

Such claims about the power of good colleagues to perfect claims and arguments figured centrally in the early-modern theory of conversation. In his famous *Civil Conversation*, Stefano Guazzo contended that social interaction tests and strengthens arguments: "And verily as the armourer can not assure him selfe of the goodnesse of a cosselet, untill such time as he hath scene it prouved with lance or harquebouse: so neither can a learned man assure him selfe of his learning, untill he meete with other learned men, and by discoursing and reasoning with them, bee acertained of his sufficiencie. Whereby it seemeth to me verie cleere, that conversation is the beginning and end of knowledge."[18] Conversation best leads to the production of truth: "while they dispute by live reasons, indevouring to get the upper hand ech of other, the perfect knowledge is come by, thereupo[n] it is commonly saide that Disputation is the sifter out of the trueth."[19] Montaigne noted that the "contradictions of judgments do not offend or upset me; they simply wake me up and exercise me."[20] Lively and penetrating discussion, he made clear, excluded mere mechanical contradiction—the sterile forms of disputation taught in the schools and universities. "We flee from correction; it is necessary to present and to produce it there, especially when it comes in the form of conversing rather than the lessons of a schoolmaster."[21] Guazzo and Montaigne characterized properly moderated and lively dispute as intellectually fruitful, personally ennobling, and politically sound.

Pascal's mathematical and natural-philosophical works in progress almost certainly circulated in Le Pailleur's group, and were discussed and improved there, just as Pascal claimed and just as theorists of conversation envisioned.[22] At one point in a consideration of figurate numbers, Pascal thanked "my friends, learned lovers of general solutions," for having pushed him to work out an appropriate level of generality when he failed to reach a far too ambitious goal. The works of others circulated as well, as du Verdus' letter to Hobbes suggests. The group discussed Fermat's attempts to restore the lost *Porisms* of Euclid, which the lawyer had sent from Toulouse. A set of such porisms—mathematical statements that the ancients judged to be neither problems nor theorems—in Fermat's hand is preserved among Pascal's papers.[23] In late 1654, Fermat communicated some important results. Pascal informed him that "all our Messieurs saw them last Saturday"—the time of Le Pailleur's academy—"and held them in esteem with all their hearts."[24] Such manuscripts circulated among members of the group; on occasion they would reach print improved, if we take Pascal at his word. Pascal's works are the outstanding printed legacy of the group.

The meetings at Le Pailleur's involved a form of intellectual civility that stressed rigor and solidity of reasoning, in contrast to more rigid forms of politeness that excluded any real challenge or contradiction. Guazzo and Montaigne presented conversation as rigorous and heated debate that should never spill into the automatic, mechanical denial or disagreement said to characterize scholastic disputation. Unlike the civility underlying the Royal Society as presented by Shapin and Schaffer or some of the courtly civilities presented by Biagioli, vigorous dispute was seen as constitutive of knowledge, and not always dangerous to it.[25] Unlike the case of the early Royal Society, mathematics figured at the center of the discussions of the Le Pailleur academy. Just as in those examples, however, a proper form of regulated and civilized discussion was understood to contribute to producing the proper sort of truth claims.

Publishing results was not the purpose of discussion for conversational theorists, who praised rather its power to form good judgment and good people. More important than discerning the truth of this or that statement was cultivating a judgment that might regularly produce true, insightful, or challenging statements in oneself and others. In his essay "On the Art of Persuasion," Pascal called for a form of judgment capable of getting behind apparently intelligent remarks to grasp the real person producing any given statement. "It is necessary therefore to sound out how this thought is lodged in its author: how, whence, how far he possesses it."[26] Any fool might produce a true and insightful statement occasionally; what mattered was the insight and judgment behind it.

A conversational circle conducive to exercising minds and creating truth had to restrict its membership. Montaigne had praised the power of "vigorous and regulated minds" to fortify the mind while denouncing the power of "low and sickly minds" to "debase" the mind.[27] Pascal accordingly counseled: "It is important therefore above all to know how to choose [conversations] well, in order to be well formed and not spoiled."[28] Although a few conversational circles were open to a wider public, most limited access to those with both the intellectual and the social skills deemed necessary for productive discussion.[29]

Not all the members of the Le Pailleur circle were good conversationalists. Some made themselves rather unwelcome elsewhere. Mathematicians were notoriously uncivil, a charge that Marin Mersenne considered in 1634. On the one hand, the contention required in mathematics "burns the blood and leads to a melancholic humor." On the other, "the naked and simple demonstrations they use render those who do not look beyond [such demonstrations] as simple as the demonstration; they are thus easily deceived." Thus, "since mathematicians distrust the rest of humanity that does not know how to use their demonstrations, they receive no instruction from civil conversation."[30] The most professional mathematician of them all, the "*geomètre tout court*," as Mersenne called him, Gilles Personne de Roberval, was considered "the most disagreeable man in the world in conversation," as Pascal later wrote (M IV:1311).[31] By insulting the leader of another conversational group, the Montmor Academy, Roberval earned the epithets "rustic" and "pedantic."[32] Years later, an eyewitness suggested that his skill in mathematics likewise reeked of the pedant, as it owed more to rules and books than to genuine mathematical discernment. Roberval was "only a studious mathematician [*geometre d'estude*], not what one calls a *mathematical mind*, as were Messieurs Huygens and Pascal." Such minds "go with a straight and solid view, having already all the principles stowed

in their heads without having to seek them out in books or torturing themselves to discover them."[33] More important than the truth of this accusa-
tion is the use of such an opposition between minds with insight, judgment, and taste and minds productive only by virtue of rules and books.[34]
Pascal sought to cultivate the former, genuinely mathematical minds, in
his mathematical works.

Publication and professional position were not necessary conditions for
admission into intellectual circles such as Le Pailleur's. In a lovely passage,
Pascal evoked thinkers who shine intellectually only when conversing:
"There are some who speak well and write poorly. The place, the audience,
excites them and pulls from their minds more than they would find there
without this heat" (S464). Much like Pascal's father, Etienne, Jacques Le
Pailleur demonstrated his mathematical acumen less through print than
through conversation, informal networks, and the limited circulation of
manuscripts.[35] The mathematical judgment of most members of these circles did not make mathematics into their métier.[36]

For Pascal, conversing about mathematics was not limited to Le
Pailleur's small group of specialists. He certainly discussed mathematics with known mathematicians such as Girard Desargues, Claude Mylon,
Fermat, Ismaël Boulliau, Etienne Pascal, and Roberval. He also pursued
mathematical questions with learned *honnêtes hommes* such as his friends
the chevalier de Méré, the duc de Roannez, and Damien Mitton.[37] With his
interest in generating the universal man capable of judging and creating
in many domains, Pascal took the interested members of a wider public
seriously, however much he recognized the limits of their understanding
and dedication. His most famous mathematical treatise addressed both
savants and more general readers; it offered exercises appropriate to the
mathematical acumen and interest of both sorts.[38]

PASCAL'S ARITHMETICAL TRIANGLE

Around the same time that du Verdus wrote to Hobbes in 1654, Pascal
wrote a speech for a mathematical academy, likely Le Pailleur's. He outlined the range of his mathematical and natural-philosophical activities and
set out his plans for future publication.[39] Before listing works he hoped to
finish soon, concerning geometry and other subjects, he mentioned several treatises, evidently complete, on various properties of numbers. These
completed treatises soon formed the most original parts of his *Treatise on
the Arithmetical Triangle*.

On 23 November 1654, Pascal had an intense mystical experience,
which he recorded in a "memorial" that he had sewn into his clothing:
"God of Abraham, God of Isaac, God of Jacob, / Not of philosophers
and savants. / Certitude, certitude, *sentiment*, joy, peace."[40] The experience

provided him with a profound certainty and repose that he had long sought in vain and led him to dedicate himself primarily to religion. He abandoned most of his scientific pursuits and distanced himself from many of his conversational partners. He later committed himself to persuading philosophers and savants to seek the certainty and peace he had found outside philosophy and mathematics.

Before his mystical experience, Pascal had numerous parts of his treatise on the triangle printed, some in French, some in Latin, some in both.[41] In general, the parts accessible to more general readers eventually appeared in French; some of the more recondite parts, more appropriate for the specialists, in Latin alone. In his speech Pascal mentioned only the advanced Latin treatises and not the more elementary French texts. He first had a coherent set of Latin texts printed, of which a single exemplar exists. He apparently then decided to revise and rewrite most of the texts in French, had them printed, and tacked on the previously printed, more advanced, Latin texts.[42] He thus decided to present his work on the triangle to a larger public of French readers, as he had his earlier works on the vacuum. After his mystical experience of November 1654, he put nearly all the printed copies away; they became generally available only after his death.

Just as a brief survey of some technical aspects of Descartes' geometry illuminates how he understood the power of mathematics to cultivate, a brief tour of Pascal's *Treatise* clarifies how he claimed mathematics could hone skills and aptitudes. In his *Treatise*, Pascal elegantly organized numerous known results about several sorts of special numbers and presented some important new results.[43] In the various parts of the *Treatise*, he illustrated a mode of mathematical practice in which he explicitly focused more on providing exercise in discovering interrelations among mathematical objects and enunciations than in claiming any kind of comprehensive or clear and distinct knowledge of the arithmetical triangle and related special numbers.

"Pascal's triangle" is well known:

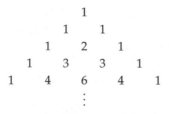

Versions of the triangle and the sets of numbers within it appear in different guises in the Chinese, Indian, Arab, Persian, and European traditions.[44] Pascal called it the arithmetical triangle, and he oriented it differently from

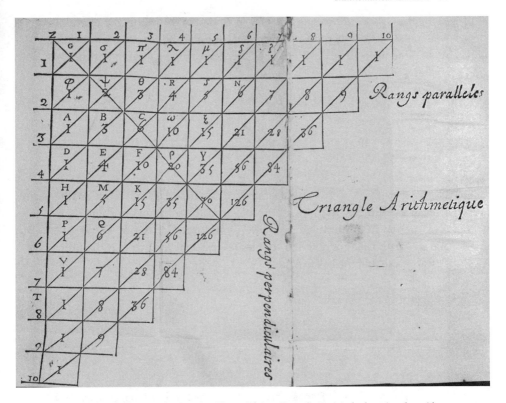

Figure 3.1. Arithmetical triangle. From Blaise Pascal, *Traité de la triangle arithmetique*... (Paris, 1665). Courtesy of the Burndy Library, Cambridge, MA.

the form opposite. His *Treatise* begins by constructing the triangle (see fig. 3.1). The number in each "cell" is produced by adding the cell in the row above it to the cell in the previous column. (I henceforth use "cell" to mean "the number in the cell.") This is the same operation as adding the numbers pair by pair along each row of the now better known form of the triangle. In the examples that follow, I refer exclusively to Pascal's form of the triangle, figure 3.1.

Having constructed the triangle, Pascal illustrated a variety of its properties in a series of propositions. For example, any cell equals all the cells in the previous column from the top of the triangle to the row of the original cell; for example, $10 = 1 + 3 + 6$. This can easily be seen in a truncated triangle:

$$
\begin{array}{cccc}
1 & 1 & \mathbf{1} & 1 \\
1 & 2 & \mathbf{3} & 4 \\
1 & 3 & \mathbf{6} & \mathbf{10} \\
1 & 4 & 10 & \ddots
\end{array}
$$

Using the Greek and Roman letters he included in the triangle, the result is $\omega = \pi + \theta + C$. The proposition holds if columns and rows are reversed in its statement, so $10 = 1 + 2 + 3 + 4$. This can easily be seen in a truncated triangle:

$$
\begin{array}{cccc}
1 & 1 & 1 & 1 \\
1 & 2 & 3 & 4 \\
1 & 3 & 6 & 10 \\
1 & 4 & 10 & \ddots
\end{array}
$$

In his brief proof of this enunciation, Pascal drew on innovative mathematical typography to gives all the stages in the proof in abridgment. By construction of the triangle, $\omega = R + C$. In turn, $C = \theta + B$, $B = \psi + A$, and so forth. Using a notation that makes evident the steps in the proof, Pascal wrote that the cell ω equals

$$
R + \underbrace{\begin{array}{c} C \\ \theta + \underbrace{\begin{array}{c} B \\ \psi + \underbrace{A}_{\varphi} \end{array}} \end{array}}
$$

so $\omega = R + \theta + \psi + \varphi$. Repeatedly applying the fundamental construction formula to itself—each cell is the sum of the two elements above it— yields a great variety of such interesting properties of the triangle. He demonstrated some propositions about proportions among the cells. Given two contiguous cells on the same diagonal, for example, the first cell is to the second cell as the number of cells from the lower end of the diagonal up to and including the first is to the number of cells from the second to the top of the diagonal. Consider E and C: 4 is to 6 as 2 is to 3. Pascal proved this through a rigorous but straightforward proof by induction.[45]

After demonstrating numerous properties of the numbers in the triangle, Pascal stopped to note: "It is a strange thing how fertile [the triangle] is in properties." He had certainly not discovered them all: "Each can exercise himself with it" in finding new ones (M II:1300). He turned then to consider a wide array of uses for this triangle. He illustrated the equivalence of the numbers in the triangle with several different sorts of special numbers useful in various parts of mathematics. He also illustrated techniques for making the triangle a fount of other discoveries. In a fine book on Pascal's triangle and its European and non-European predecessors, A. W. F. Edwards has shown that many of these relations had been discovered long before and that Pascal's more erudite contemporaries knew many, if not

Figure 3.2. First three triangular numbers: 1, 3, 6. Note that the 3rd triangular number (6) equals the 2nd triangular number (3) plus the 3rd counting number (3). In general, the nth triangular number equals the $n - 1$th triangular number plus the nth counting number.

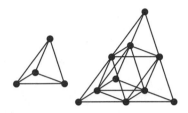

Figure 3.3. First three pyramidal numbers: 1, 4, 10. Note that the 3rd pyramidal number (10) equals the 2nd pyramidal number (4) plus the 3rd triangular number (6). In general, the nth pyramidal number equals the $n - 1$th pyramidal number plus the nth triangular number.

most, of them.[46] In elegantly synthesizing these results, Pascal presented and systematically proved most of the various known relations.

Having constructed the triangle, Pascal turned to consider a class of numbers that had fascinated mathematicians since antiquity—figurate numbers. These numbers come from counting the vertices of geometrical figures such as the triangle or the square (see fig. 3.2). A series of triangles made from a discrete number of evenly spaced points are made of 1, 3, 6, 10 points, and so on. These numbers can be extended to greater dimensions: the pyramidal numbers, 1, 4, 10, and then their equivalents in greater dimensions (see fig. 3.3).

Pascal demonstrated many of the properties and relationships among these figurate numbers.[47] One of the most straightforward can be seen in examining figures 3.2 and 3.3. Each figurate number in a certain dimension equals the number preceding it in its own dimension plus its equivalent in the next lower dimension. The third pyramidal number (10) equals the second pyramidal number (4) plus the third triangular number (6). This relationship can easily be generalized. Each figurate number can be produced from preceding figurate numbers and lower-dimensional figurate numbers. Repeating this process ultimately shows that every figurate number equals a sum of lower-dimensional figurate numbers.

This important result about the figurate numbers is, however, an alternative way of expressing the first result given above concerning the numbers in the arithmetical triangle: any cell equals all the cells in the previous column from the top to the row of the original cell: $10 = 1 + 3 + 6$. Pascal's arithmetical triangle therefore produces the figurate numbers (look at the diagonals in fig. 3.4). Considering the relationships among the numbers in the triangle can thus help make evident the relationships among the figurate numbers.[48] "This makes it known that all that has been said

		Number				
		1	2	3	4	5
	0	1	1	1	1	1
counting	1	1	2	3	4	5
triangular	2	1	3	6	10	15
pyramidal	3	1	4	10	20	35
Dimension						

Figure 3.4. Figurate number table.

about the rows and cells of the Arithmetical Triangle corresponds exactly to the [figurate numbers], and that the same equalities and proportions noted in one will be found in the other." This correspondence gives a ready means for discovery: "it will be necessary only to change the enunciations, by replacing the terms of [figurate numbers] . . . with those belonging to the arithmetical triangle" (M II:1302). Once the equivalence of two sorts of special numbers is shown, one merely needs to replace the terms in propositions concerning one sort of numbers with the equivalent terms for another set of numbers.

In subsequent treatises, Pascal demonstrated that other special numbers also appear in the triangle and that the triangle's properties belong to them as well. How many ways can one choose two things from three? Pascal showed that the number of combinations of r things taken n at a time without regard to their order is simply the $n + 1$th number of the $r + 1$th diagonal of the triangle. Calculating the combinatorial numbers is equivalent to computing the triangle: "One sees clearly how well combinations and the arithmetical triangle correspond; and thus, that the observed proportions between the rows or among the cells of the triangle extend to the reasons [*rationes*] of combinations" (M II:1242). Any enunciation proved about the figurate numbers or the numbers of the triangle has a corresponding enunciation concerning the combinatorial numbers, and vice versa. As the combinatorial numbers are central to questions of probability, for example, the triangle is useful in solving such problems. Pascal readily put the triangle to use in solving problems of fairly distributing the stakes of interrupted games of chance.[49]

Finally, the triangle is useful for finding the coefficients of binomials. A binomial is the expansion of an algebraic expression, such as $(a + b)^r$, where r is a positive integer. Although easy to compute for smaller exponents, as r gets larger the calculation quickly becomes tedious and error prone. Pascal showed that the coefficients of the expansion equal the elements along the $r + 1$th diagonal of his triangle. To find the coefficients of $(a + b)^3$, simply look at the fourth diagonal of the triangle: 1 3 3 1; for $(a + b)^4$, look at the fifth diagonal: 1 4 6 4 1; and so on, for any arbitrary whole number.[50]

PROLIFERATING ENUNCIATIONS

Beyond demonstrating propositions new and old about the various sets of numbers, Pascal illustrated how mathematicians ought to go about discovering such propositions in his *Treatise on the Arithmetical Triangle*. Like Descartes, Pascal sought to train the mind with his preferred style of mathematics. In one of the component parts of the *Treatise* he set forth his normative and pedagogical purposes. The reader will be shown the means of discovery: "in order to hide nothing about the way in which these correspondences have been pulled forth, I will demonstrate the relation of them openly [*à découvert*]" (M II:1326).[51] Most propositions in this treatise are duly followed by comments about Pascal's techniques for discovering and proving them.

Each cell in the arithmetical triangle belongs to several sets of special numbers and therefore can go by several different names. Any given cell is—at once—a number in the arithmetical triangle, a certain figurate number, a combination, and a binomial coefficient. The numbers associated with the rows, columns, and diagonals of the triangle all have distinctive significance when considered in relation to these different special numbers. Each type of description of the same number brings forth a different set of associations; each set of names is connected to a different set of propositions already proved; and yet each set of names is mathematically equivalent to the others. If we take a proved enunciation about the combinatorial numbers and replace the combinatorial names with the equivalent names used for figurate numbers, we get an equivalent, but new, enunciation. Varying the names in restating a given enunciation quickly reveals a series of new relations expressed in a new enunciation: "these are the diverse paths that reveal new consequences and that, through enunciations matched to their subjects, link propositions that seem to have no connection in the terms in which they were originally conceived" (M II:1329). Pascal gave many examples of this technique: "In place of 'the exponent of the base,' . . . one must substitute 'the exponent of the parallel row, plus the exponent of the perpendicular row minus one.' This produces the same number, and with this advantage: we know the relation these exponents" have with the figurate numbers (M II:1326–27). Because readers would have already explored the relationship between the exponents of the rows and the numerical orders, they could readily apply that knowledge to any enunciation about the "exponent of the base." Thereby they will be able to discern new propositions expressed in the language of numerical orders, and then they can turn them into propositions in the language of any of the other sorts of numbers. Pascal described these mathematical processes in a series of metaphors: the "turning about" of statements, the discovery of new paths, as well as a perspectival image: "being regarded from another side, they give other openings" (M II:1327).

In the *Treatise* Pascal presented a normative model of mathematical practice. In the first Latin version, he wrote that the finding of new paths by varying enunciations "ought to be the study of mathematicians" (M II:1203). In the French version, he was more emphatic: he "leaves it to each to exercise his genius in this research that ought to make up all the study of mathematicians" (M II:1329). Pascal's mathematical practice involved discerning relationships and liaisons among mathematical objects: "for the enunciations prepared by this art lead to many different and great theorems, in connecting those propositions that seem to be altogether different when first considered" (M II:1203). By relating apparently disparate types of mathematical objects, this form of practice helps to produce new relations *within* types of numbers and *among* types of numbers.

By showing the relations between the triangles and the kinds of numbers, a mathematician can more effectively produce propositions about each type of special number: "Thus are propositions multiplied, and not without utility: for varied enunciations, although based in the same propositions, afford various uses" (M II:1202–3). The varied enunciations within one sort of number can later be applied to another sort, and so on. Mathematical inventiveness meant constantly discerning new interconnections among mathematical objects, not reducing a variety of mathematical results to some small set of logical or arithmetical truths.[52] As Leibniz was later to remark, citing Pascal, varying expressions was an effective tool for discovery of new truths and discerning the harmony among truths already known.

Absent skill in transforming enunciations, Pascal argued, mathematics will prove difficult and tedious. "Whoever lacks this talent of turning things about will have an unpleasant practice of mathematics. But in truth this talent is not given but is assisted [*juvatur*]" (M II:1203). Practice in transforming enunciations will cultivate a talent or "mind" that will make the pursuit of mathematics easier and more productive. Although Pascal stressed that mathematical practice involves linguistic transformation, he did not espouse remaining at the level of linguistic transformation without attempting to develop a deeper understanding. Far from encouraging the reader mechanically to register the soundness of each deductive step, Pascal's effort to demonstrate his discoveries "openly" was supposed to help to develop the talent, the *ingenium*, of the reader by laying bare the insights behind the formal linguistic transformations. Changing enunciations develops a certain kind of mathematical mind, one capable of producing new relations. This practice works to develop the capacity for innovatively drawing consequences from many principles kept in mind without reducing or confusing them.

As discussed in chapter 1, the fundamental model for cultivation through exercise went back to Aristotle's *Nicomachean Ethics* and permeated early-modern accounts of how exercise could perfect habits and skills. Early-modern thinkers such as Descartes and Pascal were acutely worried that repetition of activities, like those in mathematics, might produce habituation without thereby producing an intellectual understanding accompanying the activity.[53] In "Art of Conversation," Montaigne argued that "the fruit of the experience of a Surgeon is not the history of his practices and the recalling that he has cured" a certain number of patients. "If he does not know how to pull out of his practice [*usages*] those things that will form his judgment, and if he does not know how to make us feel that he has become more Wise with the practice of his art," he cannot be considered skilled. Developing true judgment requires understanding the reasons behind experiences. "It is not enough to count experiences; they must be weighed and compared, and digested and distilled, in order to draw forth the reasons and conclusions they carry."[54] Pascal's later collaborator Pierre Nicole likewise described the "pedantic mind" as having "a mass of poorly digested precepts."[55] Pascal reworked Montaigne's point in a language of deduction:

> Those who have a discerning mind grasp how much difference there is between two similar words, according to the places and the circumstances accompanying them. Will one believe, in truth, that two people who have read and learned the same book by heart know it equally, if one comprehends it in such a way that he knows all the principles of it, the force of the consequences, the responses to the objections one might make, and the entire economy of the work; whereas for the other these are but dead words and seeds which... remain dry and fruitless in the sterile mind that has received them in vain? (M III:422–23)[56]

In openly articulating the heuristics behind his new enunciations in his mathematical treatise, Pascal explicitly worked to help the reader to profit from them and not just register their truth-value. He offered means for training a mathematical mind, not for leading the reader into the sterile world of the studious mathematician.

In several fragments of *Pensées*, Pascal considered the differences among several sorts of minds: the *esprit de géométrie*, the *esprit de justesse*, and the *esprit de finesse*—roughly, the mathematical mind, the sound mind, and the intuitive mind. One might be tempted to think that the ability to follow and to produce long proofs characterizes a mathematical mind, but such an ability characterizes Pascal's sound mind. Such a mind has "an extreme rightness [*droiture*]" and can "penetrate sharply and profoundly into the consequences of principles" (S669). Pascal claimed that the sound mind was central in sciences involving a small number of principles whose consequences are remote from their small number of premises. Hydrostatics was a chief example. Mathematics involves fundamentally many principles kept distinct in the mind.[57] Someone with a mathematical mind can contend with many principles all at once. People with this quality of mind "well draw out the consequences of things where there are many principles." Like someone thinking through the use of the arithmetical triangle, they comprehend "a great number of principles without confusing them" (S669). Such a mind may or may not be one that penetrates things to their depths—a sound mind. A mathematical mind has "amplitude," whereas a sound mind has "force and rightness."

Rigorous mathematics rests on long chains of connected arguments. Seeing only the chains fails to capture the essence of mathematical activity as Pascal saw it. While productive of formally valid chains of argument, Pascal's mathematical practice was focused primarily on discerning new relationships among mathematical entities (or principles). In his mathematical practice, he drew consequences from several sets of separate principles: the arithmetical triangle, the figurate numbers, the combinatorial numbers. In connecting them, he kept the different numbers and their uses distinct. He transferred these aspects of his practice into his account of the mathematical mind.

Pascal's friend the chevalier de Méré complained about the deleterious effects of mathematics on Pascal. "You still have a habit, which you have taken from this Science, of judging anything whatsoever only by demonstrations that are most often false." The instilled habit of only using such demonstrations imperiled Pascal's capacity for higher forms of judgment. "These long reasonings taken line by line prevent you from entering first into higher knowledges that never deceive." Such techniques of reasoning using rules belong to "little minds and the *demi-Savants*" and inhibit real judgment.[58] Far from cultivating the ability to judge well, mathematics as traditionally practiced merely develops skills in formal manipulation. It is but a craft.

With his later espousal of the powers of the intuitive mind, Pascal agreed with Méré about the limits of formal reasoning and the danger of believing that formal reasoning was capable of adjudicating all truths. Méré mistook, however, the ability to test demonstrations as the heart of real mathematical competency. Judging the soundness of long chains of reasoning characterizes a sound mind, not a mathematical one. Like far too many of his contemporaries, Méré misunderstood the quality of mind fundamentally at work in mathematical practice as Pascal conceived it. He therefore failed to gain all that one could from mathematics.

Pascal's exercise with changing enunciations did not necessarily lead one to seek some small number of fundamental basic principles of mathematical objects or to attempt to work out all the consequences of those principles. Such activity characterizes the sound mind far more than the mathematical mind. Following his mathematical practice, one aimed less at discovering some fundamental principles than at producing relations and general principles among different mathematical entities. Although the consequences produced are all certain, none of them yields comprehensive or even clear, distinct, and sufficient knowledge of these numbers and their qualities. Pascal insisted on the infinite qualities yet to be discovered. "The ways of turning about the same thing are infinite" (M II:1328). While Descartes noted the indefinite number of possible mathematical propositions to be discovered and proved, he nevertheless maintained that proper mathematical knowledge always should involve clear and distinct knowledge of the mathematical entities at issue.[59]

For Pascal, the unlimited turning about of mathematical propositions requires that mathematics be expressed in language. Pascal's technique rests on the choice of terms used in framing enunciations; its power comes from remaining at the level of linguistic transformations, not attempting to move from enunciation to clear and distinct knowledge of the imperfectly described mathematical entities. Like Descartes, Pascal recognized the power of language to aid mathematical discovery. And like Descartes, Pascal feared the collapse of mathematical reasoning into mechanical, rule-bound manipulations. Unlike Descartes, he did not call for mathematics to replace propositions expressed in language with constructions intuitively grasped all at once. Mathematical discovery required, first, the ability to keep many mathematical entities distinct in the mind and, second, skill in continually transforming linguistic expressions about these principles to find hitherto-hidden relationships among the principles.

Changing enunciations in the process of discovery is a logical and a rhetorical operation suited to human epistemic capacity. In *Pensées*, Pascal reflected on putting the same proposition into different enunciations: "Words diversely arranged yield diverse meanings. And meanings diversely arranged make different effects." Expressing equivalent propositions in diverse ways alters how human beings understand them and what they take away from them. "The same meaning changes according to the words that express it. Meanings receive their dignity from words rather than giving it to them" (S645). Fundamentally linguistic, even rhetorical, beings, humans need such different forms of expression, even in mathematics, the only certain knowledge available to them. Other forms of inquiry, Pascal insisted, had much to learn from mathematical practice.

DEFINING SOUNDLY

For Pascal and his friends, illegitimate ways of transforming statements undermined contemporary natural philosophy. A conduit essential for acquiring knowledge, language too often stood in for knowledge. Among their many occupations, Pascal's conversational circles considered questions about scientific doctrines, practice, and rhetoric: What kind of knowledge ought to be the goal of natural philosophy? What level of certainty should the sciences seek? How ought consent be created? What forms of speaking and writing are appropriate to a new science that avoids the dangers of dogmatism and incivility? Two of Pascal's friends, Le Pailleur and the poet Charles Vion Dalibray, for example, exchanged poems that debated whether human beings must remain satisfied with recording and saving appearances, or whether they can achieve some knowledge about the actual constitution of the world.[60] Dalibray called upon Pascal to help them resolve their dispute with his new experiments on the vacuum and new written forms for reporting those experiments. Dalibray instructed Pascal:

> Let us live no longer, Pascal, as slaves;
> With the cast-off skins of another let us not dress up;
> Let us write, since it is wanted that finally we write,
> But let us flee the knowledge whither the pedant
>   aspires.[61]

After all their efforts to replace substantial forms with new sorts of explanations, natural philosophers could easily fall prey to the written form of the old philosophy itself. Desiring certain knowledge of the principles of philosophy could lead one to concoct fictions and mistake them for certain knowledge. Pascal's friends called upon him to devise practices and rhetoric appropriate for a new natural philosophy. Eschewing the search for the certain principles of philosophy, this new philosophy would seek knowledge that was as certain as possible for feeble human beings.

Soon after publishing a pamphlet describing some vacuum experiments of the late 1640s, Pascal became embroiled in a dispute with the Jesuit Etienne Noël. Responding to Noël's letters and book, Pascal outlined an account of the proper role of definitions—an account he adapted before long to mathematical knowledge. In his final letter, addressed to Le Pailleur, Pascal complained that Noël had misunderstood the nature and proper use of definitions. People unaccustomed to seeing things "treated in the true order" believe that "one cannot define something without being assured of its being." They think a valid definition implies existence. In contrast, "one must always define things before seeking out whether they are possible or not. . . . for first we conceive the idea of a thing; next we give a name to this idea, that is, we define it; finally, we seek to find out whether this thing is true or false."[62] Only through further inquiry can we determine whether a defined thing exists and discover what properties the thing has: "there is no necessary liaison between the definition of a thing and the assurance of its existence; and one can just as easily define an impossible thing as a true one."[63] No matter how clear and distinct a definition may be, no definition by itself guarantees that something exists.[64] To Pascal, the error of confusing possibility and existence plagued much scholastic and Cartesian thought alike.

In his essay "Mathematical Mind," written a few years after the debate overdefinition, Pascal argued that mathematics illustrated how to use definitions correctly. "In mathematics only the definitions that logicians call nominal definitions are recognized—in other words, only the impositions of names onto things that one has clearly designated in terms that are perfectly known." Such definitions aid reasoning and thinking: "Their utility and usage is for clarifying and abridging speech, in expressing in a single name that one has imposed onto them what could otherwise only be said with numerous terms" (M III:393). The term "even," for example, abbreviates the expression "divisible by two." In providing abbreviated ways of referring to things, definitions themselves do not explicate the nature of the things they signify: "Definitions are made only for designating the things that one is naming, and not for demonstrating the nature of those things" (M III:398). Linguistic definitions involving space or time do not express the essence of empty space, any more than mathematical definitions express the essence of number, line, or space.

Definitions aid in producing statements about various entities, such as space and time. One must then demonstrate, either from first principles or from experience, whether the things do or do not possess the attributes claimed in the propositions. The numbers in the arithmetical triangle or the apparent vacuum can, for example, be shown to possess a given set of qualities expressed using defined terms. Knowledge about the apparent vacuum involves no definition of the essence of "empty space"; it comprises a set of statements that are grounded in observation and logical transformations of those observations and expressed using defined names. According to Pascal, much natural philosophy collapsed the making of nominal definitions with the subsequent proving of propositions; it conflated the proper order of inquiry and was therefore incapable of producing genuine knowledge.

Not content with nominal definitions, the Jesuit Noël attempted to define the indefinable—things Pascal claimed are known clearly, such as "space," "light," and "time." Pascal argued that names such as these "designate so naturally the things that they signify to those who know the language that the clarification that one would like to offer would provide more obscurity than instruction."[65] According to Pascal, all human beings innately have some knowledge of space and number; every person innately knows the reference of terms such as "space" and "number" in their languages; no definition of "space" could improve this knowledge of the reference of the name. That everyone shares knowledge of the connection between names and things does not imply, however, that they have innate knowledge of the nature of those things: "it is not the nature of these things that I say is common to all; it is only the connection between the name and thing" (M III:397). Attempting to define terms such as "time," "space," or "number" can only confuse, not clarify. Indeed, "there is nothing more feeble than the speech of those who want to define these primitive words" (M III:396). Noël became Pascal's stock example of the futility of attempting to define something clear in itself. "There are some who take this absurdity to the point of explaining a word by the word itself. I know one who defined 'light' in this way: 'light is a luminary movement of luminous bodies'" (M III:396). Rather than bandying about various definitions expressing the essence of light, natural philosophers should put forth propositions concerning the phenomena of light; and then they should attempt to prove or disprove whether those propositions in fact describe the characteristics of light.

Too often natural philosophers had confused definitions with propositions requiring further demonstration. A primary cause of dispute and discord in natural philosophy, this confusion precluded the work of proving propositions that had been mistaken for definitions of essence. Instead of starting by defining the essence of something, natural philosophers must realize that their so-called definitions are in fact propositions about the subject in question. They should then demonstrate that their propositions actually are the case about the thing in question, through deductions from known things and through empirical work. Rather than sparring over definitions of the essence of something, they should work, collectively and rationally, to evaluate different statements about things. Only such a rational and collective process of achieving consensus could end disputes and produce a certainty grounded not in dogmatic belief but in legitimate demonstrations.

In attempting to make philosophy attain the certainty of mathematics, too many contemporary philosophers had misunderstood the role of definitions and language in mathematics; they had therefore missed the lessons mathematical reasoning has for reforming philosophy. In his essay

"Mathematical Mind," Pascal claimed to be teaching others to recognize the lessons of mathematical reasoning: "it seems to me, from the experience I have had of the confusion of disputes, that one cannot too much enter into this sharp mind; I have produced this treatise for this reason, more than for the subject that is treated in it" (M III:399). Seeing the use of definitions at work in mathematics could help develop the sharp mind necessary to evade silly disputes over definitions and to distinguish language parading as knowledge from knowledge expressed in language.

Anyone continuing to insist on defining essences gives up on regulating philosophical language by reference to reality. In the conversational culture of Pascal's time, scholastic philosophers and rhetoricians were often decried for having permitted language to become detached from reference to the real world—for having lost touch with reality. Pascal singled out Noël. Since Pascal had refused to provide a definition of the essence of empty space, Noël imposed one on him. He delineated the best definition of "empty space" that he could glean from Pascal's writings: "a space that is neither God nor a creature nor a body nor a spirit nor a substance nor an accident, which transmits light without being transparent, which resists without resistance, which is immobile and moves itself along with the tube, which is everywhere and nowhere."[66]

Pascal responded by noting where Noël displayed great rigor and precision: "no one, after having read what I had written to him, would not laugh at the consequences that he had drawn from it. Based on the antitheses opposed with so much soundness [*justesse*], it is easy to see that he worked more to render his terms contrary one to another than to make them conform to reason and to truth."[67] Noël illegitimately transformed Pascal's enunciations about the vacuum. He regulated his language by the standards of ornate language rather than mathematical or physical reality. Language became the end, not the means. In Pascal's *Pensées*, antitheses such as Noël's exemplify figurative language that sacrifices meaning to linguistic form: "Those who make antitheses by forcing words are like those who make false windows for symmetry. Their rule is not to speak justly but to make just figures" (S466).[68] In Pascal's usage, "*justesse*" meant the hewing to the rules and standards of a specified domain, whether of reason or of words. Caught up in a logic of rhetorical figures, Noël's series of antitheses went far beyond representing Pascal's propositions about the apparent vacuum. Noël pursued entertaining ways of describing nature by appealing to rhetorical standards rather than by appealing exclusively to experience and to legitimate transformations of statements.

In the final letter of the correspondence with Noël, Pascal's father contrasted the title of Noël's book with Pascal's title *New Experiences concerning the Vacuum*: "To this simple, naive, ingenious title, without artifice and completely natural, you oppose this other title, *The Fullness of the Vacuum,* subtle, artificial, ornate, or, rather, composed of a figure called *antithesis,* if I recall correctly." In the schools, Pascal's father continued, this rhetorical figure was "not only permitted but also necessary." In contrast, "in the world [*monde*], it would not have passed, for it has no perfect sense."[69] The "world" in question is that of the conversing *honnêtes hommes,* those stalwart opponents of the pedantry of the schools, such as Pascal's family friend Jacques Le Pailleur.

Guez de Balzac, one of the foremost theorists of writing appropriate for *honnêtes hommes,* distinguished "two sorts of eloquence: one pure, free, natural; the other figurative, constrained, learnt; one of the world [*monde*], the other of the schools."[70] In a letter praising Guez de Balzac, Descartes likewise mocked those "who devote their time only to the seeking out of a few bons mots and of some games of the mind; the former usually consist in ridiculous equivocations, poetical fictions, sophistical arguments, and puerile subtleties." In contrast, Guez de Balzac's letters exhibit a "happy alliance of things with speech," with graces "so simple and so natural"; his use of rhetoric is completely different from the "deceiving and counterfeit beauties that customarily charm the people."[71] While retaining an important place for ornament and figurative language, Descartes denounced language that lacked a sufficient connection to meaning.

The *Port-Royal Logic, or The Art of Thinking* (1662) contrasted a good style with "an artificial and rhetorical style made up of false and hyperbolic thoughts and forced figures," which it declared nothing less than the "worst of all vices" (J23). The French conversational ideal of midcentury required a "habitual mastery" of rhetoric, an internalizing of the essence of good style and decorum.[72] The self-conscious, rote, and therefore artificial use of rhetoric was understood to divide craftsmen of words, the Schoolmen, from habituated users of rhetoric, whose style putatively never lost reference to the realities of the social and natural worlds. For the Schoolmen, rhetoric and dialectic were ends in themselves; for the "world," they were tools appropriate for regulating communal life and exploring the actual natural world. Or so claimed the *honnêtes hommes.*

As we saw in chapter 2, Jesuit rhetoricians stressed the power of metaphorical definitions to make descriptions become images, to appear as if they were before the eyes. Pascal and like-minded writers denounced Jesuit natural philosophers for producing false images, much as Martin Schoock had denounced Descartes. With his use of metaphor unregulated by reason or nature, Noël misused a keystone of his order's "rhetoric of paintings."[73] Whereas the conversational rhetoric of the *honnêtes hommes* supposedly moderated the passions and encouraged deliberation, consideration, and the weighing of opposing viewpoints, Jesuit rhetoric worked through emotive suasion; it offered "spectacles more true than nature" rather than nature itself.[74] In attempting to explain Pascal's vacuum experiments, Noël offered a pleasurable picture of how nature might be, rather than demonstrating from experiment and observation how it is. He transformed imaginative propositions about nature into essential definitions of its parts.

Pascal later denounced writers who produced descriptions that outstripped their true knowledge: "Eloquence is the painting of thought. And thus those who, after having painted, continue to add more make a tableau in place of a portrait" (S481). Descartes was guilty of making such a tableau: he had seen the need for a new science but had put on the discarded skin of the pedant. "The late Mr. Pascal called the Cartesian philosophy the novel of nature, similar more or less to the story of Don Quixote" (LG II:1087). For all their eloquence and beguiling charm, such metaphorical producers of tableaux do not perform valid transformations of enunciations; they do not preserve truth; they do not facilitate discovery; they do not perfect the use of language in reasoning. They substitute fictions for proved propositions and falsely claim certain knowledge of the principles of the world.

Both Descartes and Pascal saw a connection between improper forms of rhetoric and illegitimate reasoning in natural philosophy and in mathematics. Rhetoricians and scholastics alike produced a torrent of artificial techniques. Montaigne had stressed their delusive artifice: "The sciences treat things too much in an artificial way. . . . Were I of that craft [*métier*], I would naturalize the art as much as they artificialize [*artialisent*] nature."[75] Descartes and Pascal attempted the difficult task of naturalizing arts of reasoning, of producing forms of artificial reasoning that remained always linked to nature and to the aptitudes for knowing and speaking natural to human beings. They sought better linguistic techniques for producing knowledge, based, in part, on their understanding of the practice of good mathematics. For Pascal, the proper use of language could lead to certain knowledge of some aspects of mathematics and the world without ever producing clear and distinct knowledge of the entities involved. In drawing attention to mathematics as practiced, he sought means for recognizing real qualities of the world, without ever falling into a dogmatism grounded in a mistaken vision of the human epistemic condition and the power of language.

THE SURPRISING CONNECTIONS OF ADDING SUMS

With his numerous examples of the utility of the arithmetical triangle, Pascal demonstrated the power of producing new enunciations of the same proposition. As promised, many of these results were remote from any obvious implication of the triangle. One of the oldest of the parts of the treatise on the triangle considered the "summing of numerical powers." Pascal showed how to calculate the sum of any finite series of whole numbers each taken to a given exponent.[76] He wanted to show, for example, how to add the cubes of 5, 8, 11, 14 without actually performing the sum. His technique reveals, to take one case, that the sum of the squares of the first $n - 1$ whole numbers, $1^2 + 2^2 + \cdots + (n - 1)^2$, equals

$$\frac{n^3}{3} - \frac{n^2}{2} + \frac{n}{6}$$

Pascal's method for demonstrating this equality revealed an unexpected relationship between sums of powers and the arithmetical triangle.[77] His technique here illustrates how Pascal offered readers stimulating examples of connecting apparently remote enunciations; his technique is a fine example of that practice which "ought to be the study of mathematicians" (M II:1203).

First, consider the difference between two numbers $n + 1$ and n each taken to the fourth power, $(n + 1)^4 - n^4$. We saw above that the expansion of binomials such as $(n + 1)^4$ involves a series of numbers taken from a corresponding diagonal of the arithmetical triangle (the fifth). Thus, $(n + 1)^4 = n^4 + 4n^3 + 6n^2 + 4n + 1$. Substituting, we get $(n + 1)^4 - n^4 = (n^4 + 4n^3 + 6n^2 + 4n + 1) - n^4 = 4n^3 + 6n^2 + 4n + 1$. In other terms, the difference between two sequential numbers each taken to the fourth power always equals an expression with coefficients taken from the fifth diagonal on the arithmetical triangle:

```
1  1  1  1  1  1
1  2  3  4  5
1  3  6  10
1  4  10
1  5
1
```

Such a rule is true for any positive integer power.

Now write an arbitrary number taken to the fourth power as a sum of differences:

$$n^4 = n^4 + \begin{bmatrix} [(n-1)^4 - (n-1)^4] \\ + [(n-2)^4 - (n-2)^4] \\ + [(n-3)^4 - (n-3)^4] \\ + \cdots \\ + [(2)^4 - (2)^4] \\ + [(1)^4 - (1)^4] \end{bmatrix}$$

Subtracting something from itself equals zero: $n - n = 0$. Each of these differences equals zero, so nothing is added in this procedure. Now regroup these terms into a series of differences between pairs of sequential numbers:

$$n^4 = \begin{array}{l} [n^4 - (n-1)^4] \\ + [(n-1)^4 - (n-2)^4] \\ + [(n-2)^4 - (n-3)^4] \\ + \cdots \\ + [(2)^4 - (1)^4] \\ + (1)^4 \end{array}$$

Every such difference in the form $(n+1)^4 - n^4$ equals $4n^3 + 6n^2 + 4n + 1$. Replacing each difference with such an expression, we get

$$n^4 = \begin{array}{l} [n^4 - (n-1)^4] \\ + [(n-1)^4 - (n-2)^4] \\ + [(n-2)^4 - (n-3)^4] \\ | \cdots \\ + [(2)^4 - (1)^4] \\ + (1)^4 \end{array} = \begin{array}{l} 4(n-1)^3 + 6(n-1)^2 + 4(n-1) + 1 \\ + 4(n-2)^3 + 6(n-2)^2 + 4(n-2) + 1 \\ + 4(n-3)^3 + 6(n-3)^2 + 4(n-3) + 1 \\ + \cdots \\ + 4(1)^3 + 6(1)^2 + 4(1) + 1 \\ + (1)^4 \end{array}$$

Look down the parallel columns in the right-hand equation: the first coefficient 4 is multiplied first by $(n-1)^3$, then by $(n-2)^3$, all the way to 1^3. In other words, 4 is multiplied by the sum of cubes from $n - 1$ to 1, in modern notation

$$4 \sum_{i=1}^{n-1} i^3$$

Likewise, visual inspection shows that the coefficient 6 is multiplied by the sum of squares, the coefficient 4 by the sum of counting numbers. Bringing together terms with the same coefficients, we get

$$n^4 = \begin{array}{l} 4[(n-1)^3 + (n-2)^3 + \cdots + 1^3] \\ + 6[(n-1)^2 + (n-2)^2 + \cdots + 1^2] \\ + 4[(n-1) + (n-2) + \cdots + 1] \\ + (n-1) \\ + (1)^4 \end{array} = \begin{array}{l} 4[\sum \text{cubes}] \\ + 6[\sum \text{squares}] \\ + 4[\sum \text{numbers}] \\ + (n-1) \\ + 1^4 \end{array}$$

An arbitrary number taken to the fourth power equals a series of sums of numbers taken to every lower power. Modernizing we get

$$n^4 = 4\sum_{i=1}^{n-1} i^3 + 6\sum_{i=1}^{n-1} i^2 + 4\sum_{i=1}^{n-1} i + n$$

A number taken to an arbitrary power thus equals a sum involving the coefficients from the arithmetical triangle 1 4 6 4 1. In writing out the series of formulas obtained in this way, the arithmetical triangle quickly reappears in the coefficients:

$$n^2 = 2\sum_{i=1}^{n-1} i + n$$

$$n^3 = 3\sum_{i=1}^{n-1} i^2 + 3\sum_{i=1}^{n-1} i + n$$

$$n^4 = 4\sum_{i=1}^{n-1} i^3 + 6\sum_{i=1}^{n-1} i^2 + 4\sum_{i=1}^{n-1} i + n$$

and so forth, to any whole power.[78]

Recall that Pascal's purpose was to discover how to add a series of numbers taken to any whole power. These formulas provide a way. Rearranging the formula for n^4 gives us, for example,

$$\sum_{i=1}^{n-1} i^3 = \frac{n^4 - 6\sum_{i=1}^{n-1} i^2 - 4\sum_{i=1}^{n-1} i - n}{4}$$

Pascal showed that the sum of cubes equals an expression involving the sum of squares and the sum of natural numbers, multiplied by coefficients from the triangle. The same process works for any whole power, so we can always express the sum of numbers taken to a given power in terms of lower powers. Applying the formula recursively to each power, we can always find the sum we desire:

$$\sum_{i=1}^{n-1} i = \frac{n^2 - n}{2}$$

$$\sum_{i=1}^{n-1} i^2 = \frac{n^3 - 3\sum_{i=1}^{n-1} i - n}{3} = \frac{n^3 - 3\left(\frac{n^2-n}{2}\right) - n}{3} = \frac{n^3}{3} - \frac{n^2}{2} + \frac{n}{6}$$

and so forth, for any whole power.

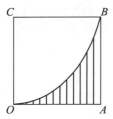

Figure 3.5. Quadrature of parabola. After Gardies 1984, p. 62.

FROM SUMS TO SPACE

The sum of the first $n - 1$ square numbers equals

$$\frac{n^3}{3} - \frac{n^2}{2} + \frac{n}{6}$$

In producing such a general result, Pascal claimed to have revealed an even more surprising connection between apparently remote things.[79] To his contemporaries, "however little versed in the theory of indivisibles," summing a series of numbers taken to any positive integral power meant one could find the area of parabolas. One could, that is, find the area for all curves such as $y = kx^2$, $y = kx^3$, $y = kx^4$, and so forth. The proof is relatively straightforward. Consider a simple parabola OAB within a rectangle $OABC$, with $OA = n$ (see fig. 3.5). If we divide the area of the parabola into n equally wide vertical strips, then the area will be approximately $1^2 k + 2^2 k + 3^2 k + \cdots + n^2 k$. By Pascal's method above, such a sum of squares (times k) equals

$$\left(\frac{n^3}{3} - \frac{n^2}{2} + \frac{n}{6}\right) k$$

If $OA = n$, then $AB = kn^2$, so the area of the rectangle $OABC$ is kn^3. The ratio of the approximate area of the parabola to the area of the rectangle is thus

$$\frac{OAB}{OABC} \approx \frac{\left(\frac{n^3}{3} - \frac{n^2}{2} + \frac{n}{6}\right) k}{n^3 k} = \frac{1}{3} - \frac{1}{2n} + \frac{1}{6n^2}$$

As n goes to infinity, the ratio of the area of the parabola to that of the rectangle becomes simply one-third.[80]

$$\frac{OAB}{OABC} = \frac{1}{3}$$

Pascal expressed the general result: "the sum of all the lines of some power is to the greatest [number] raised to the next highest power as one is to the exponential of the next highest power" (M II:1271). In other words, with his series of formulas for summing powers, Pascal had proved generally that the area of a parabola kx^r equals

$$\frac{kx^{r+1}}{r+1}$$

for integer $r > 0$.

A CONNECTION THAT CANNOT BE ADMIRED ENOUGH

Historians of mathematics see Pascal's procedures for finding the areas of parabolas—their quadratures—as a highpoint in a tradition of reducing quadratures to arithmetic.[81] Pascal saw profound significance in this demonstration, which concerned properties of the continuous qualities and used properties of the discrete counting numbers. He wished for readers to recognize "this connection that can never be admired enough, for in it nature, lover of unity, joins even the subjects that seem most remote into a single thing." In varying enunciations, Pascal had shown that things remote in appearance—the combinatorial numbers, figurate numbers, sums of powers, and arithmetical triangle—are but different ways of speaking about the same set of numbers. These special numbers all involve discrete counting numbers, whereas Pascal's quadrature of the parabolas unites the apparently far more disjoint realms of the continuous and discrete. A wondrous connection, he claimed, "appears in this example, in which the measure of a continuous quantity is conjoined with the sum of numerical powers."[82] Such a connection shows how mathematical practice investigates the properties of different sorts of objects and reveals startling and unforeseen connections among them.

In *Pensées*, Pascal returned to this "connection that can never be admired enough." In the fragment "Nature Imitates Itself," Pascal reflected on the surprising link between the counting numbers and the continuous quantity that he simply called "space."[83]

> Numbers imitate space—which are so different in nature. All is made and led by the same master: the root, the branches, the fruits, principles, consequences. (S577)

The close relationship between the apparently remote entities of the numbers and space points toward the hidden liaisons connecting other natural entities. For Pascal, investigating the numbers meant investigating the framework of the constitution of the universe. Grasping the consequences of numerical principles helped to reveal the manner in which God in fact had made nature according to number, measure, and weight.

> These three things, which encompass the entire universe, according to these words, "God made everything in number, weight, and measure" [paraphrase of Wisdom 11: 21], have a reciprocal and necessary liaison.
>
> For one cannot imagine movement without something that moves; and this thing being one, this unity is the origin of all the numbers; finally, movement being unable to exist without space, one sees the three things enclosed in the first. (M III:401)

Studying the numbers makes the intimate liaisons between space, movement, and number apparent. Human beings can have certain knowledge that these relationships exist, without thereby having knowledge of the causes behind them or the reasons for them. Such knowledge is certain without being causal, clear, distinct, or complete.

Nature's imitation of itself extends even to human activity: "Nature imitates itself: a grain, thrown into good earth, produces; a principle, thrown into a good mind, produces" (S577). In his *Treatise*, Pascal had produced his surprising connection with a method that aids in cultivating an inventive mind. Such a mind possesses the amplitude that allows it to grasp many principles all at once and keep them distinct. To use Pascal's parlance, such a mind can turn enunciations about to reveal new relationships among different sets of these principles.

In relating entities that we would otherwise think entirely disparate, this high point of Pascal's treatise on the triangle offered a key example of his normative conception of mathematics—as he said, what mathematics and mathematical practice *ought* to be. Pascal's method of varying enunciations produced a sequence of true enunciations, relations within varieties of special numbers and among varieties of numbers. In varying enunciations, human beings can find some of the true relations that God has realized in nature and has made human beings capable of knowing. They can discern some set of relations with absolute certainty; and they can discern some consequences of already-known sets of relations. Pascal did not offer a systematic method capable of producing all necessarily true enunciations. He did not claim to offer a method that could produce a comprehensive, or even a clear and distinct, idea of the mathematical objects at issue. With his method, nevertheless, one constantly produces certain connections among the objects of mathematics and of nature.

COMPREHENDING INFINITY

At some point in the early 1650s Pascal and his friend the chevalier de Méré—long-term conversational partner, source of problems in

probability, and paragon of *honnêteté*—came to disagree about the infinite divisibility of a line. Pascal mocked Méré for refusing to accept infinite
divisibility. Such an assumption was essential in proofs such as Pascal's
quadrature of the parabola. While Pascal appreciated Méré's conviction
that mathematical reasoning should not be applied in many domains, Méré
had failed to grasp where—and how—mathematics was certain. In sum-
marily rejecting infinite divisibility, he misunderstood what mathematics
could reveal about human nature; he misunderstood human nature itself.

In 1641, the Italian mathematician and natural philosopher Evangelista Torricelli showed that an infinitely long solid created by rotating a
hyperbola around its axis has a finite volume equal to a finite cylinder.
When he communicated the result in 1643, great celebrity, disbelief, and
astonishment followed.[84] Along with many other philosophers and mathematicians, Descartes had avoided questions of the mathematical infinite
primarily by appealing to the "indefinite," meaning a process continuing
without end, and by refusing to speculate about actually infinitely large
or small things.[85] In contrast, Torricelli's proof appeared to involve certain
knowledge, attainable in a finite amount of time, about an actually infinite
object.[86] (A *potentially* infinite object might be, for example, an object tracing
a line in time without ever stopping; an *actually* infinite object might be an
infinitely long line existing at a single moment.) Torricelli's result underscored how easily mathematical reasoning could lead to results violating
physical and metaphysical intuitions.

Thirty years later, the Jesuit Gaston-Ignace Pardies praised a similar finding of a finite area of an infinite space under a hyperbola in his textbook of
geometry: this knowledge is nothing less than "the most admirable thing
in the world." The proof "shows most clearly the greatness and spirituality
of our souls, since by the light alone of the mind, penetrating beyond infinity, the soul discovers things so clearly that no sensory experience could
learn it." Nothing made only of matter could measure anything infinite;
measuring the infinite requires something beyond the finite, namely, a nonmaterial soul. Since human beings can measure the infinite, they must have
a nonmaterial soul. Pardies called for using this proof against libertines and
skeptics. For Pardies, the proof showed that "infinity itself—as completely
immense and innumerable as it is—reduces nevertheless to calculation and
to measurement by geometry" so that "our mind, greater yet than [infin-
ity], is capable of *comprehending* it." He equated such comprehension with
clear and distinct knowledge. Human beings have "within us" nothing
less than "ideas and clear and distinct representations of an infinite extension." Found with human tools and without any special divine infusion of
knowledge, this extraordinary proof illustrates the true extent of human
ability: "Of all the natural knowledge that man can acquire through his

own reasoning, without doubt the most admirable is this comprehension of infinity."[87] As so often in the seventeenth century, a new mathematical result was marshaled to defend an account of the possible extent of human knowledge.

Proofs such as Torricelli's were startling; so was Pardies's claim about their significance. "Comprehensive" knowledge of something was a high standard indeed, whether from a Cartesian or a scholastic standpoint. The Jesuit Coimbra commentators explained, for example, that "[c]omprehensive cognition of a thing is the perception not only of all of its causes but also of its effects, . . . and its properties and its powers, . . . , and finally of all its perfections."[88] Comprehensive knowledge of the infinite was certainly not available to human beings. In the metaphysical picture offered by the commentators, all created things have powers and capacities proper to their place in creation. No created intellect could acquire knowledge of the infinite through its own means, for the "infinite exceeds the proportion of a finite potential." The commentators were clear that human beings, like all created intellects, including angels, cannot through their natural virtue perceive infinity, either all at once or "successively," in finite times, where the elements are understood distinctly one after another. In sharp contrast, God can "know the infinite in a perfect act," in which he knows the infinite all at once; he comprehends all its parts simultaneously and distinctly through an intuitive cognition.[89]

Descartes himself did not equate knowing something clearly and distinctly with comprehension in this scholastic sense. Only God had "entire and perfect" knowledge of all the properties of something; in contrast, human beings can know "fully" (*pleinement*), meaning that they can know something distinctly enough to recognize it as a "complete thing" (see AT IX/1:171–72). For Descartes, human beings can have neither complete (comprehensive) nor "full" knowledge of actually infinite quantities in mathematics. These fine distinctions were lost on many readers, who often conflated clarity and distinctness with comprehension.

The Jesuit Pardies held that the knowledge of the infinite provided in the squaring of the hyperbola showed that human beings had epistemic capacities far more proportionate to the universe than the Jesuit scholastics or thinkers such as Descartes had previously argued. Human epistemic capacities were closer to God's than often claimed. Some years previously, Pascal too thought that proofs such as Torricelli's revealed much about human epistemic capacity. No doubt the proof provided certain knowledge of the finite volume within an infinite surface. For Pascal, however, this proof was paradigmatic of certain knowledge that was nevertheless incomprehensible; to know the volume in no way meant one had clear and distinct knowledge or comprehension of an infinite space. This separation of certainty and comprehension is central to Pascal's account of mathematical practice. Discovering the volume meant simply discovering a relationship between an infinite space and finite numbers.[90] Two things previously remote in appearance are shown to be necessarily connected and capable of being known with certainty. In *Pensées,* Pascal alluded to Torricelli's infinite solid of finite volume: "Everything that is incomprehensible is not prevented from being. The infinite number, an infinite space equal to a finite one" (S182). Rather than testifying to human ability to comprehend the infinite, the proof testified to human ability to have certain knowledge about the infinite *without* comprehending it and *without* having clear and distinct knowledge of it. This incomprehensible knowledge could readily be applied in proving new mathematical theorems. Mathematics revealed that the human condition was far more perplexing than the Coimbra commentators, Pardies, Descartes, or Méré claimed. It seems reasonable that human beings, as finite beings, can have no knowledge of the mathematical infinite, but they do, in spades. Human beings can in fact easily gain certain knowledge about infinity—something they can never hope to comprehend.

In the essay "Mathematical Mind," probably written around 1655, Pascal considered mathematics as a source of knowledge about human ability. Mathematics offered powerful evidence for appreciating the greatness and the limitations of human beings. Unfortunately, few heeded the lessons mathematics as practiced had for philosophy. In fixing upon its outward form, logicians and philosophers had generally misunderstood the underlying "spirit" of the enterprise: "logic has perhaps borrowed the rules of mathematics, without understanding their force. And thus, in putting them pell-mell among [rules] that belong to [logic], it does not follow that they have entered into the mathematical mind" (M III:425). Like people who happen to say something insightful by chance, the logicians lacked the foundation in sound judgment—the mathematical mind—necessary to apply those rules in judging and in discovering.

In the essay, before turning to mathematics, Pascal discussed the perfect or ideal order of demonstration: "This true order, which *would* make up demonstrations of the greatest excellence, if it were possible to arrive there, *would* consist" in "defining all" and "proving all" (M III:393; my italics). Human beings simply cannot achieve such perfect demonstrations, for they would have to continue to define things without ever being able to stop. "Whence it appears that men are in a natural and unchangeable powerlessness to treat any science, whatever it be, in an absolutely accomplished order" (M III:395). This ideal rhetoric and logic is a false goal for humanity.

Human beings cannot achieve this ideal order of knowledge; if they attempt to, they will deceive themselves by inventing false principles that are masked as knowledge. Recognizing this inability ought not to inspire a fatal skepticism, for such skepticism misunderstands what knowledge is possible. The practice of mathematics offers a remarkable example of a middle path drawing on human abilities: "This [mathematical] order, the most perfect *among men*, consists not in defining or demonstrating everything, or in defining and demonstrating nothing, but in holding oneself in this middle point in not defining clear things known to all men and in defining all the rest; and in not proving things known by men and proving all the rest" (M III:395; my italics). This middle course steers clear of two errors: "Against this order sin equally those who set out to define and prove everything and [also] those who neglect to do both for things not evident in themselves" (M III:395). Traditional philosophy commits both sins: it tries to define the indefinable and it deals with things hardly evident in themselves. Duly stepping in to provide ersatz principles, the imagination masks the deficiencies of reason with convenient fictions.

The limitations and powers of human ability are built into mathematics as practiced. Pascal's account of mathematics rests on his claim that the fundamental objects of mathematics, including space, time, and motion, are known through "natural light."[91] Attempting to define these objects further is impossible. One can accept this ability to define objects or prove principles, he argued, "for this single and highly advantageous reason": they "are in an extreme natural clarity that convinces reason more powerfully than speech." This extreme natural clarity characterizes the ability to augment and to reduce numbers, spaces, and times. "For what is more evident than this truth: that a number, whatever it be, can be augmented? Can one not double it?" Likewise, "who can doubt that a number, whatever it be, can be divided in half, and its half in half again?" Indeed, Pascal held there to be "no natural knowledge at all in man which precedes these [truths] and which surpasses them in clarity." Not susceptible to proof, these truths make mathematical certainty possible. "All these truths cannot be demonstrated, and yet they are the foundations and principles of mathematics" (M III:403). Although the nature of these fundamental entities is undefined and indefinable, propositions concerning them can be known with certainty—a greater certainty than any other human knowledge.[92] This certain knowledge of undefined and indefinable things produced in mathematics testifies to the greatness of human ability. The fact that this certain knowledge rests on truths escaping all proof serves as a reminder of human inability.

The certainty of mathematics entices people to translate its logical form into other domains. In other domains, such as religion, much of natural philosophy, and anthropology, human beings simply do not have knowledge of the basic entities, known through natural light or other faculties,

with which to begin their reasoning.[93] Without something given as principles, known intuitively through natural light, a logic mimicking the mathematical style and applied to other domains can only remain unmoored
and useless, merely a speculative logic in the worst sense, incapable of
providing even nominal definitions for its basic entities. Such a logic is
condemned to deal in ersatz principles that are not evident in themselves,
as if its principles were as certain as mathematical principles. A limited human ability—mathematical reasoning—had been mistakenly generalized
into the claim that humans might gain certain knowledge of the principles
of philosophy using a logic that mimics mathematical form. For Pascal,
mathematical ability is a crucial quality of human beings; a logical ability
to grasp the principles of the world is not.

Even mathematical knowledge is sharply limited. That human beings
have some knowledge of fundamental mathematical principles does not
imply that human beings comprehend all the necessary consequences following from those principles, or even that they could comprehend them.
Human beings can prove propositions about these fundamental entities of
space and time; they can have certainty and clarity about some qualities of
these entities; but they do not comprehend these entities as wholes. Like
Descartes, Pascal maintained that the fundamental principles of mathematical reasoning are known clearly through natural light; our knowledge
of them is evident. In contrast with Descartes, Pascal maintained that the
consequences of reasoning from such naturally known principles are *not*
known evidently.[94] To conclude with certainty that something is the case
starting from evident principles does not mean that the conclusion is known
evidently. For Descartes, mathematical proofs rest on evident principles *and*
preserve evidence; for Pascal, mathematical proofs rest on evident principles but do *not* preserve evidence.[95]

Pascal separated conviction about the *existence* and certain properties
of a thing from true knowledge of its *nature*—the comprehension of that
thing in its totality or, at least, of its essence. Concerning mathematics, for
example, he claimed that we conceive that we can always go faster than
a given speed or augment a given line or remain twice as long; likewise,
we can conceive clearly that we can always cut our speed in two, shorten a
line, or remain half as long (M III:402). It is contrary to human knowledge
about these entities that one would ever have to stop these processes—
the absolute clarity of our knowledge that mathematical objects can be
augmented and diminished guarantees that one could continue doubling
or dividing them forever. For Pascal, our evident knowledge that we can
always continue to divide implies that there is an actual infinite division
in mathematical objects and in nature; and our evident knowledge that we
can always continue to double implies that there is an actual infinitely large

number (one number greater than all others) in mathematics and nature. Descartes accepted the *possibility* of dividing and augmenting without limit but rejected that this possibility implied the *reality* of infinitely small and large numbers.[96] Pascal maintained, therefore, that everyone must accept the actual existence of two infinities: one of greatness, one of smallness. Based on clear and certain knowledge of the qualities of the elements of space and time, one comes to know with certainty that these two infinities exist. One does not, however, comprehend these infinities as clear and distinct ideas. "We know that there is an infinite, and we ignore its nature, since we know that it is false that the numbers are finite. Therefore, it is true that there is a numerical infinite [*un infinie en nombre*], but we do not know what it is" (S680).[97] Human beings cannot conceive the two infinites by imagining and picturing them or by having clear and distinct ideas about them. From certain and clear knowledge of the existence and nature of the elements of space and time, they get certain and incomprehensible knowledge about the existence, but not the nature, of space and time taken as infinite wholes.

Pascal sketched three kinds of relationships that human beings have to the things they would like to know:

> We know therefore the existence and the nature of the finite because we are finite and extended as it is.
>
> We know the existence of the infinite and ignore its nature because it is extended like us but has no limits as we do.
>
> But we know neither the existence nor the nature of God, because he has neither extension nor limits. (S680)[98]

Certain but incomprehensible knowledge is at the heart of practiced mathematics, and every practicing mathematician accepts it, whatever her other philosophical commitments. Every mathematician "believes that space is divisible to infinity," although no mathematician "comprehends an infinite division"; indeed, one can no more be a mathematician without this belief than "a man [can be] without a soul" (M III:404). The chevalier de Méré, simply by doing mathematics, implicitly accepted the views he explicitly repudiated. He ought to have acknowledged how quickly human reason can illustrate its own limits.

Many philosophers and mathematicians resisted claims of knowledge about infinities because they demanded that all legitimate knowledge be comprehensible (M III:404–5). For Pascal, these incomprehensible infinities exemplified an essential sort of human knowledge about the created world. In contrast to a logic mimicking the outward form of mathematical reasoning, practiced mathematics shows that rational people who accept the human condition—anyone with good judgment—must accept the certain existence of things they can never hope to comprehend. In considering the basic entities of mathematics, human beings properly revel in its greatness and in their own greatness; they are equally led to recognize its limits and their own: "for what passes geometry surpasses us" (M III:393). Mathematics, perfectly proportioned to deal with finite objects and operations on them, forces human beings to recognize their disproportion: "those who will clearly see these truths will be able to admire the greatness of the power of nature in this double infinity that surrounds us everywhere and to learn through this marvelous consideration to know themselves" (M III:411). More than mathematical practice itself, reflecting upon mathematical thought and practice helps human beings to esteem themselves correctly and to recognize the powers and limits of human reason. Mathematical practice serves as a great source of empirical data on human abilities and inabilities.

Proofs of a finite volume in an infinite surface occasioned not just intellectual appreciation but widespread affective and cognitive wonder. Pascal sought to generalize this wonder. Marveling at human incomprehension of mathematical truths such as the two infinities leads to better knowledge of humanity. "Upon which [consideration] one can learn to figure one's true worth and to produce thoughts more valuable than all the rest of mathematics" (M III:411). Through natural philosophy, Descartes hoped to delimit astonishment, wonder so powerful that it fixes the attention only on the first surfaces perceived, and to replace it with more controlled forms of appreciation that spur further inquiry into real knowledge of the causes of things.[99] Descartes excluded the infinite as unknowable; in doing so he excluded something deserving astonishment. Pascal praised mathematical reason in no small part because it could generate affective wonder that forced human beings to acknowledge their limits, to recognize that at appropriate moments they ought to cease their investigative quest and simply gawk.[100] Mathematics ultimately can excite and not merely instruct. Seriously considering the implications of mathematical practice can *move*, can transform a human being into a state of reasonable abasement, into a different relationship with his own epistemic and affective capacities. Aping mathematical form in logical proofs could never yield such shock and affective transformation.

LOGICAL CONVERSATIONS: ARNAULD, PASCAL, AND THE YOUNG DUKE

A friendly disagreement provoked the writing of *Logic, or The Art of Thinking* (1662), the so-called *Port-Royal Logic:*

The birth of this little work was entirely due to chance and belongs rather to a species of diversion than to a serious design. A person of quality conversing with a young Seigneur, who showed great strength and penetration of mind, said to him that, when he was young, he had found a man who rendered him capable of contending with a part of logic in fifteen days. This speech gave occasion to another person present, who had no great esteem for this science, to respond while laughing that if Monsieur wanted to take the trouble, he would engage to teach him in four or five days all that is useful in logic. Once voiced, this proposition prompted some discussion, and it was resolved to make an attempt at it. . . . See then the meeting that produced this work. (J7)

The encounter described took place in spring 1657; the "person of quality" was almost certainly Pascal; the young man was Charles-Honoré d'Albert, the son of the duc de Luynes, the translator of Descartes' *Meditations* into French; the other person was the great Jansenist theologian Antoine Arnauld.[101] The pedagogical attempt eventually yielded a written account of the art of thinking, an account animated by the tensions among the interlocutors. The *Logic* brought together, not always without apparent contradiction, the views of Pascal and Arnauld, along with the later contributions of Pierre Nicole.[102]

Drawing on Descartes, the authors, primarily Arnauld and Nicole, featured mathematics prominently, as briefly discussed in chapter 1. Rather than constraining reason with the traditional rules of logic, the controlled procedures of mathematics could train reason to distinguish the true from the false. Mathematics improves the capacity and strength of the mind: "The capacity of the mind is extended and strengthened through habituation, and mathematics serves principally to this end, as do, more generally, all difficult things . . . for they give the mind a certain extent, and they exercise it in applying itself further and in holding itself more firmly in what it knows" (J16). The authors—particularly Nicole—showed little interest in mathematics detached from its proper role of helping people to pursue a moral and religious life. The sciences "are completely useless, if one considers them in themselves and for themselves" (J10). Accordingly, any parts of mathematics not conducive to the training of the mind are useless, given the purposes of human life.[103]

Nicole and Arnauld were skeptical of much contemporary mathematical practice, including techniques central for Pascal. In the ninth chapter of the fourth part of the *Logic,* the authors turn to consider "some faults that are regularly encountered in the method of mathematicians." The first fault comes from "having more care for certainty than for evidence": "it does not suffice to have a perfect knowledge of some truth to be convinced that it is true, if one does not penetrate, through reasons taken from the nature of the thing itself, *why* it is true; for, until we have reached that point, our mind is not fully satisfied at all."[104] For Arnauld, such clarification required a causal proof; he apparently equated such causal understanding with the Cartesian pursuit of evident knowledge.[105] The failure of mathematicians to insist on proofs that clarified as well as convinced was "the source of nearly all the faults" in mathematical practice (J307). In his preface to Arnauld's mathematical textbook, *New Elements of Geometry*, Nicole complained that Euclid's *Elements*, "far from being able to give the mind the idea and taste of true order, [is], in contrast, able to accustom it only to disorder and confusion."[106] Like bad poetry and bad conversation, bad mathematics could spoil the mind, no matter how certain it may be.

Having demanded evident proofs, Arnauld and Nicole turned quickly to another ordinary fault of mathematics: "proof by contradiction"— *reductio ad absurdam*. A *reductio* proof involves first assuming the opposite of what one intends to prove, then showing that such an assumption leads to a falsehood, often something such as $1 = 2$. Such proofs have been central to mathematical practice since the Greeks and have long upset philosophers, for they provide no causal demonstrations.[107] Since they afford no clear and distinct knowledge of the mathematical entities involved, Descartes refused to admit *reductio* proofs. Arnauld and Nicole likewise assailed this form of proof: "These kinds of demonstrations that show that a thing is so, not by its principles, but via some absurdity that follows if it were otherwise, are quite ordinary in Euclid. Although it can be seen that they can convince the mind, they do not clarify it whatsoever, which must be the principal fruit of this science."[108] For Arnauld, Nicole, and Descartes, producing mere conviction through a proof really did not count as knowledge capable of exercising a mind.

For Pascal, in contrast, the practice of mathematics did not always involve clarification and the production of evident knowledge, whatever philosophers, even those with whom he was somewhat sympathetic, such as Arnauld, might desire.[109] Mathematics was inferior to the perfect science (which is impossible for human beings) "in that it is less convincing, but not in that it is less certain" (M III:395). In rejecting mathematical practices such as *reductio* proofs, Arnauld and Nicole misunderstood the knowledge human beings attain through mathematics and other disciplines. This misunderstanding stemmed from a profound reluctance to admit that human beings often know things only indirectly. In "Mathematical Mind," Pascal argued, "It is a sickness natural to man that he believes that he posseses truth directly. From this comes the disposition always to deny all that is incomprehensible to him. Whereas in fact he knows naturally only falsehoods, he ought to take for true only those things whose contrary appears false to him" (M III:404).

Anyone with a true estimation of human ability must accept *reductio* proofs, above all when dealing with those things human beings simply cannot understand: "Whenever a proposition is inconceivable, it is necessary to suspend judgment about it and not to deny it on this ground, and to examine its contrary. If one finds the contrary manifestly false, one can boldly affirm the first, however incomprehensible it is" (M III:404). In rejecting *reductio* proofs, Arnauld and Nicole wrongly rejected a central means for human beings to gain knowledge. They failed to acknowledge the centrality of the incomprehensible in much certain knowledge.

In *Penseés* and "Mathematical Mind," Pascal stressed that the utility of mathematics and natural philosophy in life was less for clarifying the mind for everyday use than for revealing the gap between what could be shown to exist with certainty and our comprehension of those things. In a section largely drawn from Pascal that fits uneasily with its primary Cartesian focus, the *Logic, or The Art of Thinking* stressed the importance of accepting "things that are incomprehensible in their manner and that are certain in their existence. One cannot conceive how they could be, and it is nevertheless certain that they are." The utility of considerations of infinity is "to teach one to know the limits of our minds and to make one confess, despite what one can know about it, that some things exist even though one is incapable of comprehending them" (J278, J280). In restating Pascal's claims, Arnauld and Nicole subdued them; they largely reduced them to Descartes' prohibition of considerations of the infinite, itself largely taken from the scholastic tradition.[110]

In adopting Pascal's arguments only in part, Arnauld and Nicole missed the force of his claims. Pascal illustrated that considerations of infinity are presuppositions in *everyday* mathematical practice involving apparently finite mathematical objects. The mathematical proofs Arnauld and Nicole claimed to be evident rest on assumptions beyond human ken. Where the Jesuit Pardies saw the proof of an infinite space equal to the finite as evidence that humans could have clear and distinct knowledge of the infinite, Pascal saw in this demonstration the human ability to have certain knowledge about something we cannot comprehend—a sure sign of the incomprehensible mix of the great and the wretched in humanity. Pascal put the ubiquity of the incomprehensible at the center of his "apologetic" project sketched in *Pensées*.

HOLLOW CONVERSATIONS

For all his praise of conversation, Pascal found that of his worldly colleagues wanting, just as he ultimately found mathematics wanting. In an autobiographical fragment, Pascal described moving from scientific circles to the broader conversational circles around people like the duc de Roannez: "I had spent much time in the study of abstract sciences, and the little communication available there made them distasteful to me." The study of mathematics and the natural world did not partake enough in the sociability Pascal desired, so he abandoned them. His deepening spirituality made him recognize how little scientific subjects mattered next to the study of humanity, a study in which he felt certain that he would find colleagues and real sociability: "I believed I would find at least a number of companions in the study of man, for it is the true study that is proper to him. I was deceived: there are fewer there than those who study mathematics" (S566). Pascal expressed his belief in the superior acuity in judgment and taste of his *honnête* friends. The chasm between their heightened discernment and the vacuity of their conversational topics disturbed Pascal.

Conversation best taught one to understand and to judge people, books, and plays. It helped make reasoning rigorous. It best developed speaking and reasoning without the crutches of artificial rhetorical figures and syllogisms. Yet those steadfast conversationalists refused to turn their gaze and examine humanity in depth—they refused to consider with care the one subject that ought to be their natural study, or at the very least their self-interested study. Those people most apt to have fine discernment, those most capable of instructing and communicating their judgments, shied away from applying their skills to themselves. In his projected apology, Pascal took it upon himself to turn their discernment, their conversational acuity, their natural forms of speech and thinking, upon themselves.

It is necessary to "have these three qualities: Pyrhonnian skeptic, mathematician, Christian." Crossing this statement out, Pascal rewrote it with his definitions of these terms: "to know how to doubt when it is necessary, to assure when it is necessary, while submitting when it is necessary." Someone who only doubted or someone who was only certain misunderstood the genuine power of reason itself. Likewise, someone who only submitted "misses knowing when it is necessary to judge" (S201). Mathematics taught how to judge and how to discern when one could and could not reasonably be called upon to judge. We need not give up on knowing something of the infinite: human beings can know it exists, can even prove things about it, without ever knowing its nature. Mathematics best showed the power and the limits of humanity—the greatness and wretchedness of human beings as epistemic beings.

In his projected apology for Christianity, sketched out in the 1650s, Pascal attempted to demonstrate, in rhetorical forms adapted to real human needs, that to be truly reasonable, to truly possess the judgment of the *honnête homme* and the real mathematician, led inexorably to a vision of humanity that Christianity alone could explain. He sketched his conversational strategy:

> If he praises himself, I will abase him,
> If he abases himself, I will praise him,
> And I will contradict him always,
> Until he understands
> That he is an incomprehensible monster.
>   (S163)[111]

Only this give-and-take might lead the skeptical reader to the proper mix of esteem and disdain for human ability. Leading his readers to reflect upon mathematics and natural philosophy would be central to making evident this incomprehensible mix of the great and the base.

CHAPTER FOUR

The Anthropology of Disproportion

In a long letter of February 1697, Gottfried Wilhelm Leibniz expressed hope that his "discoveries in Mathematics, about which the public is now well apprised, ... will contribute something to the credit of my philosophico-theological meditations." He had heard "a little story" some years back about bolstering one's theological claims with one's mathematical renown: "You know that Mister Pascal, dead too soon, was at the end dedicated to establishing the truths of religion. Since he passed with reason for an excellent mathematician, those of his friends dedicated to religion ... thought that it would be advantageous to religion itself if others saw, based on his example, that strong and solid minds can be good Christians at the same time."[1]

A set of "extraordinary" mathematical discoveries about a curve called the cycloid presented a great opportunity to demonstrate that good Christians were not irrational. Initially little interested in publicizing his results, Pascal was counseled to set up a public competition to challenge the greatest mathematicians in Europe. Their struggles in contending with Pascal's difficult problems would illustrate that his turn to religion had not dimmed his mathematical acuity. In making his genius evident, another account explains, "it would take away" the normal objections of atheists "to proofs of religion, namely: that only credulous and feeble minds ... ever admit proofs supporting the truth of the Christian Religion."[2] Many skeptical contemporaries mocked the pious as irrational dupes, who were unwilling to face the absurdity of their beliefs and refused to see religion as nothing more than mere custom with no rational or revealed warrant. The province of the credulous, irrational, and the foolish, piety had no place in the life of a straight-thinking person. Demonstrating Pascal's extraordinary rationality with a mathematical competition would rebut such claims: even the most rational, the most incisive, the most skeptical, and the most mathematically brilliant could be deeply devoted to a pious Christian life. A devout geometrician, Pascal would show, was no contradiction.

In *Pensées*, his draft notes toward a prospective "apology" for Christianity, Pascal sought more than to demonstrate that living a Christian life was not irrational. Far from contradicting Christianity, he claimed, reason should point one inexorably toward it. He deployed mathematics and natural philosophy in his attempt to direct his rational, skeptical contemporaries to his rigorous and dour Christianity. In reflecting upon mathematics, Pascal wrote, "one can learn to figure one's true worth and to produce thoughts more valuable than all the rest of mathematics" (M III:411). Mathematics was powerful less in offering exercises that prepared one to judge truth and falsity in all matters, as Descartes and his followers maintained, than in revealing the limits and extent of human knowledge—of reason itself.

Pascal tried to show that reasonable people ought to reject the sundry spiritual exercises available in the mid-seventeenth century, be they the less rigorous Catholicism of the Jesuits, the promises of the Stoic, Epicurean, and Skeptical revivals, or the exercises of Descartes and his followers. In his *Provincial Letters* and *Pensées*, Pascal showed that reason could easily expose the paradoxes of the human condition, contradictions that no other philosophy could admit or explain, much less cure.[3] Honest consideration of the best human mathematical and natural-philosophical knowledge should make any reasonable person accept these contradictions, recognize the inability of other philosophies and religions to contend with them, and search in desperation for answers elsewhere.

Who did Pascal think these reasonable people were? Not the abstract *homo oeconomicus* of the rational-choice theorist. For much of his life, Pascal adapted the aesthetic and epistemic standards and rhetorical practices of the French conversational culture of the *honnêtes gens*—cultivated noble people—to different domains of inquiry. In his pseudonymous *Provincial Letters*, Pascal dramatized how the good sense of the *honnête homme*— the cultivated generalist—could easily clear away the needless obfuscations of theologians. In his *Pensées*, Pascal turned this presumptive good sense against the *honnêtes gens* themselves. He demanded that they critically examine themselves to recognize the limits of their skeptical purview and to see how they consistently failed to criticize their own actions and motivations. Cast in their own critical light, the *honnêtes hommes* would see their own hypocrisy: they advocated reason yet were remiss in applying that reason to the most important of questions, their final end. In failing to turn their critical eye on themselves, they remained blithely ignorant of their hypocrisy. This absence of conscientious self-reflection, even among the most discerning of Pascal's contemporaries, illustrated the monstrosity of human nature: the mix of perspicuity with blindness, of rational motivation with irrational and contradictory desire. No philosophy constructed by human reason could hope to explain such a contradictory being; no such philosophy, whether that of Seneca, Descartes, or Aristotle, could provide spiritual exercises necessary to console and tame such a beast. Pascal maintained that only his rigorous brand of Christianity, a Catholicism grounded in the "empirical" evidence of the Bible and the church fathers, could explain human nature and console those people who had seen and accepted their own monstrosity.[4]

Most philosophical spiritual exercises used knowledge of the natural world to comfort those who practiced them, whether to demonstrate the existence of Providence, to reveal a just creation, or to show the superiority of humanity over all other beings. In *Pensées*, Pascal used knowledge of the natural world to upset and to dismay, to shock his readers into recognizing that the natural and mathematical knowledge available to human beings rests upon presuppositions surpassing human ken. Beyond challenging traditional beliefs about nature, the new mathematical and natural-philosophical discoveries of the seventeenth century should unsettle rarely interrogated beliefs about human nature. Marveling at the limits of human mathematical and natural knowledge could lead to abasing reason without ever abandoning it.

FROM POPULARIZATION TO REASONING PUBLICLY

In 1640, the Jesuit Pierre Le Moyne published a work entitled *Moral Paintings*, in which he assembled sundry pictures, dialogues, poems, and philosophical explanations. Devised to "instruct the reader pleasurably and give him a useful and serious recreation," his moral "gallery" vividly portrayed a worldly Christian ethical life.[5] To motivate his ethical picture, he offered a popularization of the largely Aristotelian physics, psychology, and theology of the Jesuits. Using the example of the compass needle, Le Moyne portrayed the natural inclinations of all created things toward their sovereign good. With the heuristic of the magnetized needle, he offered an optimistic view of human beings, who, he believed, naturally tended to know and to do good. Through his descriptions of the natural world, as well as through his manifold poems and dialogues, Le Moyne worked to lead the reader to accept and internalize the view, supported by the Jesuit order, of pious, active Christians living integrated in their contemporary society.

Opponents of the Jesuits considered books such as Le Moyne's dangerous: these books popularized lax morals and justified non-Christian living, in this case through scientific balderdash. For more pessimistic Catholics, such as the so-called Jansenists, Le Moyne's popular books showed to what lengths the Jesuits would go to replace what the Jansenists saw as true Christian doctrine with whatever might appeal to the most people.[6] In his *Provincial Letters*, Pascal called another book by Le Moyne, *Easy Devotion*, "an entirely charming picture of devotion" before denouncing *Moral Paintings* for its savage depiction of ascetic and rigorous Christians (CF 158–60).

Some years later, Pierre Nicole reminisced that Pascal had encouraged him to work to remove the "air of harshness" from the doctrine of Saint Augustine "and to put it into a state proper for being tasted and embraced by more people." As portrayed by Nicole, Pascal seems to have recommended altering religion to fit popular taste. Although Pascal was the least likely to deviate from Saint Augustine's teaching on grace, Nicole continued, "he said nevertheless that if he were to treat this matter, he would hope to render this doctrine so plausible, and to remove from it a certain fierce air . . . so that it would be proportionate to the taste of all sorts of minds."[7] Like Jesuits such as Pierre Le Moyne, Pascal tried to make proper religion appeal to French elites. Unlike Le Moyne, Pascal sought to make them see that their current lives were radically incompatible with religion. Whereas Le Moyne popularized scholastic doctrines more or less justifying much of the current lives of his readers, Pascal publicized religious ideas calling for a retreat from most secular goals. Their different versions of making religion popular rested on profoundly different visions of religious authority and the human ability to know.[8]

With his rhetorical illustration of scholastic philosophy, Le Moyne aimed to encourage elite Frenchmen and Frenchwomen—the *honnêtes gens*—to turn to a tempered Christianity and to conform to the moderation central to Aristotelian ethics. Le Moyne illustrated that their ways of life were close to proper Christian life: that Christianity required no complete retreat to the desert, no total abandonment of the pleasures of daily life, only a tempering, a moderating of excess, and the avoidance of outright immorality and sin. The rhetorical and dialectical manuals of the Jesuits called for popularizing the truths of scholastic philosophy and theology: Jesuit writers and preachers aimed to convert authoritative theological and metaphysical knowledge into rhetorical forms capable of disseminating that knowledge and generating strong affects about it.[9] Such dissemination rested fundamentally on rhetorical decorum, the demand that a speaker or writer use a style appropriate to the audience at hand and the subject to be communicated. In the introductory material of his *Moral Paintings*, Le Moyne explicitly followed this path of decorous dissemination. He promised to present the truths of scholasticism without "the roughness and the blemishes of their birth," still too apparent in scholastic textbooks and treatises. Le Moyne would remove "all that could offend delicate souls." The true propositions of philosophy, essential to a proper ethical life, had not been given "all the proper dimensions and all the ornaments of which they are capable."[10] Disseminating proper ideas decorously meant making them, in principle, more appropriate in form and simplifying them, not changing or weakening them.[11] Le Moyne exemplified the great, and often successful, Jesuit strategy of accommodating the outward character and religious demands of the church to local conditions and needs, from Paris to courtly settings, from rural Italy to China.

Pascal publicized theology in a less disseminative vein: rather than simplifying complex doctrines for a general public, he sought to reveal the essence of those doctrines for a group of insightful nonspecialists, to tear away the jargon precluding nonspecialists from being able to understand them. In his *Provincial Letters* (1656–57), a fictional correspondence printed as pamphlets, he took the standards of judgment and commonsense reason of the *honnêtes gens* and used those standards to examine, ridicule, and condemn Jesuit theology. He flattered his readers by appealing to their sense of possessing such standards, such commonsense rationality and reasonableness, and by using their standards of judgment and of expression—or what he portrayed as their standards—to ridicule and undermine Jesuit theology and casuistry.

Scholars have long argued that Pascal wrote his *Provincial Letters* largely for a nontheologically sophisticated audience, the audience of the salon, the *mondaine*, the *honnête*—the sort of person captured in the fictional addressee, the provincial friend, as well as the fictional narrator, shocked by all he finds.[12] In the early *Provincial Letters*, the narrator, a character later named Louis de Montalte, easily and quickly comprehends matters previously discussed by specialists alone; where they insist on the need for scholastic logic, he uses only a commonsensical reason and judgment.[13] Montalte writes with the tone of the nonspecialist denouncing a pretentious academic book loaded with jargon. The letters begin "We have been misled": the letter writer and his readers have been misled about the difficulty of the theological endeavor and the necessity, therefore, for theological authorities (CF 3).

With his reason and rhetorical skill, Montalte can agreeably educate others in the most potentially complex issues of theology. In the *Letters*, he reveals the purely *formal* agreement among a variety of theologians by translating their technical language or jargon into an ordinary language of good taste and simple reason, uncorrupted by artificial subtleties of baroque rhetoric and scholastic dialectic. The theologians agree to a set of statements but disagree about the meaning of all the key terms in those statements. The *Letters* include fictional responses penned by Pascal. In one, the addressee praises Montalte for showing the superfluity of the obfuscations of the theologians. Thanks to Montalte's intrepid detective work and flat-footed common sense, the theological vocabulary—"proximate power" and "sufficient grace"—that previously had "menaced" the reading public no longer would do so. The discerning public now knows that they can grasp—and even purify—theology. They can judge the portentous blathering of professional theologians, who had too long obfuscated theology with empty scholastic artifice in order to ensure their socioprofessional and epistemic privilege. Through his characters, Pascal asserted his individual judgments against the doctrinal authority of professional theologians; and he asserted the value of the language and judgment of good taste and commonsensical reason against theological formalism and logical methods.[14] As the sequence of letters continues, the Jesuits are repeatedly shown—admittedly through Pascal's clever manipulation, distortion, and disposition of evidence—to violate common sense about nature, language, and Christianity.[15]

A brief response to the first two letters, supposedly written by the provincial addressee of the letters, illustrates the goals of the strategy: "The entire world sees [the letters]; the entire world understands them; the entire world believes them. Not only admired by theologians, they are pleasant to people of the world, and intelligible even to women" (CF 36).[16] The aesthetic power of the letters is essential to their intellectual power. Good writing is not some supplementary rhetorical flourish; producing it involves real philosophical work. The artifices of the formalistic style of the theologians produce discourse without reference to meaning.[17] In principle, in the natural style praised by Pascal, writing will be produced only with reference to meaning.

The invocation of women underscores the importance accorded nonpedantic insight in Pascal's *Provincial Letters*. Much writing on courtesy and *honnêteté* maintained that mixed-sex conversation sharpens judgment and clarifies speech. As one manuscript dialogue explains, such conversation teaches one not to confuse "the discernment of pedantry with the true knowledge [*science*] of *honnêtes gens*"; it helps to cultivate "the means of finding the middle between the affectation of the sciences and crass ignorance."[18] The provincial friend quotes a letter from an anonymous woman praising one of Montalte's letters: "It narrates without narrating; it clarifies the most convoluted affairs of the world; it jokes with skill; it instructs even those who do not know things well; and it doubles the pleasure of those who understand them" (CF 37–38). The understanding of women was believed to be an index of clarity and knowledge, not simplification, as described in the important *Honneste femme* of the 1630s: "There is nothing so true as that, when the Sciences are so well conceived . . . , women are capable of understanding them."[19]

Regardless of whether these statements report the responses of actual readers, they identify the intertwined aesthetic and epistemic values at the heart of the *Provincial Letters*. Clarification and real instruction accompany pleasure. Far from popularizing a thorny issue with simplification and rhetorical flourish, as Jesuit rhetoricians did, Montalte sweeps away the difficulties and the need for a formal class of trained theologians to contend with them. Rather than popularizing, he makes these theological questions public, that is, amenable to public reason. He brings them to his elite reading and conversing public; they appreciate and understand them; they apply a truly commonsensical reasoning to them; in doing so they, in principle, could do substantial philosophical work in purifying and clarifying them.[20]

The Jesuits recognized Pascal's attempt to deprecate formal theological work and the hierarchies supporting that work.[21] In merging two distinct domains, the serious domain of philosophy and the agreeable domain of rhetoric, the Jansenists and their lackey Montalte collapsed reason into laughter and threatened proper hierarchical organization with ecclesiastical anarchy, a concern akin to the fear of enthusiasm that so preoccupied Catholics and Protestants alike. Pascal's letters, in principle, disseminated theological and hermeneutic authority, not authoritative doctrine. They were not demotic, however. The means for securing authority moved from a formal system of theological reason possessed only by people at the top of a theological hierarchy to a putatively commonsensical reason possessed by people with discernment, those people characterized by something akin to the good sense described by the chevalier de Méré. Intelligent generalists with discernment and taste, not specialists versed in Aristotle, could ensure proper divinity.[22] The solution to the problem of proper theological knowledge required an elite, discerning reading public, not formal scholastic institutions (or individual divine illumination). Real common sense, possessed by discerning elites, not philosophical theology, would secure the order of the church and society.

Yet the worldly ideals of *honnêteté* of the seventeenth century were far distant from the rigorous Christianity of Pascal and his Jansenist colleagues. In *Pensées*, Pascal took the standards of judgment, modes of rhetoric, and proper use of reason that he claimed to use against the Jesuits and turned them back onto the *honnêtes gens* themselves. Were they clearly to express the truth about themselves, they would recognize the irrationality of their largely secular lives within their own secular framework. Proper reason, possessed not by scholastic philosophers but by *honnêtes gens*, called for a rethinking of everyday life.

Le Moyne used the natural-philosophical example of a compass needle to help his worldly readers recognize the striving of all things, including human beings, for their sovereign good. Pascal turned his analysis to natural-philosophical and mathematical knowledge to help his worldly readers recognize the paradoxical greatness and wretchedness of humanity and the monstrosity of human motivations. Just as making theology public made Jesuit casuistry absurd, making public a clear account of the nature of mathematical and natural-philosophical knowledge made overestimating human cognitive ability absurd.

FROM MATHEMATICS TO NATURAL KNOWLEDGE

In *Pensées*, Pascal drew on his philosophical and rhetorical account of mathematics to develop a new apologetic role for knowledge of nature. "How many beings have [microscopes] discovered for us that simply were not for the philosophers of the past?" (S645). Despite the humility that such discoveries ought to have engendered, Pascal's contemporaries failed to draw the proper epistemic morals from them:

> "There are grasses on the earth; we see them; on the moon, we would not see them. And on these grasses, hairs, and on the hairs, little animals, but after that, nothing more?" Oh, presumptuous one!
> "Mixtures are made of elements, and the elements not?" Oh, presumptuous one! (S645)

New natural philosophies, using new forms of observation, had undermined traditional views of the natural world. In all but a few observers, these new philosophies had perversely failed to undermine the self-satisfaction and deceptive certainty accompanying those traditional views. New technologies like the microscope and telescope had revealed without cease things never envisioned in traditional natural philosophy. Most philosophers reflecting on the significance of these new technologies had used them to extol the ability of humanity, not to show its perverse overconfidence in its current state of knowledge. New, unexpected things ought to cause *cognitive* shock about confidence in old ideas; they ought to cause wonder; they ought to challenge the sovereignty of reason, not to instill undue epistemic confidence. They have caused little doubt and given little pause—or, in a few cases, generated far too much doubt. Pascal's contemporaries, by and large, still maintained that they could grasp the fundamental principles of nature by turning their experiences and reasoning into principles of nature.

Pascal claimed that dispassionate, rational analysis of geometric practice and the knowledge it provides clarified the limited access to mathematical truths that logical form provides human beings. Rationally considering the new natural-philosophical knowledge was supposed to temper a rampant overconfidence in the extent of human epistemic powers. In the case of mathematics, the certain knowledge that one can always halve or double a number or line implies than one can never achieve some indivisible or largest quantity. It is irrational not to believe in the actual existence of the infinitely small and infinitely large; it is likewise irrational to think human beings can comprehend such entities. In the case of natural philosophy, certain knowledge of the possibility of ever more minute or giant structures reveals the impossibility of ever becoming certain that one possesses the real principles of knowledge of this world. To think that one can stop examining nature at ever smaller and larger scales is irrational and imprudent. Yet Pascal's contemporaries stopped. They systematically misunderstood the implications of real mathematics; they ignored the evident, reasonable consequences of changing natural knowledge. Under the guise of defending rationality and scientific knowledge, philosophers were simply being irrational and imprudent.

"Disproportion of Man," S230, one of the most accomplished pieces of writing in *Pensées,* leads the reader to see and feel how natural knowledge ought to undermine hubris. With the term "disproportion," Pascal swiped at a key doctrine of Jesuit scholastics such as Francisco Suárez: that human beings naturally have cognitive abilities "proportionate" to knowing nature through the senses with certainty, before *and* after the Fall. Suárez argued that "every natural truth seen in the order of nature and acquired by itself, that is, without special revelation but by the aid of the senses, . . . is proportionate to the light of the natural intellect." After attacking a view that the Fall "wounded" human intellectual ability, making it disproportionate to natural truths, for example, Suárez forcefully contended that "all objects proportionate to human intellect before [Original] Sin are equally proportionate after it."[23] For Jesuits such as Suárez, human knowledge of nature showed how fit human beings were for living in and knowing their universe; this fitness revealed the fundamental continuity between the pure pagan nature understood by Aristotle and the postlapsarian human condition—the continuity upon which Jesuit theology, pastoral practices, and spiritual exercises rested to no small extent.

In the draft of his apology, Pascal set forth a plan to use the best current natural knowledge to illustrate the falsity of this happy picture of human proportionality. A crossed-out preliminary note reads: "See where natural knowledges lead us. If they are not true, there is no truth in man; and if they are, he finds in them a great subject of humiliation that forces him to bring himself down in one way or another (S230; all struck out)."[24] In the fragment in which he fulfilled this plan, Pascal offered a series of visualizations about the natural world from what we might call the philosophy of science to bring alive (in the technical rhetorical sense—that is, to illustrate or to make evident) his point. His consideration of human knowledge about the world was intended, not just to inform readers about human cognitive limits and good epistemological beliefs, but to move readers, to transform them intellectually and emotionally into accepting human limitation.

The fragment comprises a series of exercises in reasoning, visualizing, and imagining. In the following, I describe the effects of the passages on the "reader," for simplicity's sake. The form of rationality Pascal assumed is examined below in some detail. Like an Ignatian spiritual exercise, Pascal's fragment seems to be written in part as a set of instructions for a spiritual director leading someone through a series of exercises. Through ordinary human sight, the reader is to envision a vast universe astonishing in itself: "Let man contemplate then the entirety of nature in its high and full majesty . . . let him view this brilliant light placed as an eternal lamp for illuminating the universe; let the earth appear to him as a mere point compared to the vast circuit that this star describes; and let him marvel that this vast circuit is but a delicate point in relation to the circuit followed by the stars in the firmament" (S230). Vision leads one to recognize the minuteness of the earth in a large universe: the earth seems a tiny, but proportionate, part of the universe.

Microscopes and telescopes had expanded the range of human vision. They readily indicated the dangers of taking the currently visible as the extent of the real; they underlined the danger of taking the phenomena of ordinary experience as the actual way the world works; they also underscored the danger of taking the phenomena they revealed as the only ones remaining to be found. It is irrational and imprudent to assume that the limits of human instruments are calibrated to the limits of the world. Pascal had his reader imagine possibilities beyond those instrumental and visual limits: "if our vision stops there, let imagination go beyond" to possible beings and structures of creation. Imagining such possibilities should force the reader to abandon any pretense of grasping the entirety of the universe: "The entire visible world is but an imperceptible feature in the ample breast of nature; no idea approaches it" (S230). As in mathematics, our human faculties of rational knowing push us to recognize the possibility of that which we cannot comprehend. It is irrational and imprudent not to acknowledge the possibility of structures not only beyond the visible but also beyond human imagination.[25] All of this is not to abandon science but to recognize its genuine implications. Rather than stopping, satisfied that all possibilities have been given, readers must recognize their inability to convert the infinitely many imaginable accounts of the structure of the universe into confirmed and certain knowledge about its actual metaphysical and physical constitution.

Having imagined the ever larger, the reader should envision the ever smaller: "to present to him another prodigy just as astonishing, let him search in what he knows of the most delicate things that a maggot offers him, in the smallness of its body, parts incomparably smaller, legs with joints, veins in its legs, blood in its veins, humors in its blood, drops in its humors, vapors in its drops, so that continuing to divide these last things he exhausts his forces in these conceptions.... He will think perhaps that it is there, the smallest thing of nature" (S230). Pascal shifted from the images provided by vision to those conceived solely through the imagination. "I want to make him see a new abyss within all that; I want to paint for him not just the visible universe but the immensity of nature that one *can* conceive of in the confines of this abridgement of an atom" (S230; my italics). With his imagination, the reader is to generate different possible microstructures of the parts of an atom, not to decide which among them is in fact true, but to make him concretely visualize those things beyond his epistemic capacities to know and to choose among.

With this exercise in *enargeia* (the emotionally powerful narration of things so that they appear as if before the eyes), Pascal envisioned recursive pictures within pictures: "Let him see there an infinity of universes, each having its firmament, its planets, its earth, in the same proportions as in the visible world, and, on this earth, animals and, finally, maggots again. In them he will find anew what the first maggots revealed, and finding yet in the new ones the same thing, without stopping or resting, so that he loses himself in marvels just as astonishing in their smallness as the previous were in their extent" (S230). In its recursive construction of possibilities, the imagination acts like mathematics to produce a series of finite things that human beings can conceive and understand and that do not surpass human understanding. This sequence of understandable things, proportionate to human senses, imagination, and reason, in turn should force human beings to accept the potential existence of things that they can neither conceive nor understand, that do surpass them, that are disproportionate to their abilities. Rationally, the reader has recognized that the universe need not be fitted to any of his epistemic capacities: sight, sight with instruments, reason, reason with instructions, or imagination. The reader must accept the possibility that nature has attributes, properties, and principles not even imaginable. Of these, only "the author of these marvels comprehends them. All others cannot" (S230).

Pascal emphasized that contemplating the natural world and universe had humbled few; he took this lack of humility as an important empirical fact about human beings. Meditation upon nature had long been a staple of contemplative and philosophical literatures, usually to extol humanity rather than to humble it. In Guillaume Du Vair's widely read *Holy Philosophy* (1584), contemplation of the natural world leads to the recognition that nothing in nature can equal or contain the greatness of the human soul. The soul "finds nothing there that can hold it back or contain it." The human soul transcends all that it sees: "For, more magnificent than all that, it encompasses heaven and earth, surrounds the world, pierces the depths of abysses, knows all things, moves and handles itself, and is so beautiful that if we kept it in its natural beauty, everything else in this low world would seem to us in comparison both ugly and deformed."[26] Human understanding and ability to pierce through to the realities of all created things are so apparent that readers should recognize the superiority of the human soul and grasp its value, which is greater than that of all other created things. In Du Vair's work, knowledge of the natural world leads to the affective recognition of human superiority to it, and the raising of one's soul toward knowledge of God.

Du Vair's book exemplifies a widely disseminated genre. The raising of oneself to higher knowledge through reflection upon nature featured prominently in Marsilio Ficino's much-imitated symposium commentary and in a wide range of popular philosophical books of the sixteenth and seventeenth centuries drawing on both Ficino and earlier sources such as Saint Bonaventure and pseudo-Dionysius.[27] Such views were hardly confined to philosophers and erudite readers; Jesuit rhetoricians widely promoted the consideration of nature as a pathway to proper knowledge of God and the universe in their efforts to attract and reform the French courtly and legal/bureaucratic (robe) nobility.[28] Scholastic authors such as the Jesuit Robert Bellarmine, with his popular Latin work *The Ascension of the Mind to God through the Ladder of Created Things* (1614), offered powerful models, but Francophone Jesuits made it into a widely disseminated art.[29] The rhetorical manual writer Etienne Binet's enormously popular *Essay on the Marvels of Nature . . . Very Necessary to All Who Profess to Eloquence* (1624) provided a stock set of examples for preachers and writers.[30] Jesuit writers such as Caussin, Le Moyne, and Joseph Filère used these examples to great effect, transforming the affective experiences of Ignatius's *Spiritual Exercises* into forms appropriate for French courtly and *honnête* readers.[31] Jesuit writers, aiming to transform *honnêtes gens* and courtly readers through verbal displays of nature, linked wonder to the grasping of the certain order of the world. Real affective change was to accompany the ascension attendant upon greater understanding.

In traditional contemplative literature, such as the examples above, the reader comes to ever greater cognition of deeper truths, ever more totalizing viewpoints, ever greater comprehension, an ever greater sense of human capacity, finally reaching a state of inner certainty and knowledge of God and his creation. Through contemplating nature, he or she becomes a better likeness and image of God. In one of Descartes' most important exercises for Elisabeth, he stressed the gap between human cognitive ability and a potentially infinite world. Descartes, however, turned this human inability to know the world into the grounds for ever more creatively imagining the perfections of all beings in creation, including human beings.[32] In contemplating nature with Pascal's guidance, the reader comes to ever less knowledge, ever less totalizing viewpoints, an ever greater lack of comprehension, and finally recognition of the limits of human knowledge.[33] The authority of nature leads, not to recognizing the divine and to cultivating human kinship to the divine, but to recognizing human incapacity before even what little of nature human beings can see or imagine. Contemplation does not make one more godlike; it does not reveal God; it shows human cognitive and interpretive limits.

Human *inability* to know the world, not an ability to comprehend it in its totality, ought to cause wonder. Pascal took the finest examples of human intellectual ability and humbled reason by producing wonder at its own incomprehensible and necessary products. "Nothing conforms so well to reason as this disavowal of it" (S213; see also S220). With both mathematics and natural knowledge, the searcher takes human knowledge to its limits and marvels. Only then is one forced to recognize that man "surpasses man," that the abilities of human beings can make their real limits apparent.

Recognizing their lack of privileged epistemic access should force human beings to account for their value in the universe: "Let man, having returned to himself, consider what he is in comparison to what is; let him look at himself as tossed into this lost district of nature; and let him learn, from this little prison cell where he finds himself lodged—by this I mean the universe, let him learn to value the world, kingdoms, cities, and himself—to learn his true value [*prix*]" (S230). Du Vair and his like would have man value himself above the rest of nature; Pascal would have him place little value upon himself. Given the scope of the infinitely large and small, rationally considering his true value should engender emotional uproar. Upon seeing himself suspended "between these two abysses of infinity and nothingness, he will tremble in viewing these marvels, and I believe that, his curiosity changing into admiration, he will be more disposed to contemplate them in silence than to investigate them with presumption" (S230). With this series of imaginative exercises in what we would call the philosophy of science, Pascal worked to effect real transformation in the reader's attitude toward the world and actions in it.

Having proceeded through these exercises, the reader has achieved a form of rational humility. "See then our true state. It is what renders us incapable of knowing certainly or ignoring absolutely" (S230). Consideration of nature leads less to knowledge about nature than to knowledge about human beings. Human beings can recognize *that* the universe is incomprehensibly vast and incomprehensibly minutely detailed without ever being able to understand it in its totality. They can recognize that human beings have no special place, no special epistemic access, without knowing why they lack such a place or access.[34]

Considering the two infinities of the ever larger and the ever smaller revealed in considering natural philosophy works above all else to destroy human overconfidence: "Having failed to contemplate these infinities, men have turned with temerity to the study of nature, as if they have some proportion with it" (S230). Contemplating infinities is an essential curative technique for human beings filled with presumption. Pascal's two-infinities argument offers experience in recognizing the truth of things that reason cannot comprehend. Although the argument offers no experience, cognition, or mystical union with God, it provides affectively powerful experience of the force of certain truths that are not comprehended. Considering carefully human knowledge and ignorance of nature provides experience in recognizing the power of reason to illustrate its own limits; this study can effect real emotional and epistemic transformation by leading one to accept a limited but crucial role for reason in organizing human life.

HUMAN PRESUMPTION AND THE MISTAKEN
MIMICRY OF MATHEMATICS

By reflecting upon mathematics, Pascal invited his readers to distinguish practiced mathematics from logic that mimicked mathematics. Under what conditions could so many people so mistake the applicability of geometric reasoning that they believe that nongeometrical knowledge might consist in demonstrations based on principles known with certainty? How could almost everyone, including the most learned philosophers, make this trivial mistake? "It is a strange thing that they wanted to understand the principles of things, and from there to come to the point where they think they know everything, by a presumption as infinite as their object. For it is without doubt that one could not envisage this design without either presumption or an infinite capacity like that of nature" (S230). This nearly ubiquitous misunderstanding of the nature and applicability of mathematics and natural philosophy offers powerful empirical evidence about the motivations and qualities of human beings.

People desiring to know the principles of things have, in the name of improving reason, detached themselves from actual human knowledge of nature: "when they lack true objects, they must attach themselves to the false" (S544). Their presumption leads them to generate ersatz principles and artificial rules. Together these principles are a pervasive, mistaken mimicry of real geometrical knowledge, a façade of reason and logic offering a fine front to a ramshackle edifice of imaginary principles and arbitrary rules. The faculty of the imagination duly provides any missing principles to those who seek them: "For reason has been obligated to cede itself [to imagination]. The wisest take for their principles what the imagination of men has introduced with temerity in each place" (S78).[35] The mistaken presumption that human beings can have knowledge of certain principles, which is grounded on a false vision of the exact sciences, fuels the production of ersatz, imaginative principles. Real geometrical ability is wrongly taken to signal a general ability to know the principles of the world through a logic formally similar to geometrical reasoning.

This presumption of making other knowledge look like mathematics necessarily involves ignoring the rules and limitations of practiced mathematics, that is, the rules and limitations imposed by actual human nature, in favor of self-imposed artificial ones.[36] Pascal and his father portrayed the Jesuits as besotted by the imaginative principles, the suppositions from nothing, of their philosophy and rhetoric. The logician belongs in an unhappy—and to Pascal an unsavory—collection of producers of social, natural, and religious artifice: "It is a pleasant thing to think about—all those people in the world who, having renounced all the laws of God and of nature, make laws up themselves, laws which they obey exactly, as, for example, the soldiers of Mohammed, thieves, heretics. *And also the logicians*" (S647; my italics).[37] Much as Pascal used slurs against Judaism to capture hermeneutic error, he used slurs against Islam to capture mistaking philosophy for revealed religion. For Pascal, Islam was an invented religion, its foundation simply the prophet Mohammed's say-so, without sanction from God, without proofs or evidence, without fulfilled prophecies or external authority.[38] Just as following their self-imposed law supposedly does nothing for Muslims or heretics but leads them into misplaced confidence in salvation, following self-imposed syllogistic laws leads logicians and philosophers into perverse overconfidence in their beliefs and the possibility of their salvation.[39] In criticizing philosophical artifice, Pascal echoed the complaints of his Jansenist friends about Jesuit theological innovation. Following Cornelius Jansen, Antoine Arnauld described the account of grace of the great Jesuit theologian Luis de Molina as "completely philosophical and completely human," jury-rigged from old heretical notions and the "useless imaginations of a man who distrusted the entire tradition of the Church."[40] In Pascal's view, the Jesuits' promises of salvation, like the promises offered in Islam, stemmed from imaginative constructions bolstered by an empty logical architecture, detached from actual humanity, scripture, and tradition.

Like the Jansenists, the *honnête homme*, the discerning generalist, could see the dangers of speculative logical pursuits. In criticizing the limits of logic, Pascal echoed Méré's admonishment to Pascal to seek means of knowledge other than mathematics, means more appropriate for knowing the things and people of this world.[41] Although Méré misunderstood the real strengths of mathematics in his general critique of formal reasoning, Pascal concurred with him on the need for alternative paths to knowledge. Possessing a nonmathematical form of judgment distinguished someone truly knowledgeable in mathematics from the sterile and deceived logician or the mere procedure-following calculator. Only with such alternative ways of knowing could one avoid the dangers of formalism and grasp the true "spirit of these rules" regulating mathematics. In the essay "Art of Persuasion," which complements "Mathematical Mind," Pascal called for a discernment capable of getting behind mere form to grasp the essence, the intention, the thoughts behind statements, books, and things. "One must therefore determine how this thought is lodged in its author: how, whence, how far he possesses it" (M III:423). Only with such a form of discernment, deployed to consider human nature in depth, could anyone hope to make evident the obscurities of human nature and to set about finding a cure proper to that nature.

HONNÊTETÉ AND COMMUNITY-MINDED RATIONALITY

Pascal's arguments just presented rest on a range of assumptions about how rational people ought to act. The great currency of such questions about rational actors in contemporary social sciences and philosophy threatens to obscure Pascal's thought by imputing an anachronistic account of rationality to him. Pascal described rationality as including more than maximizing self-interest. The arguments reconstructed here depend on a historical vision of rational action and decision making.[42]

Nevertheless, one may attribute a germ of decision theory to Pascal. The chevalier de Méré is best known for asking Pascal how to divide properly the stakes of a game of chance stopped midway through. The problem: Two players are playing a game of chance requiring three points to win, with 64 pistoles at stake. Player A has two points, B one point, when the game breaks off. How ought the winnings to be fairly divided? What is the reasonable expectation of each player?[43] In a series of letters, Pascal and Pierre de Fermat offered different means for answering the question and, together, legend has it, founded the probability calculus and decision theory.[44] Both mathematicians found that players A's expectation was 48 pistoles from the 64; that was his *just* portion.[45]

Pascal met and came to know well the chevalier de Méré and his fellow theorist of *honnêteté* Damien Mitton in the circle of the duc de Roannez.

Pursuing more than simply gentlemanly conversation and diversion, Roan-nez and his friends undertook a series of collaborative business ventures, from the draining of swamps to the creation of the first regularly sched-uled carriage system in Paris.[46] Business, mathematics, polite conversa-tion, economics, and gentlemanly behavior were not isolated domains of interest.

As Lorraine Daston has argued, Pascal phrased his answer to Méré's question about games of chance in the language of contract law. He framed the answer as giving each person what was due to him *"en justice."*[47] The question at hand, then, was not only how a rational actor, in the modern sense, ought to act but also how a reasonable being ought to act so as to render justice to all parties, so that all parties get their just due. In early-modern Europe, such considerations about the justice due to members of a community were often taken as integral to rationality itself and were not understood merely as a secondary result derived from the rational consideration of self-interest.[48] The communal justice central to this form of rationality had deep roots in Roman civil law of contracts. Seventeenth-century writers on law such as Hugo Grotius and Pascal's friend Jean Domat highlighted the communal demands on contracts in their accounts of how communities stay cohesive and peaceful.[49] Contracts with unfair distributions were inherently invalid, antisocial, and grounds for legal action. Many quarters expressed great interest in determining what just dis-tributions should be, particularly in cases involving uncertainty and risk.[50]

Concerns about the rationality appropriate to communal life extended beyond deeply held legal beliefs. Community-minded rationality was at the heart of the obscure subject of *honnêteté*, a central concern of those in the Roannez circle, as well illustrated in the brief account of the subject written by Pascal's friend Damien Mitton: "*Honnêteté* therefore must be considered as the desire to be happy, but in such a way that others are as well." Producing such happiness rested fundamentally on aptness of judgment, acuity of mind, and the will. "To have this *honnêteté* in the greatest degree, one must have an excellent mind and a well-made heart, and the two must work together in concert." Mitton defined the greatness of these capacities in terms of both the reasonable and the just: "Through greatness of mind, one knows what is most just and most reasonable to say and to do; and through the goodness of the heart, one never lacks wanting to do and to say what is most reasonable and most just."[51] Only with a fine sense of judgment and a strong, habitual will to act justly can one act and speak to produce collective happiness, including one's own happiness.[52] According to Mitton, *honnêteté* involves a rational being who recognizes that pleasing others is integral to the practice of reasoning, and who habitually pleases others and produces community. This habit of discernment tempers, in fact, the pursuit of self-interest: "It is this management of happiness for us and for others that one ought to call *honnêteté*, which is only, to grasp it correctly, well-regulated *amour-propre*."[53]

In his considerations of human rationality, Pascal discussed both a communal vision of habitual rational action and a simple, more familiar account of human beings driven by self-interest, simple *amour-propre*. He rejected both as adequate depictions of human motivation. Pascal repeatedly attempted to show that when predictions were made according to what was "just," whether in mathematics, natural philosophy, or human activity, the predictions jarred with how people actually act—the empirical evidence of human nature. Any cursory study of human beings quickly would produce such evidence, and yet Pascal's friends Mitton and Méré ignored it. These *honnêtes hommes*, those most apt to have fine discernment, those most capable of instructing and communicating their judgments, shied away from turning their acuity to consider their own thinking and actions.

By focusing on his friend Mitton, Pascal personalized these criticisms. A series of fragments reproach Mitton and his account of *honnêteté*, not just for ignoring human faults but for covering them up: "The self is hateful. You, Mitton, cover it up; you do not thereby remove it. Thus, you are always hateful" (S494).[54] Mitton's careful balancing between communal action and self-interest was legerdemain. Pascal had Mitton reply that one ought simply to diminish the baleful effects of self-interest, thereby removing the cause for hatred. Pascal counterblasted that Mitton has underestimated human corruption: "In a word, the self has two qualities: it is unjust in itself, in that it makes itself the center of everything; it is inconvenient to others, in that it wants to make use of them, for each self is the enemy and would like to be a tyrant over all others. You remove the inconveniences of it, but not the injustice of it" (S494). Mitton's *honnêteté* was supposed to produce the sociability foreign to self-interested human beings. In his words, the *honnête homme* "is not interested; but since he knows the needs of life, his conduct is always regulated, and he never lives in disorder."[55]

Mitton's misunderstanding of human sickness undermines the suitability of his cure for human ills more generally. His cure merely changes appearances, not internal states; it never moves the patient to real change. *Honnêteté*, perhaps the best social art and epistemic practice available to human beings, can serve as no real spiritual exercise, for it insists on the ability to regulate oneself through reason. In notes to himself, Pascal told himself to "reproach Mitton for not being moved" (S433), although he well knew that "nature is corrupt and that men are contrary to *honnêteté*" (S529 bis). In its further consideration of human foibles, *Pensées* aimed to force intended readers—people such as Méré, Mitton, and Le Pailleur—to study humanity with care, to bring their real discernment and rhetorical skills to such a study. Pascal showed his readers the consequences to which their good judgment ought to have been directing them all along, but had not for want of effort and serious application.

FROM DISCERNMENT TO SELF-REFLECTION

In *Pensées*, Pascal portrayed the detached everyday skeptics of his time, the elite men and women who cast a disdainful eye upon royalty, peasants, Christianity, and Aristotelianism alike. Armed with a more critical than constructive form of reason, they make up Pascal's category *"demi-habiles,"* the half-clever ones, a category including the *honnêtes gens* just discussed. Their straight thinking and their reason uncorrupted by formal systems allowed them easily to undermine the metaphysical nonchalance with which most people high and low in early-modern France lived.[56] In Pascal's picture, they possess the discernment to recognize the artificiality of much of the social world, the discrimination to reject the more puerile aspects of religion, and the distinction to abandon the widespread Aristotelian accounts of the natural, social, and supernatural worlds.

To clarify his account of the hierarchy of knowers, Pascal considered how different groups understand the nature of political and social hierarchy. "Gradation. The people honor those of great birth. The *demi-habiles* distrust them, saying that birth is not an advantage of a person, but of chance" (S124). Cutting through the artifice of contemporary beliefs and practices, the *demi-habiles* recognize and mock the mistaken essentializations of the common people, who wrongly believe, for example, that royalty and nobility are essential qualities. In Pascal's estimation, their judgments about the opinions of the people are only half-right—hence his term *demi-habiles*. If they pushed their criticisms further, they would recognize the value in customary beliefs. Although true nobility is but a fiction, it is a fiction upon which peace rests. To seek a political order grounded on something besides a customary fiction would simply yield incessant civil war, the worst of secular evils. The people, not the semiclever *demi-habiles*, are right, but for the wrong reasons: "The truth is very much in their opinions, but not from where they see it" (S126). Pascal's highest sort of knowers, the *habiles*, the truly clever, recognize these customary truths, for they acknowledge the limits of human reason, the permanence of the *amour-propre* and imagination, and the dangers of upsetting custom.

Why do the *demi-habiles* fail to recognize the value of these customary truths? Whether Stoic philosophers, Skeptics, or *honnêtes hommes*, they all wrongly estimate human ability in their critique of the status quo. They fail to interrogate themselves enough, to sound the depths of human nature. As we have seen, in *Provincial Letters*, Pascal drew on the presumption of the elite reading public of *honnêtes gens* that they possessed a commonsense rationality and qualities of judgment able to examine and ridicule Jesuit theology, casuistry, and pastoral practice. Applying common sense, rather than the perverted reasoning of the philosophers, to the world and to human beings should lead the clearest-thinking people of his day toward a largely negative view of the greatness and wretchedness of human nature. Yet they have rarely taken this self-reflexive step, despite their skepticism of worldly affairs; they have hardly ever turned their skeptical, probing reason on themselves.

Pascal did more than chastise his potential readers. He used their failure to reflect upon themselves to illustrate human nature. Pascal claimed that the most helpful way to illuminate misguided people is to acknowledge the coherence of their beliefs given their position: "When one wants to correct with utility and demonstrate to another that he is deceived, it is necessary to observe from which side [*côté*] he envisions the thing, for it is ordinarily true from that side, and to confess to him that truth but to reveal to him the side from which it is false. He will be happy with that, for he will see that he did not deceive himself; he simply lacked seeing from every side" (S579). Central to Pascal's effort was showing these *demi-habiles* that they were as delusional, as taken in by imaginary constructions and presuppositions, as the pedants and people they daily mocked. The *demi-habiles* had no more knowledge of the conditions systematically producing their beliefs than the people they mocked did. Prompting his readers to examine their own systematic failure to try to know themselves, then, could help them truly to know themselves.

DENATURING MAN: THE MONSTROSITY OF HUMANITY

Like Pascal, the more philosophical of the *honnêtes hommes* tended to reject an overestimation of human ability; some descended into outright skepticism. Unlike dogmatic philosophers, persons with discernment recognize that "they do not find in themselves the enlightenment" that can structure a good life, one proper to human beings. Perversely, this epistemic incapacity gives them an excuse for not searching elsewhere for that true good. They simply stop, rather than looking "elsewhere and examining in depth if this opinion is among those that the people received in a credulous simplicity" or one having a "very solid and unshakable foundation." Despite their abilities, their critical acumen, they fail to examine things to their depth. Like Mitton, they hide human nature rather than investigating it. Even in their own terms of self-interest, this absence of critical attention makes little sense: "This negligence in a matter where it is a question of themselves, of their eternity, of their totality, irritates me more than it moves me" (S681). They ignore a pursuit clearly in their own secular self-interest.

In a series of projects, Pascal worked to subvert a vision of human beings as naturally suited to community and prone to act for the common good. Insofar as they do act for a common good, he argued, they act largely to preserve their own self-interest (see S244, among others).[57] Human beings are more bizarre creatures than merely self-interested beings using reason to pursue their own good: they evidently do not pursue even their own self-interest consistently.

Reasoning based on self-interest ought to lead toward choosing the greater good for the self. In his famous wager, in S680, Pascal worked to show that all people who reason according to their own self-interest ought prudently to act as if they were true Christian believers. The expectation of an infinite good, no matter how small its probability, still trumps any finite secular good.[58] Pascal's examination of self-interest does not rest on the wager alone. Ought not a purely self-interested being be quite concerned about its final fate: to know whether its soul is immortal, whether a god exists, and so forth? Indeed, should the magnitude of such questions, whose answers after all must dictate the proper way of living, not weigh more heavily upon such rational and self-interested beings? "Thus, our first interest and our first duty is to enlighten ourselves about this subject, upon which depends all of our conduct" (S681). Given the evident difficulties in answering such questions, should not purely self-interested beings, riddled with terror before their ignorance, be earnestly pursuing real answers with all their intelligence?

Many people, including the discerning *demi-habiles* and Pascal's *honnêtes* friends like Méré and Mitton, were not terrified by their lack of knowledge of their fate, however assiduously they otherwise pursued their self-interest in worldly endeavors. This widespread lack of terror made Pascal marvel. "It is a monstrous thing to see in the same heart and at the same time this sensibility toward the least thing and this strange insensibility toward the greatest" (S681; see also S682). When it comes to considering their ultimate fate, human beings fail to pursue their self-interest, which they skillfully do in everyday affairs when operating under similar conditions of uncertainty. Pascal stressed that the warrant for believing in this account of human monstrosity comes not from piety or the dictates of scripture: "I do not say this from the pious zeal of spiritual devotion." His wonder and argument derived from secular, not pious, convictions: "I mean on the contrary that one ought to have this opinion through a principle of human interest and an interest of *amour-propre*." But many human beings, even the clearest-thinking ones who doubt the easy answers handed to the crowd, simply do not have such an opinion (*sentiment*). "And thus he who doubts and who does not search is altogether both truly unhappy and truly unjust" (S681).

Any adequate description of humanity must account for this systematic and widespread failure to pursue self-interest consistently. Reason combined with self-interest predicts models of human actions that jar incongruously with the monstrous reality illustrated by how most human beings live. To account for this failure, aspects of evident human nature that contradict reason and self-interest must be discovered, acknowledged, and explained.

In *Pensées*, Pascal used "monster" and "monstrous" several times to describe human beings and their choices: he claimed he would continue to contradict his interlocutor until the latter concedes that "he is an incomprehensible monster" (S163). In early-modern Europe, the term "monstrous" technically referred to an ill- or disproportioned jumble of parts from various animals or to a collection of accidents. Speaking technically as a natural philosopher, F. Liceti defined "monster" as "anything that is born among animals with a disposition and arrangement of limbs completely different and entirely contrary to the nature of those who have engendered it, such as, for example, a baby without feet, a girl with two heads, a baby with the head of a dog, a Centaur and so forth."[59]

Calling human beings and their motivations monstrous thus underlined that human beings are a mix of ill-fitting parts, not a unified whole. In declaring that humanity is monstrous, Pascal suggested that it lacks a unified "nature" whence spring human actions. Human nature is thus unnatural; its unnatural essence must be something produced either preternaturally or supernaturally, in the categories of Pascal's time. Any serious study of such a beast must seek causes beyond the ordinary course of nature.[60] A preternatural being such as a two-headed lamb can come about occasionally through a chance confluence of natural causes; but human monstrousness is too ubiquitous to be preternatural. Pascal maintained that no account of human nature based simply on natural causes, whether regular ones or obscure causal chains, can explain this ubiquitous monstrosity.[61] Only some causation beyond nature can possibly explain human nature. Only a form of causation beyond nature can explain human blindness to this monstrosity to be seen everywhere: "It is an incomprehensible enchantment, and a supernatural slumber, that shows an all-powerful force that causes it" (S681). Reason, capable of studying many of the regularities of nature, cannot hope to comprehend such a thing—this "supernatural enchantment" that most people *must* be under since they blithely ignore their real state.

Rather than forcing the reader to accede to the existence of infinity and to wonder at it, reason here demands one to acknowledge and then to marvel at the irrationality of much human motivation. Earlier Pascal made his reader wonder at the objects of mathematics and the cosmos that reason can convince one of without providing understanding; now the reader is to marvel at the monstrous condition of everyday humanity, a jumble of reason, self-interest, and some other, still-mysterious tendencies. Reason again demands acceding to the existence of that which certainly exists but is incomprehensible.

The enchantment goes deeper, for not only do human beings act monstrously but they systematically ignore their monstrosity. Monsters were an early-modern vogue, in learned and popular culture alike. Monstrous births violating the laws and ordinary ways of nature regularly provoked terror, wonder, and repugnance in the sixteenth and seventeenth centuries; they encouraged transformations in natural philosophy. For all this concentration on finding exotic monsters, for all the publications and public displays of the monstrous, Europeans neglected those always present: themselves. Monstrous human life was everywhere visible; it negated continuously the laws of Aristotelian natural and moral philosophy but typically went unnoticed by even the most incisive commentators upon the human condition. This constant ignorance ought to shock one into accepting human monstrousness even more: "That a thing as visible as the vanity of the world be so little known ... *that* excites wonder" (S50). The lack of self-knowledge of so-called rational man further undermines the suitability of any account of human nature that rests upon self-interest. "What astonishes me most is to see that all the world is not astonished by its feebleness" (S67). The monstrousness of humanity is surprising; humanity's lack of knowledge of such an obvious and ubiquitous fact is even more so.

Much as he had used the widespread misunderstanding of mathematics to illustrate human presumption, Pascal used the widespread misunderstanding of obvious human monstrousness to underscore its hold on human activities and motivations. So-called rational man is blind *and* usually blind to his blindness, subject to motivations he does not see. Just as with infinity, the power of reason to prove with certainty does not imply that one can comprehend the causes of human blindness and presumption through reason.

Both the general absence of self-reflexivity and the possibility of such self-reflection offer insight into human nature. *That* human beings, even the most acute philosophers and *honnêtes hommes,* do not recognize their infirmity, and do not even bother really to try, reveals important aspects of human wretchedness. That human beings *can* recognize their sickness when prompted, when challenged by wondrous unknowable things, as in Pascal's invocation of the two infinities, captures the height of human greatness. The blindness of thinking, especially by the philosophers self-designated as its practitioners, illustrates human hubris. The ability compassionately to understand that blindness illustrates proper human dignity.

The incomprehensible monstrosity of humanity springs from this naturally inexplicable mixture of wretchedness and greatness, which Pascal aimed to illustrate to his reader: "If he praises himself, I'll abase him, / If he abases himself, I'll praise him / ... / Until he understands / That he is an incomprehensible monster" (S163).

When Pascal said that truly to philosophize is to mock philosophy (S671), in large part he meant that a true self-reflexive philosophy, conscious of the limits of reason and the conditions of reasoning, ought indeed to mock philosophy. Pascal's critique, we should note, was more than a proto-Kantian examination of the limits of reason in itself. His critique went further: all self-satisfied critical spirits—the *honnêtes gens* Pascal had so long cultivated and reviled—equally need mockery. This mockery, however, must never stop at refuting the mistaken doctrines and hidden beliefs of the philosophers and the *honnêtes gens*. True philosophy must demonstrate to them the legitimacy of those doctrines and beliefs from their points of view, as well as the legitimacy of the doctrines of the "people" ridiculed by the self-satisfied and the smug.

CONTRARY TO NATURE, NOT TO REASON

In analyzing the blindness of human beings to their own self-interest, Pascal showed his readers that human beings are contrary to nature, as it was generally understood, but not contrary to reason. The unnatural nature of man does not contradict reason or logic itself. What might it mean to have a nature contrary to nature? How could one have an unnatural nature? Is not whatever a thing is in essence precisely its nature? Pascal used the term "nature" to capture what reasonable beings would think the world must be like, given a reasonable understanding of, or perhaps reasonable suppositions about, the good, God, causation, and so forth. Using such reasoning, philosophers had long provided detailed accounts of how nature—including human beings and God—simply must be. The course of events usually observed in nature and society suggests and appears to confirm such reasonable accounts: "When we see an effect always happen in the same way, we conclude from this a natural necessity, to the effect that it will be the same the next day and so forth." We are thereby deceived: "But often nature contradicts us and does not subject itself to its own rules." These mistaken judgments indicate much about human nature: "The mind believes naturally, and the will loves naturally [i.e., following nature]. In this way, when they lack true objects, they must attach themselves to the false" (S544). Nature appears not to permit vacuums; nature thus has a horror of vacuums. Human beings certainly seem to be social and political animals; human beings thus are by nature social and political. Hard work yields profits; good works on earth correspondingly will bring salvation.

The term "natural" in *Pensées* captures such a series of "naïve" inductions and hypostatizations, based, perhaps, on the widespread Aristotelian account of acquiring knowledge through everyday experience.[62]

With his example of the compass needle tending to its good, the Jesuit Pierre Le Moyne illustrated a commonsensical, "natural" created nature. The happy, proportioned world described in Jesuit popularizations brought together the interconnected set of natural, moral, political, and spiritual doctrines that Pascal called "natural" and "sweet." Jesuit philosophers and theologians had been at the forefront of articulating the notion of "pure nature," a tool for imagining what the world would have been like had God chosen for there to be no Fall and no revelation, where "human nature is not ordained to a supernatural end and received no other gifts of supernatural grace."[63] In other words, pure nature was a tool for imagining what nature would be like were there only natural goals and natural means to achieve them, and no supplemental supernatural goals and means, such as heaven and Christ's crucifixion.

In practice this tool for hypothetical thinking often functioned as an analytical tool for understanding postlapsarian human beings. The relative autonomy of the natural realm from the supernatural in this view emphasized human power to contend with and know natural things; this relative autonomy extended to an account of the innateness of sociability and natural law. Supernatural attributes, such as higher ends for humanity like true beatitude, appeared simply to be added to the purely natural form of human beings along with their abilities to know the world. After the Fall, by virtue of divine concurrence, human beings remain capable of knowing and pursuing natural good on their own, but that pursuit requires greater effort. Human beings are not corrupt through and through, even though their access to the supernatural is highly limited. Catholic rigorists, above all Cornelius Jansen in his *Augustinus*, assailed the exponents of "pure nature" for positing too autonomous a natural, moral realm and for ignoring the necessarily supernatural elements in human nature after *and* before the Fall.[64] Pascal's work on the vacuum, political order, and human qualities undermined the naturalness of the order upon which the Jesuit view rested.[65]

Natural things, then, are those things whose supposed existence is derived from logic (or reason) plus a number of assumptions that are not analytical truths in that logic but are based instead on commonsense inductions.[66] So universally were these inductions held that they were regularly taken as analytical aspects of reason itself. According to Pascal, the universality of such error led to the custom of mistaking the rational, the most apparently likely, for the reality. Such customs produce ersatz principles that human beings mistakenly but profitably naturalize and use to justify their claims to knowledge with the certainty of mathematics.[67]

Just as they mistake dressing well for true nobility, they mistake apparently rational human beings for beings capable of being ruled by their reason.

Numerous possibilities other than these "natural" positions about the world, humanity, and God are not logically inconceivable or contradictory. According to Pascal, human beings are little inclined to hold, or even to imagine, that these possibilities might be the way the fallen world in fact is. Most human beings mistake the pleasant possibilities for necessary essences. Exploring the created world beyond the most common empirical experiences, as in the "Disproportion of Man" fragment, Pascal's political thought and his experimental narratives quickly reveal countless unnatural natures filling the world. Vacuums exist. Man is a dreadful monster of disproportionate parts: his is an almost entirely unnatural nature, ill-equipped to know the world and to regulate himself. Even certain geometric knowledge is a monstrous thing, for it rests ultimately on the body, not the mind. Insofar as human beings have any true principles of knowledge, Pascal maintained, human beings have them only through the faculty of the *coeur*—that is, they have the highest principles of knowledge through the body, not the mind (S142). They have but a minute fraction of the necessary fundamental principles of knowledge—only those concerning time, space, and movement, none concerning nature more generally and certainly none about moral and political natures (S94).

For Pascal, the universe is filled with things that are contrary to an intuitive common sense of how the world ought to be but that are not contrary to reason. Pascal's geometrical and natural-philosophical exercises, followed by his exercises of reflecting on human nature, lead his readers to break with commonsense inductions about the world. They work to make them accept the utter reasonableness of an odd sort of thing, everywhere in the created world:

> Incomprehensible.
>
> Everything that is incomprehensible is not prevented from being. The infinite number, an infinite space equal to a finite one. (S182)

Any discerning reader is supposed to accept the possible existence of incomprehensible things, those contrary to nature but not contrary to logic. Entities of this type are essential to understanding, even partially, the *actually created* natural, social, and divine worlds. Infinities, vacuums, monstrous man are all incomprehensible things, contrary to nature but not to logic or reason. They are the seemingly unlikely, or "unnatural," possibilities that have in fact been realized. The "natural" sweet possibilities, like sociability or the horror of the vacuum, ought not to be mistaken for necessary essences, because they have not in fact been realized in nature, as careful empirical work readily illustrates. Human nature, and much of nature, is simply unnatural. Like infinity or the vacuum, which are unnatural aspects of nature, disproportionate man transcends comprehension. The explanation of disproportion is not "within yourselves" (S182).

For Pascal, Christianity, revealed through the often apparently contradictory evidence of scripture, alone can explain the ubiquity of the unnatural and the monstrous. The biblical Fall corrupted human beings and nature alike. Without the Fall, rational natural philosophy, rational moral and political philosophy, and natural theology, all drawing on epistemic procedures such as "naïve" induction, could very well produce true knowledge. The monstrousness of humanity, for Christians, is a product of the fall from grace, a fall of infinite distance. "Adam's sin . . . is so great that, although one cannot conceive its extent, it suffices to say that, to pay for it, it was necessary that God be incarnated and suffer up through death to make understood the extent of evil in measuring it through the extent of the remedy."[68] Making the extent of the Fall *comprehensible* required nothing less than the incomprehensible incarnation of God himself. According to Pascal, Christ's suffering was the key epistemological moment for the grasping of the scope of the sin of humanity.

At this point in his apology, however, Pascal had not led his reader to ponder and to take seriously the explanations offered by Christianity. Pascal was still producing the evidence necessary to make religion plausible, reasonable, because it is not natural. "Men are wary of religion. . . . Curing this requires beginning by demonstrating that religion is in no way contrary to reason" (S46). With examples from mathematics, natural philosophy, and human nature itself, Pascal prepared the reader not to brush aside Christianity as irrational and therefore impossible. Irrational, not at all: "If the principles of reason are shocked, our religion will be absurd and ridiculous" (S204). Natural, however, Christianity is not: "the Christian religion [is] so contrary to nature" (S717). Neither are humanity, infinity, or the vacuum natural: creation itself is an unnatural nature. Christianity, for all its bizarre beliefs, should remain a possibility for *honnêtes hommes*. Like the two infinities and the vacuum, the peculiar doctrines of the biblical Fall, the Incarnation, and transubstantiation are logically possible but seemingly unlikely things. Like those examples from the sciences, these Catholic doctrines are not contrary to reason, but they do contradict a commonsensical view of what nature must be like.

EMPIRICAL AND AFFECTIVE FAILURE
OF OTHER SPIRITUAL EXERCISES

In a general overview of ancient spiritual exercises, Pierre Hadot contends: "Thus, all spiritual exercises are, fundamentally, a return to the self, in which the self is liberated from the state of alienation into which it has been plunged by worries, passions and desires. The 'self' liberated in this way is no longer merely our egoistic, passionistic individuality: it is our *moral* person, open to universality and objectivity, and participating in universal nature or thought."[69] Pascal's exercises using mathematics and natural philosophy involved a return to the self but brought no liberation from worry, passion, or desire. According to Pascal, philosophies—Stoic, Aristotelian, Cartesian—offer the *illusion* of movement into the universal and objective, and the *illusion* of overcoming the passions obscuring the true self and its freedom. Such a liberated true self is but an illusion because it mistakes how human beings ought to have been made, according to natural criteria, for how they in fact are. The liberation effected through human philosophy involves, not truly knowing oneself, but rather preventing oneself ever from assenting to the monstrous reality of the human self by covering that reality with a deceptive façade.[70] Pascal insisted that most philosophy is recondite diversion offering pompous cover stories. In contrast, proper reason, that involved in practiced mathematics, not speculative logic, does not liberate: the limited but very real access to the universal and objective should force the self to recognize the intractability of its egoistic, passionistic, subjective individuality. Properly deployed, reason reveals desires; it causes wonder; it stokes the worry and despair that human beings ought to have. The philosophical liberation of the self is an illusion, one whose persistence ought to serve as a constant reminder of its own falsity but monstrously does not.

"One must know oneself. While this would not serve for finding the truth, at least it serves to regulate one's life. And nothing is more correct" (S106). Although Pascal's own conviction about this fundamental claim stemmed from his Augustinian Christianity, the first half of his apology for Christianity attempts to illustrate it secularly, based on the effects of reason on this self and the empirical observation of others. The cures for human unhappiness come, not from attempts to overcome that individuality, but from honestly recognizing and attempting to explain its monstrous, divided nature.

In a fragment explicitly labeled to come "[a]fter having explained incomprehensibility," Pascal laid out the demands any religion or philosophy must fulfill to be empirically and affectively adequate:

> The greatness and the wretchedness of man are so visible that the true religion must teach us both that there is some great principle of greatness in man and that there is some great principle of wretchedness.
>
>   It must further explain to us these astonishing contrarieties.
>
>   . . . .

> It must teach us the remedies for these inabilities and the means for
> obtaining these remedies. (S182; see also S237)

The true spiritual exercise—the proper philosophy or religion—must offer convincing descriptions, explanations, and practical remedies for the
greatness and wretchedness, for the monstrousness, of humankind. For
Descartes, contending forms of exercise almost all failed because they distracted the attention of the mind toward disunified particulars. For Pascal,
contending forms of exercise failed because they were deceptive, empirically inadequate, insufficiently explanatory, and above all failed to provide
any remedies that could temper human disquiet and despair.

Who were these deluded philosophers? Stoics, Epicureans, Skeptics, and
perhaps Cartesians.[71] When Pascal used the term "philosophers" in *Pensées*,
he included almost exclusively thinkers concerned with ways of life, with
modes of caring for the self.[72] He divided the schools of philosophical
spiritual exercises into two categories, the dogmatists and the skeptics,
both dependent on the power of logical demonstrations to affect human
conviction. These schools failed at describing human nature. They failed
at explaining that nature. Even more signally, they failed to persuade in a
manner appropriate to that human nature. Essentially flawed in failing to
describe unnatural man, they could offer no peace or comfort.

The Jesuit Pierre Le Moyne attacked the Stoics for failing to recognize
that the human body is a unified republic, a well-organized state with a
hierarchy of elements, each having a proper function, each proportionate
for this function. They wanted to crush the passions rather than properly
order them into an Aristotelian moderation. "Nature will never suffer the
smothering of the passions that it has made. . . . The mind must make use
of them and not smother them."[73] The overly austere Stoic ways of life
rested on a fundamentally mistaken account of human nature; therefore,
they could only fail, as nature would always subvert them. Their ways of
life and accompanying exercises had to be rejected as evidentially false and
offering no real remedy.

Whereas Le Moyne rejected the Stoics and others for failing to describe
human proportionality, Pascal rejected them for failing to describe human
disproportion. The philosophers' varied descriptions of human nature captured at best only some unified set of its disjoint parts. They wrongly
turned occasional qualities of individuals into essential qualities of human
nature.[74] Stoics saw only the ability to reason and control; Epicureans saw
only the inability and uselessness of doing so. Dogmatists saw only the
ability to know; Skeptics, only that of doubting. None had captured the
monstrous but evident mix of good and bad, knowledge and ignorance: "if
they know the excellence of man, they are ignorant of his corruption . . . and

if they recognize the infirmity of [his] nature, they are ignorant of his dignity" (S240). Unable to describe humanity properly, philosophy can hardly
succor any human. The exercises prescribed by the philosophers aim too
high or too low.

> The philosophers did not prescribe opinions proportionate
>   to the two states.
> They inspired movements of pure greatness, and that is
>   not the state of man.
> They inspired movements of pure baseness, and that is
>   not the state of man. (S17)

Neither a Stoic life of reasoned control and direction of the passions nor
an Epicurean life of passionate abandon is sustainable. The passions will
inevitably upset the life of reason; reason and its concerns will unfailingly
upset the repose of enjoying corporeal pleasures (S29).

Too feeble to tame a monstrous human being, reason cannot produce
the hierarchy of mind over passions and concupiscence.[75] One ought not
to abandon the use of reason to organize one's life. Pascal maintained that
reason is too powerful to let human beings fall into a life of base hedonism
and fleshly pleasure; reason cannot abandon its attempt to know, regulate,
and control. Nevertheless, any spiritual exercise resting on reason alone
fundamentally mistakes human nature—makes it a natural nature—not
the actual, unnatural human nature Pascal worked to force his readers to
recognize. Human psychology precludes demonstrative proof from providing any sustained, real spiritual comfort. Although Pascal's own account stemmed from his beliefs about the Fall, his argument for the failure
of proofs rested—in principle—only upon the observable aspects of actual
human behavior that he displayed to his readers.

Pascal emphasized the ridiculousness of demonstrative proofs as "remedies," as modes of practical consolation for humankind's doubts and passions. Even if fallen man *could* demonstrate the existence or nonexistence of
God, such proofs would not convince or transform other men (S222).[76] Even
if one could generate impeccant proofs mimicking geometric form, human
beings would begin to doubt moments after assenting to them. Although a
proof may convince us of something, the human constitution often makes
us relapse after having experienced the proof. To keep proofs constantly
in mind is too difficult, and hardly appropriate for the desperate seeker
after truth. "For to have always proofs present is too much trouble. One
must acquire an easier belief, that of habit" (S661). Even should there exist
perfectly good metaphysical proofs of the existence of God, they would not
satisfy the human seeker. Having experienced such a perfectly good proof,

one would quickly return to the state of disquiet and worry the proof was to cure.[77]

More popular than metaphysical proofs about God were ones based on the natural world. Pascal was blunt about their legitimacy and utility: "David and Solomon never said: 'There is no vacuum; therefore there is a God'" (S702).[78] Hugo Grotius, in his enormously influential *True Religion Explained and Defended*, used the horror of the vacuum to illustrate the providential qualities of nature. The horror shows that all things are not only "ordained for their *proper ends*, but also for the good and benefit of the whole *Universe*." Such an organization simply demands a deity: "Now it cannot possibly bee, that this common end [should be] thus intended together with an inclination of things thereunto, but by the power and purpose of some *inteligent nature*, whereunto the whole *Universe* is in subjection."[79] Pascal's own experiments with the vacuum had shown that such a vision of the universe was hardly evident; it mistook some appearances for essences. To rest on such uncertain and changing knowledge offers little real proof: "to say to those that they need only see the least things that surround them, that they will there see God displayed openly, and to give to them as every proof of this great and important subject the course of the moon and the planets, and to claim that one has achieved its proof with such a speech, [all of this] would be to give them grounds to believe that the proofs of our religion are quite weak. And I see through reason and experience that nothing is more capable of producing distrust" (S644).[80] Proofs in natural philosophy or in metaphysics are easily and often upset, so arguments taken from nature hardly prove to be epistemically or psychologically satisfying. The continued attempt to use them yields suspicion, not belief.

In undermining the "dogmatists" and their rational or empirical proofs of God's existence and nature, Pascal equally undermined the skeptics, for their attacks rely on demonstrative proofs. While skeptical attacks on reason do show the powerlessness of reason to achieve certainty about the basic principles of knowledge, they do not therefore eliminate certainty that there *are* some certain principles of some domains of knowledge: "I pause at the unique strength of the dogmatists, namely: speaking in good faith and sincerely, one does not doubt natural principles" (S164; see also S142). Human beings, according to Pascal, can never sincerely and truthfully say that they really believe that they do not exist, that time does not exist, or that there are not principles for mathematics and physics. Skeptical proofs suffer the same ills as all rational proofs. Rational proofs, whether for or against knowledge, can convince only briefly. Just as the certainty engendered by a demonstrative proof of God quickly fades, the uncertainty promoted by a skeptical attack on knowledge quickly disappears. The skeptic truly disbelieves only briefly and always returns to more basic beliefs concerning human ability to know. So-called skeptics speak without sincerity; their true beliefs, statements, and actions diverge. Skepticism leads only to evanescent peace followed by sustained disquiet. Just like dogmatic proof, it fails utterly as a spiritual exercise, given the way human beings are in fact made up.

To insist on rational proofs, whether to prove or to disprove, is to misunderstand how human beings come to be certain and how they remain satisfied in that certainty: "we must not misrecognize ourselves: we are automaton as much as mind. . . . Proofs convince only the mind." More than minds, never capable of rationally self-organizing themselves, human beings need considerably stronger means for transforming themselves, means that can "incline the automaton, which carries along the mind without it thinking about it" (S661). This something, for the "people," is the nonchalant following of custom; but that is far too suspect, if not vulgar, for Pascal's intended audience of elite gentlemen and gentlewomen, his *demi-habiles*. How to affect them?

In his essay "On the Art of Persuasion," Pascal argued that there are "two entries through which opinions are received in the soul": the understanding and the will. The understanding is "the most natural"; but "the most ordinary, although contrary to nature, is that of the will." Beings contrary to nature do not gain conviction through proof; they are genuinely brought to conviction "through *agrément*"—the fundamental skill of the *honnête homme* (M III:413). *Sentiment*, affective change, not intellect alone, can yield conviction and peace.[81]

For Pascal, reason is not powerless to produce this affective transformation. Much of this chapter and the last have illustrated how Pascal made reason into a means for *moving* his contemporaries. Pursuing the implications of mathematics leads mathematicians to wonder at infinities. Reason forces them to recognize that infinities exist; reason forces them to admit their incomprehension before them; reason forces them to accept their limitations. Wonder before infinity can be far more than a spur to further inquiry: wonder should be a permanent affective state supervening upon recognition of the boundaries of human ability. Pascal repeated this model with natural philosophy and with the diagnosis of the monstrosity of humanity itself. Reason can prevent *honnêtes gens* from misrecognizing themselves; reason can make the incomprehensible, contradictory elements that compose human beings apparent. Reason can do more than appeal to the mind; it shocks and surprises by revealing a terrible, surprising monster calling out for some explanation. Having shown the "corruption" of human nature "by nature itself," Pascal argued, only one creed has always demanded its believers to accept a beast so contrary to nature: Christianity.

FROM EMPIRICAL MONSTROSITY TO SCRIPTURE

Prosopopoeia

"O men! Vainly do you seek in yourself the remedy for your miseries! All your light can only lead you to know that it is not in yourselves that you will find the truth or the good.

The philosophers promised it to you, and they could not deliver.

They know neither your true good nor your true state."

Pascal, *Pensées*, Fragment 182

The philosophers enjoin all to know themselves, and therein to find the cures for their miseries. Pascal did not disparage completely the attempt at philosophical spiritual exercises, the use of reason to regulate one's way of life. The honest search for knowledge of the self reveals much greatness and much wretchedness, in a perplexing jumble beyond human ken. Proper philosophy, the true philosophy that mocks philosophy, recognizes its inability to comprehend and explain the monstrous nature of human beings. Reason brings to light the salient empirical features of human motivation and activity; reason moves one to wonder at these features; proper reason leads one to look outside the self and outside reason for answers. The discernment of the *honnêtes gens* applied to themselves ought always to lead them away from themselves as the source of their own consolation.

Mathematics and natural philosophy offered the model for this movement. Pascal's two-infinities argument was meant to shock and terrify the reader. It shows how reason—the very best mathematical and natural-philosophical reason human beings have—leads to the knowledge that things it cannot understand exist, and that human knowledge always rests on things far from reason's comprehension. Descartes' mathematics offered experience with the clear and distinct. Pascal's two infinities and similar arguments offered experience in reasoning with the incomprehensible in delimiting reason and in recognizing its true power. Admiring and admitting yield no true understanding of the principles of things, but rather a humbling recognition of the monstrousness of humanity. Pascal's mathematical practice was preparation for the proper unbiased search for human created nature and for a proper diagnosis and effective cure.

Truly caring for the self means trying to know the self through intense concentration and investigation; it entails always failing when confronted with the contradictions of human action, motivation, and epistemic potential. Wonder before these contradictions makes apparent the necessity of turning outside the self for understanding and spiritual help. Only an outside resource might be able to explain human contradictions; only an outside resource might succor the suffering of the monstrous self left to its devices. Reasonable beings must look outward. Where might it be reasonable to seek answers?

Pascal's exercises in coming to accept the monstrousness of humanity were supposed to involve only secular arguments and considerations from empirical evidence of nature and human nature.[82] He brought his readers to the basic Augustinian picture of human beings after the Fall, without requiring them to assent to any doctrines grounded in Christian authority alone. In the second part of his projected apology, Pascal demanded that his readers turn their reason to another empirical domain, the texts of the Hebrew Bible and the New Testament. In an organizational fragment, he sketched these two parts of his apology, their endpoints, and fundamental grounds for proving those goals: "First part: That [human] nature is corrupt, through nature itself. / Second part: That there is a Repairer, through Scripture" (S40). Remarkably, as Pierre Force has illustrated at length, Pascal's hermeneutic project rested on the secular reading of the Bible and examination of the historical grounds for its authority.[83] Pascal did not assume that his readers accepted the authority of the Bible any more than he assumed that they accepted his Augustinian anthropology. Belief in the authority of the Bible need not precede the reading; rather, as Force has shown, the text itself, properly interpreted, was to offer ample ground for accepting its authority. The reading practices that the *honnête homme* brings to any text were to drive this proper biblical interpretation and construction of the authority of the book. The apparent contradictions in the Hebrew Bible can be resolved with any satisfaction, Pascal contended, only through a partially spiritual interpretation of key prophecies, the Christian one of the New Testament. Only with the careful sorting through of the historical and predictive data in the Bible is the monstrousness of humanity described; only therein does it come to be explained.

To Pascal, the Bible exemplifies rhetoric tuned to the disproportion of humanity. As Philippe Sellier has stressed, the Bible offers, in Pascal's eyes, a panoply of rhetorical forms, a diversity geared to the eternal human need for diversion. A diverse order serves fallen humanity far better than a geometrical or logical order in questions of belief. The biblical order works to "excite, not to instruct" (S329). Mere instruction aims at the wrong result: intellective knowledge, not spiritual transformation. Affective change comes through the art of *agrément*, the art of the *honnête homme*. Christ's speech embodies the ultimate *honnêteté*. "Jesus Christ said great things so simply that it seems that he had not thought them, and so sharply that one sees that he thought them. This clarity joined to this naturalness is admirable" (S340).[84] According to Pascal, the Bible offers narratives well tuned to the actual constitution of human beings, their combined wretchedness and greatness, their need for intellectual and affective nourishment.[85] Only the Bible and the tradition and rituals of the church based on it might provide effective means of consolation. They, he boldly contended, are the only resources any semirational being should settle for; hard work with them provides the necessary, but not sufficient, conditions for receiving a satisfying and truthful sentiment—a truly consoling faith. On 23 November 1654, Pascal had a tremendous mystical experience, which he recorded in his "memorial." Not nature, not philosophy, but such mystical experience alone might provide the *sentiment* capable of bringing true peace:

> God of Abraham, God of Isaac, God of Jacob,
> Not of philosophers and savants.
> Certitude, certitude, *sentiment*, joy, peace.
>   (M III:50)

PART III Leibniz

CHAPTER FIVE

Forms of Expression

"The use of geometry rests in application," Gottfried Wilhelm Leibniz explained to the famous jurist Hermann Conring—above all, the application of "exercising the intelligence." Leibniz could attest that geometry "serves to accustom one to the solid and the certain," for he had pursued mathematics "in France, more perhaps than is necessary for the stretching of the mind." He had no regrets: "for, from that time in which I attended more diligently to geometry, I began to judge all things a bit more carefully."[1] While Leibniz did not explicitly invoke Descartes, the Frenchman, like Plato, had clearly been right about the power of mathematics to exercise the mind.

From the manifold specimens of his mathematical work, Leibniz told Conring of a key preliminary to the invention of his calculus: his quadrature (finding the area) of the circle. In slightly anachronistic terms, Leibniz had demonstrated that

$$\frac{\pi}{4} = \frac{1}{1} - \frac{1}{3} + \frac{1}{5} - \frac{1}{7} + \cdots$$

The magnitude of the circle, he wrote, "can most simply be expressed by this series, that is, the aggregate of fractions alternately added and subtracted." While the series was useful for finding numerical approximates to the area, Leibniz ended his comments by emphasizing, "but this, as I said, is to be considered primarily for the exercising of the intelligence."[2]

Leibniz developed the infinite series expressing the area of a circle early in his famously productive sojourn in Paris from 1672 to 1676. He suggested that such written expressions offer the only exact knowledge of the quadrature of the circle available to embodied human beings without divine intervention. In a subsequent letter to Conring, he noted that the series provides the sort of knowledge mathematicians should desire, but not the knowledge they in fact desire.[3] Leibniz maintained that, although rarely offering constructive, intuitive, or evident knowledge, symbolic expressions can offer entry to a much wider range and depth of knowledge than many contemporaries thought possible. Although not producing clear and distinct intuitions, such expressions can provide written substitutes for evident knowledge, substitutes that are literally visible all at once. Not to accept such expressions would be to neglect a fundamental means for acquiring better knowledge, ending disputes, and building a solid foundation for the good life. Leibniz wrote in 1677 that in mathematics we reason, not on "the thing itself, but on the characters that we have substituted in place of the thing."[4] Only by reasoning with new characters might human beings make their best effort to answer the pressing questions of metaphysics, natural philosophy, and ethics. Only by stretching the mind with such mathematics and such characters might human beings best perfect themselves and others.

In this chapter and the next I consider how Leibniz came to maintain that symbolic expression could constitute legitimate knowledge; I describe some of the techniques and doctrines—philosophical, mathematical, and practical—that he devised given such a belief. In these two chapters, I focus on a cross section of Leibniz's earlier works, written from around 1670 to the mid-1680s.[5] (While the early views and practices discussed here likely have implications for interpreting the later writings of his "mature" philosophy, I do not make any strong claims about such implications.) In this chapter I discuss how Leibniz produced his quadrature of the circle, the terms in which he defended his solution as legitimate knowledge, and some of the considerations about many larger philosophical and practical questions that he drew from it. His great mathematical discovery of the quadrature and his defense of symbolic expression as legitimate mathematical knowledge, I suggest, became possible in part because of his practical attempts to create new symbolic and optical technologies that would permit human beings to see many things all at once. By tracking his interest in these concrete techniques, we can better reconstruct how Leibniz developed some central concepts and practices in his mathematics and early philosophy, and we can understand less anachronistically the importance he attached to them. I document the constitutive importance of his optical descriptions, metaphors, and practices for Leibniz. I track his use of perspectival heuristics, from creating amusing spectacles to offering a key mathematical insight, from motivating the invention of new techniques of painting with mirrors to describing the creation of variety in the world. In reworking and further articulating his philosophical positions, he drew

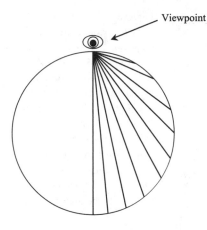

Viewpoint

Figure 5.1. Perspectival intuition behind Leibniz's quadrature.

on his optical, notational, and mathematical practices. Drawing in part upon his mathematical solution of the quadrature and his arguments that this solution really was mathematical knowledge, Leibniz came to argue that bringing the soul and mind closer to God required a sophisticated deployment and involvement in the *material* processes of notation. Perfecting ourselves means making refined use of corporeal limitation and corruption in order best to know our world and our duties within it. In the next chapter I will show how Leibniz further developed his writing technologies and put them at the center of his practices aimed at producing a society of habitually charitable people.

LEIBNIZ'S QUADRATURE OF THE CIRCLE

Finding a geometric quadrature of the circle means constructing a square whose area is equal (or proportional) to the area of a given circle. Quadratures of the circle often involved using a series of bands formed by parallel lines to divide the circle. In contrast, Leibniz's quadrature of 1673 used a simple intuition: divide the circle from a single point of view to produce an infinite number of triangles (see fig. 5.1).

The proof has several stages:

- Divide the circle into triangles intersecting at a point on the circle (A) (see fig. 5.2).
- For each curvilinear triangle ($AFE, AED, ADC, \ldots$), construct a rectangular area ($JKP\alpha, IJO\beta, HIN\gamma, \ldots$) using tangents drawn from the circle. The quarter of the circle has area $AFK +$ (all the triangles), which approximately equals $AFK + 1/2$ (all the rectangles). The sector of the circle is thereby "transmuted" to another curve defined by the corners of the rectangles ($\gamma, \beta, \alpha, \ldots$), when the number of triangles becomes

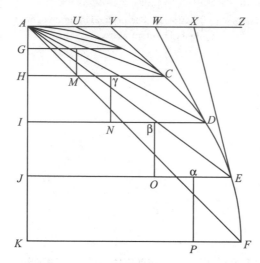

Figure 5.2. Transmutation of circle (not to scale). The rectangles are formed by copying the distances AX, AW, AV, AU, ... (formed by the tangents), onto the segments KF and JE, JE and ID, ..., to form KP and $J\alpha$, JO and $I\beta$, ... (Each rectangle has twice the area of its corresponding triangle.) Thus, the circle is transmuted into a curve defined using the tangents to the circle.

infinite. The area of the quarter circle equals $AFK + 1/2$ (the area under the curve formed by $\gamma\beta\alpha\ldots$).

- Find the area of this new curve, described by the equation

$$x = \frac{2az^2}{a^2 + z^2}$$

where x is a variable corresponding to the sequence of values AG, AH, AI, ..., and z is a variable corresponding to AU, AV, AW, ...

Following a procedure of Nicolaus Mercator, Leibniz performed long division on this equation and produced an equivalent equation for the curve in the form of an infinite sum. He could compute the area under the curve described by such an infinite series. His result[6] was that a quarter of a circle with radius 1 has an area equal to

$$\frac{1}{1} - \frac{1}{3} + \frac{1}{5} - \frac{1}{7} + \cdots$$

Leibniz insisted that displaying how he had discovered the proof was far more important than giving the proof itself.[7] Commentators have likewise noted that the actual quadrature was less important than how Leibniz translated the circle into a more tractable curve. Instead of resolving figures with parallel lines into rectangles, Leibniz noted, "I have found a general method of usefully resolving every figure into an infinity of little triangles ending at a single point, by means of convergent ordinates."[8] This procedure, he claimed, would open up a "great field of new inventions" (K36). He dubbed this procedure for converting curves "transmutation."[9]

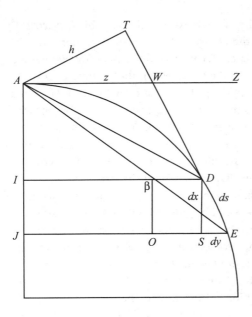

Figure 5.3. Characteristic triangle. The characteristic triangle allows one to find different ways of characterizing a curve using the infinitely small triangle *DSE*. One can find, for example, multiple expressions of the summations of areas. The triangle *DSE* is similar to triangle *ATW*, so $DS/DE = AT/AW$ and thus $DE \cdot AT = DS \cdot AW$. (In more modern terms we might say $h \cdot ds = z \cdot dx$, so that $\int hds = \int zdx$.) With these relationships, the transmutation theorem easily follows. The area of the triangle *AED* is $1/2(base \cdot height) = 1/2(DE \cdot AT) = 1/2(DS \cdot AW)$. Since $AW = I\beta$, the triangle is equal to $1/2(DS \cdot I\beta)$. So the sum of such triangles equals one-half the sum of the rectangles. After Guicciardini 1999, p. 140; Hofmann 1974, p. 55.

At this stage in his research, he judged this procedure to be among "the most general and most useful that exist in geometry"; he brashly maintained it would work to convert "any curve, whether laid down by chance or arbitrarily without a certain law" (K35–36).

The transmutation theorem builds on the "characteristic triangle" that Leibniz had developed in 1673 from procedures in Pascal's *Letters to Dettonville*.[10] The characteristic triangle associates a very small triangle *dx, dy, ds* with a finite triangle including either a tangent or normal to the curve (see fig. 5.3).[11] The similarity of the triangles reveals the equivalence of several ways of determining the area of the curve and helps in converting difficult curves into more tractable ones.[12] Drawing upon this characteristic triangle, the transmutation theorem uses the tangents of one curve to construct an equivalent one.[13] Using tangents to find quadratures was a central focus of Leibniz's mathematical work; the full implications and power of the approach would become apparent to Leibniz in late 1675 and 1676, when he made the key breakthroughs in his differential and integral calculus.

Right around the time he had discovered his quadrature, Leibniz touted the tremendous significance of the transmutation theorem and characteristic triangle in a capsule history of mathematics entitled "The Ends of Geometry." He condemned Descartes and his followers. "Common is the sin of those who have vowed obedience to Descartes' words."[14] As Leibniz put it, Descartes "believed that a curve could not be found equal to a line by any human art. He skillfully enough expressed this in his *Geometry.*"[15] Descartes was wrong about the ability of human art to understand the relation between curves and lines. Techniques unforeseen by his method showed his error. By developing the treatment of the arithmetic of series and the manipulation of irrational equations, mathematicians had gained entry into the arithmetic of the infinite. Following upon Viscount Brouncker's quadrature of the hyperbola, as well as the work of John Wallis, Leibniz's quadrature of the circle was an early success, and one that rested on a more rigorous foundation than the interpolations of the English.[16] With his general transmutation theorem, Leibniz claimed, he could methodically transform figures whose equations include intractable quantities so that the properties of those figures could be grasped with infinite series of rational numbers. Previously, such transformations had been discovered only by chance, he claimed; now they could be methodically performed though the easy transmutations procedure Leibniz had suggested. He "dared" to suggest that he had opened the "source" (*fontes*) of mathematics sought since Archimedes and Apollonius.[17] Too many mathematicians refused to accept the need for such a form of mathematics, drawing on written expressions. Too many had trusted the word of Descartes and other overly restrictive mathematical authorities.

DEFENDING INFINITE SERIES

Responding to Leibniz's announcement of his result, Christiaan Huygens, his mathematical mentor in Paris, tempered his praise. Commending the procedure as "very beautiful," he held that it offered "a new way which seems to give some hope of coming to the real solution."[18] Huygens did not consider Leibniz's infinite series as providing genuine mathematical knowledge of the area of the circle, just as many contemporaries rejected the results of Wallis and Brouncker.[19] Much as Descartes had to contend that his new machines resulted in genuine knowledge, Leibniz had to defend his enrichment of the domain of truly knowable mathematical objects and the means for knowing them.[20]

Like all problems in traditional Euclidean geometry, the problem of squaring the circle demanded construction of a solution, that is, *constructing* a square with area equal (more rigorously, proportionate) to a given circle using only a compass and straightedge. Even in Descartes' account, algebraic formulas, in principle, provided knowledge only insofar as they ultimately resulted in geometric constructions—Descartes merely expanded the legitimate means for constructing them. Calling a quadrature permitting such a construction a "perfect quadrature," Leibniz maintained that such perfection likely exceeded human power.[21] A taxonomy of quadratures that he drew up includes both geometrical construction and giving a value "through expression" under the category of "exact quadrature."[22] His result was such an expression, one as exact as one desired.

From his first tentative announcement of his result to Henry Oldenburg in 1674, he carefully chose his language for describing this new mathematical knowledge. "I can *exhibit* the relation of the diameter to the circumference exactly, not as a certain number to another number . . . , but by a ratio of a number to a whole certain, simple, and regular series of rational numbers."[23] Mathematical knowledge envisioned as *exhibition* meant the production and then proof of a simple rule for the continuation of a series.

Using a language of human limitation in a more developed defense of his quadrature, Leibniz anticipated an objection: "You inquire if the magnitude sought cannot thus be exhibited, since it is not *in our power* to progress into the infinite. I confess it is so" (K79; my italics).[24] Like his imagined critic, Leibniz maintained that human capacity prevented any attempt actually to comprehend all the terms, to comprehend the infinite in its totality.[25] Despite or, rather, because of this confession, he insisted on a new path forward. As he put it in the most elaborate treatment of his quadrature, written in 1675–76: "I do not promise to exhibit [the area] by a certain geometric construction, but rather by an Arithmetical or Analytical *expression* of a series, for such is permitted, as long as the *ratio* of the progression is apparent" (K79; my italics).[26] An expression meant certain knowledge of the rule (Leibniz often used the vague Latin term *ratio*) producing a sum, not a complete and simultaneous comprehension of all its terms. First systematically used in his mathematical papers of late 1674, the term "expression" quickly came to play a central role in Leibniz's account of human knowledge and in his account of the interrelatedness of everything in the world, as discussed below.[27]

For Leibniz, such an expression could constitute exact knowledge, even though it provided neither a construction nor a quantity. "A value can be expressed exactly, either by a quantity or by a progression of quantities whose nature and way of continuing are known" (GM V:96). As Marc Parmentier has noted, not only "can an infinite series be equal to a finite number, but one can affirm that this equality is rigorous, even though the series can be neither written nor completely surveyed."[28] Leibniz's series was no approximation: using the rule for his series, one can compute a partial sum with an error smaller than any finite quantity; the series can provide a sum as exact as we might desire.[29] Leibniz stressed that knowing the rule behind the progression constituted knowledge, in contrast to blindly computing terms or simply knowing more of them: "For the nature of a series, even an infinite one, can be understood, even only having perceived a few terms, so long as the *ratio* of the progression has appeared. Once discovered, we continue in vain when it is a question of illuminating the mind rather than trying to achieve some mechanical operation" (K79).[30] The quadrature could indeed help such mechanical operations, but discovering the rule behind a progression—not blind computation using that rule—provided knowledge and illumination.[31] This striking drive of Leibniz's to avoid needless computation framed his quest for a general method of discovery of the *rationes*—the "reasons"—underlying complex phenomena, such as the quadrature of the circle. His quadrature, far from merely providing an approximation, permitted "the mind as if with a single blow to penetrate" and to grasp the value (K80).[32] Since they did not provide geometric constructions, Leibniz's expressions could not result in the clear, distinct, and simultaneous knowledge Descartes demanded. These expressions provided, however, an ersatz form of simultaneous knowledge, appropriate to human beings.

In demanding traditional geometric or algebraic answers, his contemporaries blocked the discovery of additional supplementary practices within human power. Or so Leibniz argued. Much as Pascal had maintained, the mistaken desire for divine knowledge often held back the search for more appropriate techniques. Leibniz put it: "How we wish for many things that we have, and how often do we ignore those things which are in our power to know."[33] The quadrature could teach an important lesson about what was within human power to accomplish, about what was in fact proportionate to fallen human epistemic capacities, and what means were needed to supplement those capacities.

Recognizing the proper limits to mathematical knowledge meant coming to accept the proper "ends" of mathematical inquiry. Leibniz wrote to Conring that the quadrature "is not that which mathematicians commonly desire, but that which they ought to desire; for it is impossible to explicate the ratio between a circle and a square with a single number; therefore, it is necessary to produce a series of numbers extending to infinity."[34] He hinted at a natural basis in human faculties for his new mathematical epistemology: "where *equations* are lacking, nature has furnished us another means for relating problems to numbers, which is that of *progressions* of numbers."[35] Anyone daring to reject the use of infinite series Leibniz viewed as either ignorant or ungracious, for their unwillingness to use those faculties God had granted to created beings (K69).

For Descartes, as illustrated above, symbols were a crucial but worrisome propaedeutic to intuiting a complex mathematical situation clearly, distinctly, and simultaneously. A solution to a geometrical problem with algebraic help should result in a geometrical construction, and not merely in a formula. In contrast, Leibniz stressed the inability to intuit simultaneously many complex truths except through expressions—collections of appropriate characters. In order to maximize human capacity, mathematicians must allow the introduction and use of helpful algebraic-like characters *and* permit expressions involving those characters to be final results. Such expressions must be allowed if one wished to have the maximum amount of partial but certain (mathematical) knowledge.

Much like Descartes, Pascal, Pardies, or Arnauld, Leibniz used mathematical questions to probe human capacity. In early-modern Europe, the quadrature of the circle figured in debates about the limitations of human knowledge. One Jesuit mathematical treatise, for example, explicitly connected the quadrature to the question of the proportionality of the human intellect to natural truths after the Fall: "although the quadrature of the circle is possible in its nature, theologians have nevertheless thought fit to inquire whether in these days, that is, after the Fall of Adam, man can attain knowledge of this matter without the aid of special divine grace."[36] Leibniz's private philosophical, mathematical, ethical, and practical notes of the 1670s and early 1680s reveal him concentrating on augmenting human deficiency through human technologies and symbolic techniques.

THE TRANSMUTATIVE HEURISTIC

From early in his life, "harmony" had been central to Leibniz's metaphysical, as well as aesthetic, views. He defined it often as "diversity compensated by unity."[37] He used the language of harmony to explain achievements in geometry. Girard Desargues and Blaise Pascal, for example, "have accomplished the universal demonstration about the conics, and by this means, the harmony of the conic sections and their common properties appeared." Where the ancients had proved analogous theorems about the various conic sections separately, Desargues and Pascal proved theorems about them all taken together.[38] Unfortunately, most discoveries of such harmonies, even in mathematics, had been made more by chance than by method.

Drawing up his recent mathematical achievements, Leibniz called for a "method of universality" capable of systematically discovering harmony—unity in diversity. A general method requires, in the first place, "the reduction of many different cases to the same formula, rule, equation, or construction; and in the second place, the reduction of different figures to a certain harmony, in order to demonstrate or to resolve universally a quantity of problems or theorems about them."[39] The text calls for improving techniques for quadrature as a prime example of this process: "if with time the Geometry of infinities can be made a bit more susceptible to analysis, then the problems of quadratures, centers [of gravity], and the dimensions of curves will be resolved by equations. There are grounds for hoping that—although Mr. Descartes dared not to aspire to it—we will draw forth a great advantage from the harmony of figures for finding the quadrature of one as well as the others."[40]

Leibniz's quadrature of the circle illustrated both forms of reduction. Through his transmutation theorem, the circle was shown to be "harmonious" with another tractable curve; his characteristic triangle, work with series, and other techniques, he hoped, would allow many more curves to be understood as harmonious with tractable ones.[41] The expression of the infinite series of the areas would permit certain knowledge about a wider array of curves—the knowledge Descartes thought impossible. Leibniz's quadrature of the circle, then, exemplified a general heuristic for systematically discerning interrelationships among objects in mathematics and in the world. First, it illustrated the power of transforming things—transmuting them—in order to discern the "harmony" among apparently disparate things or objects of thought, such as the harmony between the circle and the curve produced in his proof. Second, it illustrated the power of an appropriate form of written expression to help discern, or in many cases to permit one to discern, such harmony.

In various writings, Leibniz cited a perspectival intuition of Pascal's and Desargues' as a central stimulus for his quadrature and mathematical method more generally. When he read Pascal's now-lost manuscript on conic sections, Leibniz duly copied out (or summarized) a key statement of method: "In geometry, every method for discovery by the means of situation, and therefore without calculation, consists in grasping many objects simultaneously in the same situation. This is done, sometimes by means of a figure that includes many [different objects], sometimes by motion, sometimes by mutation."[42] Pascal thought that changing viewpoints could obviate the necessity for calculation and expression in symbols. Leibniz demurred. While they had the insight of requiring multiple viewpoints and attempting to connect them, Pascal and Desargues failed to provide a helpful system of notation to make relations apparent. Like many of their contemporaries, they resisted the further development of symbolic notation in mathematics. They remained only at the level of images, whose manipulation, Leibniz complained, fatigued the mind and impeded progress.[43] Such mutations were but one-half of Leibniz's heuristic. (To be sure, Pascal had praised the transmutation of enunciations; but most of his enunciations were written out in ordinary language, not in algebraic symbols.) Desargues and Pascal had failed to provide symbolic techniques to aid the imagination.

Leibniz often credited the key intuition behind his quadrature—the dividing of the circle into triangles from a single point—to the work of Pascal and Desargues: "Messieurs Desargues and Pascal very well took" standard parallel lines for a species of convergent lines, whose meeting point is infinitely distant.[44] Once alerted to the work of Desargues, Leibniz sought out Desargues' extremely rare books and Pascal's unpublished manuscripts but had difficulty in gaining access to them. Toward the end of his Parisian years, as already mentioned, he managed to copy parts of Pascal's now-lost manuscript on conic sections.[45] In the period of the quadrature, around 1673, Leibniz apparently knew the geometrical insights of Pascal and Desargues exclusively through rumor and the descriptions of others.

The simple use of their insights in Leibniz's transmutation theorem hardly required the full details of the geometry of Pascal and Desargues. Many years later, in a letter to Bernard Le Bovier de Fontenelle, Leibniz noted that nonspecialists have a habit of offering "more singular thoughts—slightly vague and wandering concepts." Someone "not at all a mathematician" had inspired his arithmetical quadrature; this person had "printed something of geometry."[46] Such a description could hardly have applied to Pascal, but Leibniz could have meant the engraver Abraham Bosse, a disciple of Desargues whose work Leibniz read. Bosse noted that Desargues wrote "that parallels are the same as those lines that end in a point, and that they do not differ from them"[47]—a key intuition behind the transmutation of the quadrature. For most readers, Bosse's telegraphic remarks would have meant little. By the time he arrived in Paris, Leibniz was well versed in perspective and optics, however little up to date he was in mathematics more generally. In the years immediately prior to leaving for Paris, he had already worked hard to create new optical and perspectival devices and drew heavily on optical and catoptrical metaphors. In his notes on Pascal's methodological remarks, Leibniz wrote: "Furthermore, from motions and mutations, it seems useful to consider the mutations of appearances, that is, the transformations of optical figures; it is to be seen whether by this means we might go beyond the cone to greater things" (M II:1127).[48] At the heart of the novel projective geometry of Pascal and Desargues, such mutating of optical figures, known as anamorphosis, was a central preoccupation of a much wider culture of perspective.

To convey the importance of perspective and other optical techniques in Leibniz's mathematics and then in his philosophical practice, the next few sections discuss his intense involvement in the varied cultures concerned with perspective in the mid-seventeenth century, just before and during his Parisian trip. In perspective and optics, as understood and practiced during his time, Leibniz found potent models for both parts of his transmutative heuristic: he found and developed a model of the hidden unities in the apparent diversity of nature; he likewise found and developed important technical practices for improving human ability to see many things distinctly and simultaneously. The next section considers the wider culture of seventeenth-century perspective understood as a form of natural magic.

The subsequent section considers how Leibniz worked to invent new optical and symbolic techniques to improve human ability to see all at once. In his quadrature, he brilliantly united these two strands of his work.

PERSPECTIVE, ANAMORPHOSIS, AND DIRECTING THE ATTENTION

One evening late in his four-year sojourn in Paris from 1672 to 1676, Leibniz witnessed a public display of a machine for walking on water. Much taken with this display, he produced an extended textual reverie, which he labeled "Funny Thought" ("Drôle de pensée").[49] His subject was the profit, personal and societal, that spectacles such as demonstrating new machines could bring. He envisioned sites for the public display of new curiosities to stoke interest and investment in invention and technologies. His funny thought soon took another turn: he dreamt of "academies of games," devious sites where vices would be turned to virtue: "gambling would be the most beautiful pretext in the world for beginning a thing as useful to the public as [these displays of useful devices]. For it is necessary to make the world fall right into a trap, to profit from its feebleness, and to deceive it in order to cure it. Is there anything so just, then, as to make extravagance serve the establishment of wisdom? It is truly . . . to make an antidote from a poison."[50]

The quixotic and delightful "Funny Thought" offers a tour through the whole range of late-seventeenth-century amusements that Leibniz judged salutary for the advancement of knowledge, productive industry, and morals: extraordinary animals, fake wars, experiences with the vacuum, rare instruments, calculating machines, magnets, the camera obscura, fire-eaters, telescopes and microscopes, pleasant disputes, comedies of all nations, windmills, instruments that play themselves—in the end, "all sorts of things."[51] Like so many other early-modern figures, Leibniz was much taken with all manner of seemingly magic transformations, metamorphoses, and transmutations of natural and artificial things. Like many contemporaries, he wished to channel interest in such spectacles into epistemic and moral reform.

Among his laundry list of diversions and spectacles, Leibniz referred to the witty painting techniques and the games of light and shadow associated with the rules of perspective: "Optical curiosities would hardly cost much and they would make up a great portion of these curiosities."[52] For example: "Paintings that one sees from one side in only a certain manner, and in an altogether different manner from the other side."[53] Often used to hide the sexually explicit, such drawings and paintings—called anamorphoses— were an early-modern rage and took many forms, often involving mirrors (see figs. 5.4–5.5).[54] In the seventeenth century such "depraved

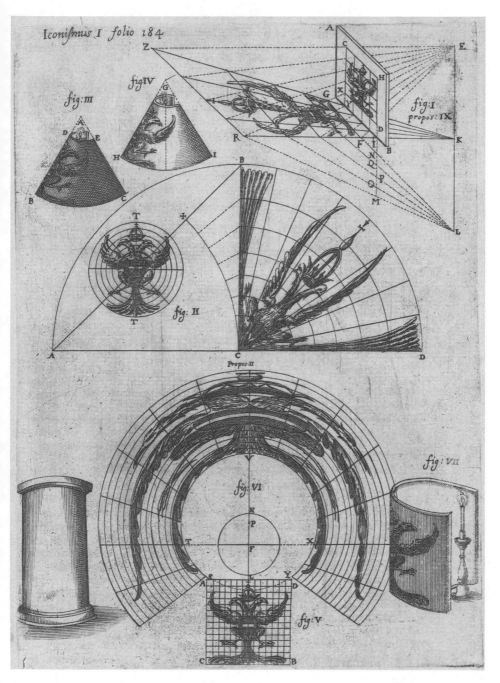

Figure 5.4. Example anamorphoses. From Athanasius Kircher, *Ars magna lucis et umbrae, in X. libros digesta*...(Rome, 1646), following p. 184. Courtesy of Rare Book and Manuscript Library, Columbia University.

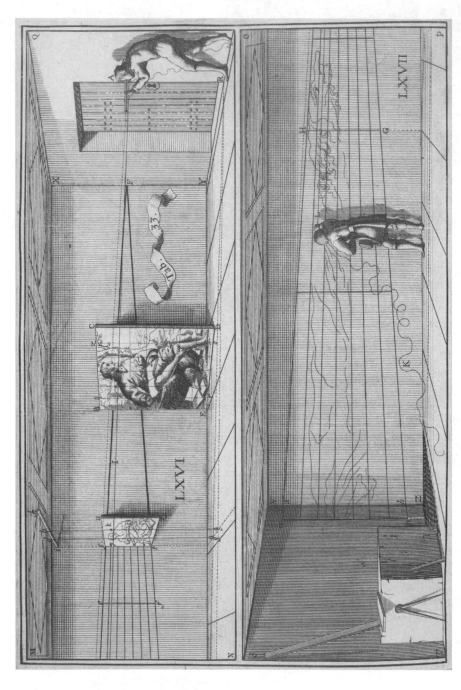

Figure 5.5. Method for the creation of the fresco *Saint Jean l'Evangéliste à Patmos* in anamorphosis. From Jean François Niceron, *Thaumaturgus opticus* (Paris, 1646), plate 33. Courtesy of Rare Book and Manuscript Library, Columbia University.

perspectives" had become both highly theorized and put to sundry religious and political uses.[55]

In Paris after 1645, the discerning tourist could see two astonishing frescoes at the house of the Minims, the religious order of Marin Mersenne. According to the 1698 edition of Germain Brice's *Description of the City of Paris:* "The figures of St. John and the Madeleine that one can only perceive in a certain position and regard from a marked point . . . occupy the entire length of two galleries."[56] The two frescoes, among the greatest achievements of the Minim Jean François Niceron, exemplified the technique of anamorphosis described in his *Curious Perspective, or The Artificial Magic of Marvelous Effects* (1638). Niceron's great frescoes appeared as smudges when looked at straight on, but astonishingly became portraits of sanctity when viewed from the end of the hallway (see fig. 5.5).

The rational transformations associated with anamorphoses, like other modish perspectival tricks, belong to the tradition known in the Renaissance and seventeenth century as natural and artificial magic.[57] With these arts, one could excite and direct the passions, as suggested by Niceron's subtitle, "artificial magic of marvelous effects." Seventeenth-century masters of spectacle enhanced every sort of authority with anamorphoses and other similar optical displays. Niceron used them to inspire piety. Jesuits such as Athanasius Kircher, Gaspar Schott, and Mario Bettini produced long treatises and sundry anamorphoses, including "gardens of the passion," replete with Jesuit symbols and instruments of Christ's Passion—all tools in the service of the Catholic Reformation.

When considering perspective as something more than a technique in painting, modern readers tend to think immediately of metaphors of point of view or of worldview. For Leibniz's contemporaries, perspective and associated practices such as anamorphoses were understood as powerful techniques for instructing, intriguing, and delighting. Perspective tricks were often held to be potentially more instructive than many other natural and artificial curiosities. The art historian Celeste Brusati has shown that perspectival boxes and other forms of perspectival artifice, including anamorphoses, were understood to do far more than merely delight and astonish. While they certainly provoked wonder, they were supposed "to stimulate the intellectual curiosity of the beholder and to incite discovery of the secrets of nature revealed in their making."[58] Early in Niceron's influential *Curious Perspective,* for example, he detailed an "Optical Experience that perfectly teaches perspective."[59] Perspectival demonstrations and tricks with light directed curiosity to salutary inquiry into the rules of nature—hence their place in Leibniz's "Funny Thought."[60]

By the time of his arrival in Paris, Leibniz knew the technical practices and the perceived utility of such perspectival tricks well. He closely followed technical developments in optical theory and in instrumental practice; he knew and cited the accounts of using perspective and image-casting as a natural magic useful for bolstering the faith and encouraging greater reflection. The natural-magic works of the Jesuits and Niceron had been duly digested into German in the great *Physical-Mathematical Delights* (see fig. 5.6). First published by Daniel Schwenter and then greatly ex-panded by the poet Georg Philipp Harsdörffer, these volumes offered a Lutheranized version of the *propagatio fidei per scientias* of the Catholics—the propagation of the faith through natural philosophy and natural magic, most famously used in China.[61] In additional to its numerous other delights involving number, weight, and measure, the work contains the whole range of optical and catoptrical devices useful to provoke wonder and amaze-ment: using a mirror to make ghosts appear, using a cylindrical mirror to make an image appear in the air, and the making of sundry anamorphoses more generally.[62]

Leibniz knew the technical discussions and pedagogical techniques of the *Physical-Mathematical Delights* well. Its first volume provided the young Leibniz with much of the inspiration and the details of the simple math-ematics of his "Dissertation on the Combinatorial Art" (1666).[63] Drawing upon texts such as Harsdörffer's, the "Funny Thought" of 1675 offered an extremely up-to-date overview of the whole range of techniques for using natural magic and the sciences to effect moral and religious transfor-mation. Before and after Leibniz arrived in Paris, he avidly read many of the latest works on optics, vision, and optical machines and attempted to offer radical new machines to improve vision and to help in the creation of perspectival representations. In Paris, his reading continued to include optical treatises using perspective, technical treatises on perspective, and accounts of the uses of perspectival and optical tricks.[64] All three subjects came together when he wrote down the funny thoughts prompted by the machine for walking on water.

In a mode of cynical realism rarely associated with Leibniz, he explained how curiosities would attract the crowd to be educated: "All *honnêtes gens* would want to have seen these curiosities so to be able to speak about them. Even women of quality would want to be taken there, and more than once. Thereby we would always be encouraged to push things further."[65] Leibniz should be understood as one of the many proponents of the propagation of the faith through natural philosophy and natural magic.[66] The need for something to lead people to self-reflection appeared in one of Leibniz's most important dialogues of the early 1670s, the *Philosopher's Confession*. People desire to perform sinful acts because their passions have clouded their intellects. Fortunately, "in the middle of the shadow, like a gleam of light through a crack, the way of evading it is in our power, so long as we

Figure 5.6. Title page of volume 1 of Georg Philipp Harsdörffer and Daniel Schwen-
ter, *Deliciae physico-mathematicae, oder Mathematische und philosophische Erquickstunden*
... (Nuremburg, 1651–53). Courtesy of Rare Book and Manuscript Library, Columbia
University.

will ourselves to do so." While human beings indeed possess the power to will the good, unfortunately mere arguments often cannot lead them to do so. It "never enters the mind of many who do not so will." These deluded ones exist "without *reflection* or attention, like those who look and look but do not see, or listen and listen but do not hear." How to get them to hear, to see?[67]

Like Pascal, Leibniz knew "*habiles gens* who will not take the trouble to read any demonstration"; he did not want "the severity of demonstrations" to keep them away from serious reflection.[68] More decorous practices were needed to capture their attention and lead them to proper philosophical reflection and transformation. To get people to examine their intellects and, therefore, to will the good requires manipulating their passions to ferment an irrational desire to pursue the rational. Although he was clear that a true ethical transformation demanded knowledge carefully considered by the intellect, Leibniz stressed the potency of moving the emotions to lead people to pursue a path toward such serious philosophical engagement.[69] While not proper for true philosophical transformation, elegant speech, for example, was useful for securing the attention, moving the soul, and imprinting things on the memory.[70] The "power of persuading consists as much in explicating reasons as in moving the emotions, as . . . in the art . . . of grabbing the attention." For the attention "is nothing other than reflection."[71] Such reflection involves, not simply making a rational choice once to pursue the good life, but a fundamental transformation through a reordering of our affects and understanding of the good.[72]

For the Jesuits, the natural magic of the propagation of the faith through science was but one of many rhetorical means for moving people to religious and ethical change and to self-reflection. From his youth, Leibniz held out the Jesuits as exemplary in using science, arts, and medicine to gain the favor and attention of the people.[73] In his early work on teaching and learning jurisprudence, he praised the power of play to encourage learning and good habits.[74] The rhetorical practices of getting human beings to reflect included the educational casino and wonder-producing displays: "it is necessary to make the world fall right into a trap to profit from its feebleness and to deceive it, in order to cure it. Is there anything so just then as to make extravagance serve the establishment of wisdom?"[75] In the late 1670s Leibniz urged his duke in Hanover to establish centers to set such traps and lead people to true wisdom.[76] As we will see in the next chapter, he dramatized the power of such deceptive traps to convert the blasé *honnête homme* from modish skepticism and worldly detachment to a reflective life of scientific knowledge and genuine tranquillity.

In the final section of his "Funny Thought," Leibniz described an illuminating spectacle to surprise, enlighten, and improve the public. Leibniz proposed drawing upon shadow-casting and the rules of perspective to heighten the power of the spectacle.[77]

> One could still add shadows there. Let there be a theater with the spectators at one end. There will be light and little figures in wood, shuffling, which throw their shadow against a transparent paper, behind which there will be light as well; this will make the shadows on the paper appear extremely brilliant and huge. But, so that the figures of the shadows do not appear all in the same plane, perspective can remedy it by the diminishing size of the shadows. They will come from the edge toward the middle, and this will appear as if they were coming from backstage to the front. They will grow in size, by means of their distance from the light; this will be really easy and simple. There will be sudden marvelous metamorphoses, perilous jumps, flights. The Magician Circe who changes form; devils that appear. After all this, all of a sudden, everything will be made dark; the same wall will serve; all the light cut out, except a single one close to movable little figures of wood. This remaining light, with the help of a magic lantern, will throw against the wall moving and admirably beautiful figures, which follow the same laws of perspective. (A4,1:567–68)

Examples such as shadow theaters and anamorphic paintings put the controlled rational transformation of light and vision in the service of intellectual and moral transformation. In "Funny Thought," Leibniz sought not just to generate affect but to spur knowledge, in order to produce more profound intellectual political and moral effects. Shadows and perspective surprised and produced emotions of wonder. Such wonder entailed a cognitive and emotional shock; it shattered one's certainty about how the world operates. Cognitive shock could motivate seeking an order behind appearances. It could motivate seeking knowledge. However, as in so many *Kunstkammer* or other collections or simple demonstrations of phenomena, it could yield what Descartes characterized as mere astonishment—the freezing of thought, not the pursuit of principles of order, of better knowledge.[78]

Much as Niceron sought to teach perspective through optical experiences, the artist and theorist Samuel van Hoogstraten staged shadow plays to teach optical principles to art students by moving figures closer and farther from light sources (see fig. 5.7).[79] With his own version of a theater of shadows, Leibniz suggested that wonder could be channeled into the desire to know the underlying natural laws and rules. The shadow theater deceived through marvelous appearances but suggested that a small number of laws (the *rationes*) produced the shifting shadows. A model of unveiling, in which the laws at work behind complex appearances are

Figure 5.7. Dance of the shadows. From Samuel van Hoogstraten, *Inleyding tot de hooge schoole der schilderkonst* ... (Rotterdam, 1678), p. 260. Courtesy of the Beinecke Library, Yale University.

revealed, appears prominently in Leibniz's own accounts of mathematical and natural-philosophical practice.

In his writings after 1676, the interrelationships among diverse appearances made evident though such anamorphic perspectives figured prominently both within Leibniz's technical practices and as a general metaphor for scientific activity itself. A series of undated manuscripts demonstrates his long-standing interest in the theory of perspective understood along these anamorphic lines.[80] One important manuscript, probably written after 1695, offers an extremely general theoretical treatment of the problem of projecting any sort of image onto any kind of surface—a highly technical version of the sort of problem at the heart of anamorphic texts. Niceron, for example, offered the following problem: "On a proposed plane and at a given distance and height of the eye, to put into perspective all sorts of objects with the universal perspectival instrument."[81] Leibniz defined the entire science of perspective as a field solving precisely such problems in great generality: "Perspectival science is the art of exhibiting the appearance of an object on a panel [*tabula*], that is, given any panel [*tabula*] and object and given the placement among them, and the placement and shape of the light, to draw on the surface lines representing the outlines of the objects and the surfaces expressing the light and shadow representing the light and shadow of the object."[82] According to Javier Echevarría, this document offered a massive generalization of the panel (*tabula*) to any shape and permitted the eye to be infinitely near or close. Leibniz worked, then, to transform the theory and art of anamorphosis into a major theoretical area of study.[83] For all his great mathematical sophistication, he retained the terms, concerns, and goals of the natural-magic culture of anamorphosis and depraved projections. The harmony to be discovered among different anamorphic views, moreover, long served Leibniz as a general metaphor for the unity underlying diversity. The harmony among the various conic sections was but one crucial example.

In his more methodological texts, Leibniz presented the discerning of the geometry underlying diverse views as a paradigmatic example of discovering underlying laws of nature and mathematics amid complicated phenomena. In a document of the 1680s calling for the creation of a "general science," Leibniz used perspective and shadow play to exemplify how a small number of mathematical and physical principles could unite and explain diverse phenomena. Given an understanding of the geometry in a given situation, one no longer needed to retain diverse appearances. The reasons behind perspective and shadow play provide, he said, an "abridgement," one that metaphorically makes all the diverse views available at once "in the blink of the eye."[84] Perspective was paradigmatic for the sciences generally, all of which need techniques to combine diverse appearances and to grasp their underlying reasons and causes.

We are here far from the "optimistic" Leibniz satirized by Voltaire's *Candide*. In the 1670s, Leibniz earnestly, if often naïvely, sought means to improve society, knowledge, and individuals through the creation of affects and the subsequent spurring of rational inquiry to grasp the order behind appearances. Leibniz's quadrature rested on a fundamental perspectival intuition; he understood it as contributing to the discovery of harmony among mathematical and natural things; his rich epistemic and affective understanding of perspective helped Leibniz develop and articulate his views and techniques of the methodic discovery and the potent display of truths. With its use of powerful notations, his mathematics suggested a formal way of discovering the harmony among apparently diverse and realized objects, that is, of finding the reasons—the *rationes*—behind them.

OPTICAL TECHNIQUES, WRITTEN REPRESENTATIONS, AND SEEING ALL AT ONCE

In 1671, the year before he went to Paris, Leibniz struggled in his metaphysical and epistemological writings to work out his account of the relationship between appearances and their underlying reality.[85] In articulating his views, he drew often upon metaphors and analogies with mirrors and lenses. Alongside this philosophical work, he envisioned practical tools to help produce superior knowledge of appearances and the harmony underlying them. Mirrors and lenses figured as key components. In the years immediately before his trip to Paris, Leibniz sought to invent optical devices and notational techniques to allow human beings to see many things distinctly and all at once. To understand his mathematical achievements and the philosophical consequences he drew from them in the later 1670s, we must step back into a period when Leibniz had little idea what his mathematics might achieve.

In 1671 Leibniz was working on a series of devices intended to allow one to perceive the world distinctly, not confusedly, and to perceive many things all at once. In a letter to his future patron and employer, Duke Johann Friedrich of Brunswick-Lüneburg, he briefly outlined several new inventions:

> In Optics, I have discovered: first, a certain Type of Tubes or Lenses that I call *Pandoche* [i.e., all-receiving] because they grasp the entire object uniformly, and collect distinctly lines off the axis as well as those on it. . . . Second, cata-dioptric tubes: which combine a mirror and perspective with one another. . . . Third, a means, which has hitherto been vainly sought, to measure with perspective from a single place.[86]

Leibniz felt strongly enough about his inventions to have a short description published in 1671: "Notice of an Optical Advance."[87] He sent off this description to several important figures, including the optician and philosopher Baruch Spinoza and the influential Frenchman Pierre Carcavy; he also mentioned his interests to the great Jansenist philosopher Arnauld in his important long letter of 1671. Spinoza, Carcavy, and another correspondent gently chided him for failing to explain himself sufficiently.[88] It is as difficult for us as for them to reconstruct what Leibniz's devices were supposed to be, even though we have some of Leibniz's working papers, published nearly a hundred years ago.[89]

More important here than the details, real utility, or practical application of these schemes are Leibniz's goals, the terms in which he expressed them, and his sources of inspiration. In an evidently contemporaneous document Leibniz explained succinctly: "The end of the optical tube is to perfect vision. Excellence of vision is seeing (1) the thing larger, (2) clearly, and (3) multiple. Excellence of the telescope is thus that it magnifies a thing, illuminates it, uncovers much of it all at once" (Ger 91). Along with citing Johannes Ott's treatise on using the laws of mechanics to understand ocular problems, Leibniz provided several examples of tools for perfecting human vision: the binoculars of the Capuchin Antonius Maria Schyrleus de Rheita, which were discussed in his wonderfully entitled *The Eye of Enoch and Elias* (1645) and were useful "so that one uncovers more all at once"; and finally the suggestions of one Lana (Ger 91).[90]

Today Francesco de Lana Terzi, of the Society of Jesus, is best known for his proposals for flying balloon ships, but from around 1671 to 1674 his work interested Leibniz greatly. Of a kind with other Jesuit polymaths such as Kircher and Schott, Lana Terzi worked with Kircher in the Collegio Romano before publishing his remarkable compilation of natural magic in 1670, the *Forerunner (Prodromo)*.[91] Lana Terzi described natural magic as the "cognition of natural and hidden causes" and of the works "of extraordinary and marvelous effects."[92] In addition to considering aeronautics, means for perfecting painting, perpetual motion, the philosopher's stone, and some surprising qualities of numbers, Lana Terzi included an important treatise on designing optical instruments. While certainly skeptical of some of Lana Terzi's brash claims about subjects such as alchemic transmutation, Leibniz read much of the *Forerunner* with care, especially the optical sections, on which he took notes, and praised it widely.[93]

In considering how to improve telescopes, Lana Terzi noted that "the perfection of the telescope is to magnify the object, and to make it appear clearly and to present it distinctly, or more precisely, without confusion or the blinding flash of light, and to see in one vision many objects."[94] In his optical projects and writings, published and unpublished, Leibniz seems to have drawn heavily from Lana's techniques, his goals, and his vocabulary.[95] In a note about his "all-seeing" lenses, Leibniz explained that the "*pandoche* lenses collect all the rays in a single line, into whichever plane represents the object distinctly." By moving a plane within the mechanism, one could work to make the various points of the object appear distinctly (Ger 97). He likewise praised concave mirrors, popular, if problematic, magnification devices in the seventeenth century: "mirrors no less than lenses can magnify to infinity." He hoped they could be used to make images "free from all confusion" (Ger 97–98).[96] The goal of increasing distinctiveness and decreasing confusion framed his search for techniques proper to human knowledge and human perfection.[97] In his various schemes, Leibniz hoped to use mirrors and lenses to magnify and resolve objects more distinctly.

Magnification is not the only desideratum for microscopes or telescopes; as important is keeping distinct points of a magnified object from becoming a confused smudge. The more an instrument can keep points of the magnified object distinct, the greater its capacity to resolve. A large number of phenomena sharply limit this resolution.[98] Like almost all of his contemporaries, Leibniz's primary concern was spherical aberration. Rays at a distance from the central axis refracted through a spherical lens focus along a line, not at a single spot. Spherical lenses make distinct objects become bigger but confused. While Leibniz and his contemporaries knew hyperbolic or elliptical lenses and mirrors could overcome these difficulties in theory, such lenses and mirrors remained technically unfeasible. Making the aperture smaller limited the aberration, but only at the cost of darkening the image.[99]

In 1671 Leibniz thought that the perfection of optical tools was of the utmost importance for human perfection. Leibniz was writing in the brief glory years of seventeenth-century microscopy, before a long, fallow period of little progress in overcoming the technical difficulties of greater magnification, illumination, and resolution largely caused by the inability to produce better mirrors and lenses. By the eighteenth century, microscopes came to be seen more as toys than serious natural-philosophical equipment.[100] Around 1671, Leibniz clearly argued that perfecting techniques for focusing to one ray would mean a great advance in human ability to know and then to use nature. He had great hope that the engineering and manufacturing obstacles would soon be overcome: "All this can be put into practice. And progress will be given in magnifying objects to infinity. . . . [better lenses] can be used in microscopes for the illumination of nature and the perfection of medical things" (Ger 92). Even if his optical speculations yielded no practical devices, they were good to think with. As he noted a few years later, "ingenious and beautiful, albeit useless, inventions ought not to be less valued, for they are . . . examples of the art of discovery and provide new viewpoints for advancing this art."[101]

Notations for Seeing All at Once

Around the same time that Leibniz was pursuing optical techniques to allow human beings to see many things distinctly, he was exploring representational and notational techniques to permit one to represent many things all at once in one visual field. In a remarkable draft "Exhibiting All Books in One Figure," written around 1670–71, he listed a series of written signs and figural techniques to make up for human cognitive limits. He advocated drawing upon algebra, shorthand, mapmaking, and perspectival drawing.[102] Such compendia and figures would be a "great help in uncovering new things and the unconsidered, as all things can be seen at once in an overview, compared among themselves, and gaps revealed." These techniques would permit, for example, "every optical principle" to be "explicated in a large figure." Likewise, "All of geometry can be contained in a large figure, one with the demonstration of all theorems in shorthand in the margin." Such facility in teaching did not mean imparting superficial knowledge, for these improved expressive techniques could allow one to consider entire demonstrations all at once: "Everyone will examine in a single moment the foundation of demonstrations, and the reasons of the reasons all the way to the first principles [*primas*]."[103] With a mathematical or philosophical proof surveyable in this way, one could literally see all the grounds for the conclusion at once; the grounds for the truth at issue were to become *visible*. The centrality of these goals remains evident in his early mathematical practices in the Paris period. In a remarkable consideration written while working on harmonic progressions, Leibniz extolled the power of progressions to help find a method capable of "representing all of Euclid's *Elements* with its demonstrations in a table, by the power of a universal writing." He adds: "Similar tables can comprehend all the theorems of Archimedes, Apollonius, Pappus, etc."[104]

Among the various technical means useful in these endeavors, Leibniz included numerous varieties of alphanumerical signs and the apt use of color. These new forms of abbreviating ought to draw on the best of representational practices, whether from painters, mathematicians, accountants, Ramist table crafters, or mapmakers.[105] Leibniz envisioned that the expressive power of these varied representational techniques would augment human ability in every important domain of inquiry; they would aid in teaching current knowledge, in recognizing the interconnections among extant knowledge, and in making new discoveries. In the 1670s, Leibniz constantly used the term "compendium" to describe abridgements, shortcuts, abbreviations, and compilations of all sorts, including diagrams, figures, symbolic expressions, techniques of calculating, and even theorems.

Another contemporary document focused more closely on the power of such compendious means of representation. The use of representational techniques such as signs in reasoning, he argued, was already ubiquitous, though his contemporaries often failed to admit their dependence on them. Leibniz stressed the necessity of using such signs for human beings to reason at all in many subjects, such as mathematics: "no one can reckon, especially with large numbers, without names or numerical signs;...For who, without the life of a Methuselah, will imagine distinctly the unities that are in 1,000,000,000,000." Moving from this mathematical example, he offered a far more general principle: "no one can survey prolonged reasonings for long with the mind, unless they are known through certain signs, that is, names." Far from being a negative innovation, the use of such signs is ubiquitous, if often unacknowledged. "And I am accustomed to calling these sorts of *thoughts blind*—nothing is more frequent or more necessary for human beings." The symbols of recent mathematical analysis, for example, whatever naysayers such as Hobbes may have thought, are thus "of such great use for quickly and securely reasoning."[106] Once one recognized this human dependence on signs, it was all the more imperative to improve them.

Such signs were especially suited to representing many things in one visual field. Through the power of names and signs "a great number of things may be comprehended compendiously, so that many things may be surveyed quickly." Mathematical theorems, in fact, ought to be understood as a form of compendium: "There is no other use of a theorem than to say many things compendiously."[107] A theorem, in other words, was above all a means for us to be able to cognize and to reason about the complex properties of the objects of mathematics and other domains.

Long before the call for distinctness and grasping many things all at once came to play such a central role in Leibniz's mathematics and his theory of definition, these goals were central for framing his efforts to develop practical techniques—optical, written, and otherwise—to improve human cognition, to help grasp the underlying harmonies among books, statements, curves, and planetary bodies. Far from a mere metaphor for clarifying philosophical questions, his constant invocation of seeing complex things all at once and distinctly was central to his technological and natural-philosophical practice. When he produced an innovative quadrature of the circle that yielded a written expression, not a geometric construction, Leibniz had a powerful array of beliefs and goals to draw upon in defending the legitimacy of expression as genuine mathematical knowledge.

If we jump from his writings on compendious signs just before the Paris period to those written soon after the achievement of the quadrature, we can see the subtle shifts of emphasis and language his new mathematical competence and achievements helped to produce. He reiterated his claim that every "theorem is a shorthand, that is, a compendium of thinking." Prior to his mathematical discoveries of 1673–74, his primary examples of compendia (or shortcuts) included figures, tables, and charts abbreviating arguments, as well as arithmetic and algebra. From 1674, his foremost example was an infinite mathematical progression subject to a finite rule, such as found in his quadrature. The "glory [*laus*] of all the abstract sciences consists in speaking and writing in compendious signs."[108] The infinite series of his quadrature fulfilled his hopes for a potent form of representation capable of improving human knowledge. It is hardly surprising that he was able to accept a quadrature that produced only an expression and not a geometrical construction. Pursuing techniques proper to human knowledge and cognitive abilities helped Leibniz to recognize, to deem legitimate, and to justify his real mathematical achievements. Those achievements in turn helped him to further articulate and develop his criteria for the tools necessary for improving human knowledge and human life. They helped transform his early, fairly vague hopes for powerful compendia into the series of concrete techniques for seeing many things at once, to be examined in the next chapter. Rather than merely viewing Leibniz's praise of mathematical and logical symbols as an obvious truth that he naturally grasped, we need to understand the framework in which he came to praise them far beyond any of his contemporaries. He did so under the aegis of optical tools and written compendia appropriate to allow human beings to see much all at once.

Tools for Mathematics

That Leibniz sought notational means to improve mathematics is well known. Less well known are his sundry efforts at advocating and inventing other technologies for advancing mathematics. Before his discoveries of 1672–76 revealed the power of infinite series, he sought out new instruments and machines to use in mathematics, both for solving problems exactly and for producing sufficiently accurate approximations. Leibniz's techniques included more than his well-known efforts at creating mechanical calculating machines.[109] He thought that optics offered a potentially fruitful source of machines and techniques for use in mathematics, both for constructing curves and in measuring objects: "our reason is so instituted that we do not estrange practical geometry from optics but rather work so that the parallels among them are revealed." Optics offered hope for solving central outstanding problems in geometry, such as the discovery of mean proportionals.[110] Again Leibniz referred to Lana Terzi—hardly a great mathematician: "A method can perhaps be found by us to find practically, not just two, but many more proportional means. See what Lana says in his *Forerunner* where he puts the microscope to certain geometrical uses" (Ger 107). The Jesuit Lana Terzi had indeed ended his *Forerunner* by speculating on the potential uses of microscopes and other optical devices in mathematics, including the discovery of mean proportions and other techniques useful for quadrature.[111] By invoking the mathematical potential of optical devices, Leibniz placed himself firmly within this early-modern pursuit of mechanical and technological solutions to problems in mathematics. Galileo's compass and the curve-producing machines in Descartes' geometry perhaps best exemplify the centrality of machines for thinking about mathematics and solving problems.[112] Just as Descartes had expanded the class of acceptable curves to include those created by machines with one source of motion, many others sought to find machines to solve problems involving other curves. Optics could perhaps produce curves, Leibniz argued, much as Descartes' machines had.[113] Leibniz noted in particular that optical means might help with the higher-order curves preoccupying his contemporaries. He asked "whether the Logarithmic line [i.e., curve] can be described by optical means—on this line [curve], see Renaldini, James Gregory, and Father Pardies. Perhaps, even, something can best be achieved by optics" (Ger 107–8).[114] A few years later, he cited the same authorities, but now in the context of his written expressions capable of solving problems about such higher-order curves.[115] The technologies needed to contend with such curves turned out to be largely algebraic and symbolic, not optical or mechanical—but Leibniz could not be sure of that in 1672 and early 1673.

In discussing optical techniques for improving mathematics, Leibniz likewise referred to the tools for copying figures and representing nature that Niceron and Harsdörffer and Jesuits such as Christoph Scheiner, Mario Bettini, Christoph Grienberger, and Lana Terzi considered at such length.[116] Kircher's *pantometrum*—or universal measurer—was to be used for measuring objects, such as cities or mountains, to make accurate plans and depictions of them, and for transforming geometrical objects into other geometrical objects.[117] Using the examples of Scheiner's pantograph and Kircher's *pantometrum*, Leibniz specifically commented on how useful such machines might be for providing "the perfect proportion of the circle" (Ger 107–8). Leibniz hoped that some form of the pantograph, now best known as a children's toy, could somehow become a practical tool for measuring the circle (see fig. 5.8).[118]

Although the telegraphic and fragmentary quality of Leibniz's remarks makes discerning the mechanisms of the tools he envisioned difficult, his comments nevertheless illustrate his goals in looking for such tools and his hope in their power. Leibniz shared the interests of Jesuits such as Scheiner, Bettini, and Kircher for solving problems in mathematics through such tools. Throughout the 1670s Leibniz envisioned numerous instruments and machines to advance mathematics.[119] Taking a cue from Lana Terzi and others, Leibniz hoped to apply instruments, perhaps including optical instruments, to the hardest nuts in mathematics: "And it could be applied also to the analysis of suitable numbers using figures, as in the proportion by the circle and the arithmetical line and others." He hoped to find "compendia" for dealing with equations involving roots of numbers: to do so, he said, "truly is art" (Ger 108).[120]

While Leibniz sought to develop various instruments to aid in solving mathematical problems, his work from 1672 onward focused above all on algebraic and arithmetical signs, in which he was becoming ever more proficient. At the end of a study of using instruments to aid the solution of problems in algebra and geometry, written in spring or summer 1673, he

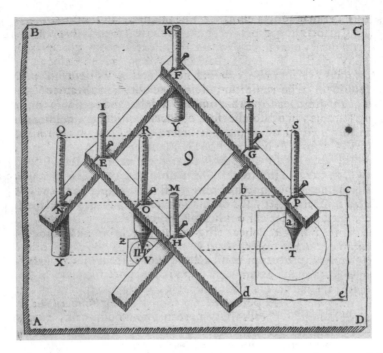

Figure 5.8. Pantograph. Still sold as a toy for children, the pantograph is a straightforward device for tracing an image and easily magnifying or reducing it. From Christoph Scheiner, *Pantographice seu ars delineandi* ... (Rome, 1631), p. 29. Courtesy of Rare Book and Manuscript Library, Columbia University.

remarked that perfecting the "science of progressions" was to be "worked on," for that study, of series and sequences, promised a key path forward in finding compendia for reasoning.[121] In a contemporaneous document, written in the months before he discovered his quadrature of the circle, he spoke ever more optimistically about the power of numerical progressions: if a method of using progressions to contend with fractions and roots in general were to be developed, "every figure could be squared, . . . and every median proportion could be found, and geometry could be perfected." At the end of the document he advocated perfecting geometry by combining the science of progressions with "motion through every case"—motion to transform one curve into another, likely using instruments, perhaps pantographs or compasses, or geometrical constructions associated with those tools, or optical and projective transformations.[122]

His quadrature—of late that summer or fall—depended on just such a combination of advances in the study of progressions and an innovative transmutation of the circle (see figs. 5.2 and 5.3 above).[123] Rather than using a mechanical or optical instrument, he used the transmutation theorem to transform a curve into a more tractable one through a geometric construction. A few years later, he remarked that he had "made an enumeration of a number of metamorphoses" before coming upon the idea of dividing the circle into an infinity of triangles intersecting at a point on the circle.[124] In discovering and then accepting his form of the quadrature of the circle, Leibniz brought together numerous resources: first, his projective heuristic for transmuting curves, drawn in part from discussions around perspective; second, his efforts to improve the science of progressions, drawn in part from the mathematics of Wallis, Mercator, and others; and third, his sense that symbolic expressions could be legitimate, even necessary, tools for much human knowledge, drawn in part from his drive to make compendia useful for every domain of human endeavor. Both parts of his quadrature—the transformation from a point of view on the circle and the progression seeable all at once—were made possible by his work on practical techniques for improving human epistemic abilities. His genius was to join these resources in a rigorous mathematical manner and to contend that the progressions produced constituted a new form of exact, and not just approximate, reasoning appropriate for human beings.

Leibniz never lost his interest in creating practical techniques, be they optical, mechanical, or symbolic, to aid embodied human beings in overcoming confusion and in seeing things more distinctly and all at once. The science of progressions proved nearly as powerful as he hoped. As the vast number of studies contained in the recently published volume of his work with series illustrates, Leibniz worked hard to perfect the study of progressions, often with brilliant results.[125] His success informed his philosophical theories and practical efforts alike.

Like the Jesuit natural magicians and their German followers, Leibniz thought that the illustration and discernment of causes through marvel-producing machinery could encourage moral and religious transformation. With his theater of shadows, his educational casino, and his dialogues, Leibniz sought practical means for leading people to reflect upon the structures of nature and their significance. Such reflection could transform one's ethical life by leading to a better appreciation of the natural and human world. All of this was no less true of his more technical endeavors in optics and in forms of writing. His projected symbolic compendia, he claimed, would be capable of solving many problems: "for multiple things expressed compendiously and simultaneously can be compared easily in thinking, and run through, and organized, all to one end, namely, the solving of problems, and of the greatest of problems: *obtaining happiness*."[126] Perfecting tools for improving human knowledge could play a signal role in self-cultivation.

EXPRESSIONS, SERIES, AND THE CALCULUS

Leibniz was right: compendious notations and techniques for working with series and sequences quickly transformed the study of curves and numbers. He developed the core of his version of the differential and integral calculus in late 1675–76, including the notations $\int$ and d, and began publishing it in the mid-1680s.[127] Leibniz created his calculus in large part by extrapolating his procedures for *finite* differences between a series of numbers into procedures for the very small or infinitesimal differences between a series of values along continuous quantities.[128] In a breakthrough paper of October 1675, he wrote "$\int$ signifies a sum; d a difference. From a given y an $a \cdot dy$ can be found ... that is, the differences of those y's. Henceforth, an equation can be changed [*mutaris*] into another."[129] He characterized differentiation as finding the differences between elements within a series and summation as finding sums of such differences between elements.[130]

Using sums of differences to solve problems figured centrally in Leibniz's practice with numerical series and sequences. Much as Pascal did in his *Treatise on the Arithmetical Triangle*, Leibniz rewrote sums of numbers as sums of their differences to help solve problems of infinite and finite summations. Suppose one wanted to find the sum $b_1 + b_2 + b_3 + \cdots + b_n$. If there is some a_i such that $b_i = a_i - a_{i+1}$, then $b_1 + b_2 + b_3 + \cdots + b_n = (a_1 - a_2) + (a_2 - a_3) + (a_3 - a_4) + \cdots + (a_n - a_{n+1}) = a_1 - a_{n+1}$. For example, consider an infinite sequence of fractions with the "triangular" numbers in the denominators:

$$\frac{1}{1}, \frac{1}{3}, \frac{1}{6}, \frac{1}{10}, \frac{1}{15}, \ldots$$

These inverted triangular numbers can be expressed as

$$b_i = \frac{2}{i^2 + i}$$

which equals the difference

$$\frac{2}{i} - \frac{2}{i + 1}$$

Setting

$$a_i = \frac{2}{i}$$

the sum $b_1 + b_2 + b_3 + \cdots + b_n$ equals

$$a_1 - a_{n+1} = \frac{2}{1} - \frac{2}{n + 1}$$

If n goes to infinity, then the sum $b_1 + b_2 + b_3 + \cdots$ to infinite terms equals 2.[131] By changing the expression of the series, in other words, the sum is found easily.

Pascal investigated the triangular and other figurate numbers; Leibniz investigated their inverses. Pascal sought general rules underlying the properties of the various figurate numbers, their sums, and their differences; Leibniz sought the same for the inverses.[132] To complement Pascal's arithmetical triangle, Leibniz constructed a "harmonic" table—one drawing upon the harmonic sequence $1, \frac{1}{2}, \frac{1}{3}, \frac{1}{4}, \frac{1}{5}, \ldots$:

$$
\begin{array}{cccccc}
1 & \dfrac{1}{2} & \dfrac{1}{3} & \dfrac{1}{4} & \dfrac{1}{5} & \dfrac{1}{6} \\[2ex]
\dfrac{1}{2} & \dfrac{1}{6} & \dfrac{1}{12} & \dfrac{1}{20} & \dfrac{1}{30} \\[2ex]
\dfrac{1}{3} & \dfrac{1}{12} & \dfrac{1}{30} & \dfrac{1}{60} \\[2ex]
\dfrac{1}{4} & \dfrac{1}{20} & \dfrac{1}{60} \\[2ex]
\dfrac{1}{5} & \dfrac{1}{30}
\end{array}
$$

In Leibniz's harmonic triangle, each element comes from finding the difference between two elements of the previous row. Each row is a difference sequence of the one preceding it and is a sum sequence of the one following it. From this table, one can easily discover numerous properties of sequences of these numbers, such as

$$
\frac{1}{2} = \frac{1}{3} + \frac{1}{12} + \frac{1}{30} + \frac{1}{60} + \cdots
$$

Leibniz drew methodological morals from his successes. Putting series into tables such as the arithmetical and harmonic triangles allowed one to recognize them as series of series and thus aided in discovering more general unifying principles—the unity underlying their diversity.[133] Most series could be understood far better, he claimed, if only we worked out the series whose differences they were. In 1674 Leibniz called for techniques to speed the discovery of such differences for geometric curves and sequences of numbers alike. He long hoped to use results like the transmutation theorem to aid discovery about sequences of numbers, and new findings concerning numbers to aid discovery in geometry.[134]

In developing his calculus, Leibniz translated these techniques and ways of thinking about differences and sums of numbers to problems involving continuous quantities, such as finding tangents to curves, determining their

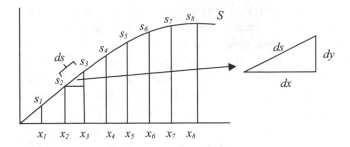

Figure 5.9. Dividing a curve. The y_i's are excluded for clarity.

length, and calculating the areas under them. Leibniz conceived of curves as comprising collections of very small differences.[135] Consider a curve S as a polygon with infinitely many sides (see fig. 5.9).[136] These sides have endpoints $s_1, s_2, s_3, \ldots$, with coordinates $x_1, x_2, x_3, \ldots$ on the x-axis and $y_1, y_2, y_3, \ldots$ on the y-axis. Viewing a curve as an infinitely sided polygon permits understanding it as a sequence of differences: each side of the polygon has a length $ds = s_{i+1} - s_i$; each ds has a corresponding $dx = x_{i+1} - x_i$ and $dy = y_{i+1} - y_i$. There are many ways to divide up the curve by reference to various differences, which need not be constant; Leibniz soon put these different ways of dividing a curve to use.[137]

The fundamental insight of the calculus comes in recognizing that finding tangents and finding quadratures, that is, differentiation and summation, are inverse operations. In explaining the essence of his calculus, Leibniz underscored the continuity between his work on series and sequences and his efforts with continuous quantities:[138] "Foundation of the calculus: Differences and sums are reciprocal to one another, that is, the sum of the differences of a series is the term of the series, and the difference of the sums of a series is also the term of the series. The first I denote $\int dx = x$; the second $d \int x = x$." To illustrate the relationships between these procedures, he constructed a table of sequences connected by addition and subtraction:

Let the differences of a series, the series itself, and the sum of the series be, let us say,

Diffs.		1	2	3	4	5		...	dx
Series	0	1	3	6	10	15		...	x
Sums	0	1	4	10	20	35	...		$\int x$

Then the terms of the series are the sums of the differences, or $x = \int dx$; thus, $3 = 1 + 2$, and $6 = 1 + 2 + 3$; on the other hand, the differences of the sums of the series are terms of the series, or $d \int x = x$; thus, $3 = 4 - 1$, $6 = 10 - 4$.[139]

Leibniz's calculus did not involve derivation and integration, operations on functions that produce functions. As H. J. M. Bos underscored some years ago, for Leibniz, differentiation and summation are operations on a sequence of variables that produce another sequence of variables.[140] With his calculus, Leibniz created a set of operations upon series of differences that helped one move easily and quickly among different expressions of the same series, in order to solve problems involving tangents and quadratures.

Given this understanding, "Finding the tangents to curves is reduced to the following problem: to find the differences of series."[141] Why? The segments *ds* of the infinitely sided polygon are tangent to the curve; if the segment is extended, a line tangent to the curve is produced, with the slope

$$\frac{dy}{dx}$$

Finding the tangent, then, involves simply being able to compute the differences *dy* and *dx*, for which Leibniz set out straightforward rules, first published in 1684; over time he extended these procedures to ever more mathematical objects.[142] His technique for finding tangents helped avoid the laborious calculations and reductions plaguing the standard methods of his day.

As for quadrature, Leibniz explained: "finding the areas of figures is reduced to this: given a series, to find sums, or (to explain this better) given a series, to find another one whose differences coincide with the terms of the given series."[143] For Leibniz, $\int y dx$ meant $y_n(x_{n+1} - x_n) + y_{n-1}(x_n - x_{n-1}) + \cdots + y_0(x_1 - x_0)$ for infinitely small divisions of x. Just as finding the sum of a series of numbers could be achieved by expressing it as another sum, one comprising differences, finding an area involved expressing it as a series of differences whose sum is already known or could be more easily computed. Techniques such as the characteristic triangle helped find such equivalent sums (see fig. 5.3); so did the algorithmic techniques that Leibniz and his followers eventually developed. In the meantime, he freely moved back and forth between geometry and algorithms to discover useful series of differences.

In his early work on summation, Leibniz freely mixed new analytical techniques with his geometrical techniques for transmuting areas and volumes into rational figures and, often, into infinite series. He long underscored the power of the transmutation theorem of 1673, his fairly general means of using properties of the tangent to transmute one curve into a more tractable curve whose area could more easily be computed and expressed. Much to the glee of his English rivals in the bitter priority dispute that followed, he made brash claims about the generality of this procedure, as in his study entitled "Expression of a Quantity through a Series."[144]

In a paper entitled "General Rule for Expressing the Areas of Figures through Infinite Series" (mid-1676), he set out the techniques and goals of his early summations well: "*To discover a general method by means of which* it will always be possible to find a *rational figure homologous to a given one,* and indeed an expression of the proposed curve through an infinite series." His first example was finding the area of part of the circle, which he expressed as $\int \overline{y\,dx}$ (Leibniz often used bars to group terms where we would use parentheses). Computing this summation involved finding a way to express ydx as a "rational figure," which he did using the equation for the auxiliary curve produced with his old transmutation theorem.[145] He worked for many years to generalize such geometrical transmutations into more general algorithms and eventually came to understand the transmutation theorem as equivalent to $\int y\,dx = xy - \int x\,dy$.[146] For Leibniz, this equation—a form of integration by parts—showed the equivalence between two ways of expressing series as sums of differences.

The new calculus was the greatest of Leibniz's many efforts to find and manipulate expressions useful in mathematics. In fulfilling many older goals, the notations and methods of his calculus confirmed, amplified, and deepened his sense that expressions were essential for extending human reasoning. Written expressions and techniques for manipulating them came to figure ever more centrally in Leibniz's account of the knowledge available to fallen human beings.

FROM MATHEMATICAL PRACTICE TO PHILOSOPHY

Not long after discovering his quadrature, while still in Paris, Leibniz began sketching a series of private, wide-ranging reflections on metaphysics, theology, and epistemology, today collected under the title *De summa rerum*.[147] Perspective and expression occupy central generative roles in these incredibly dense, rich, and at times contradictory notes. In December 1675, at the moment he was elaborating and exploring the insights behind his new calculus, Leibniz took his new account of the circle, known in part through an expression of an infinite series, and used it to set forth considerably more general considerations on the limitations and possibilities of human knowledge: "We think about the circle; we demonstrate about a circle; we recognize a circle; its essence is known to us, but only part by part. If we could think of the total essence of the circle all at once, we would have the idea of the circle. But it is to God alone to have ideas of composite things."[148] Leibniz's language here suggests a term-by-term refutation of intuition as understood by Descartes, who demanded that "two things are required for intuition: first, the proposition intuited must be understood clearly and distinctly; next, it must be understood all at once, and not bit by bit."[149] Leibniz evidently thought that neither of Descartes' demands characterized human knowledge of circles, or of anything more complex.

Before 1672, before his quadrature and extensive work on series, Leibniz still sought something like clear and distinct knowledge of ideas themselves. By late 1675 he attributed the ability to have such simultaneously grasped intuitions of all but the simplest truths to God alone. This marked movement can be tracked minutely through his mathematical and philosophical papers. Just before his trip to Paris, Leibniz explained that a "definition is explication by voice" and "definition is a signified idea." A definition is capable of signifying the idea itself. Reasoning is a "chain of ideas."[150] Another paper from the same period notes that something true is "sensible clearly and distinctly."[151] Like many of his contemporaries, Leibniz moved freely between Cartesian definitions using standards of clarity and distinctness and more traditional scholastic definitions, often depending on audience. In any case, he clearly maintained that explanations could produce simultaneously cognized ideas about things and that reasoning moved from idea to idea, even if at times mediated through "blind thinking" using symbols.

An important letter of late 1672, written after he had begun his serious study of numerical progressions, shows the impact of Leibniz's increasing mathematical sophistication on his philosophical considerations. Much of the letter sounds like the passages just quoted: propositions come either from the senses—observations and experiences—or from "the clear and distinct imagination, that is, from ideas or, if you prefer, from definitions." A definition is "nothing other than the signification of an idea, as are the theorems of arithmetic and geometry."[152] In his further remarks, he demonstrated his concern with producing practical compendia to aid human cognition and his increasing familiarity with recent mathematics. He showed a greater sense of the variety of possible linguistic definitions and their distinct uses: "The same ideas can be expressed by varied definitions. . . . I recall Pascal's praise somewhere, where he commended the variation of enunciations of the same theorems, and in which he said that [such variation] ought to be considered the whole of the study of mathematicians. For thus the way to the new and the untouched is opened."[153] While making the reference sound offhand, Leibniz very nearly quoted the key passage on discovery from Pascal's *Treatise on the Arithmetical Triangle*, considered in chapter 3 above.[154] At the time he wrote this letter in the second half of 1672, Leibniz was working hard to develop Pascal's techniques for dealing with numerical progressions.[155] In the years following, he repeatedly underlined the power and necessity of changing expressions to encourage discovery of underlying harmonies among mathematical objects.

Nevertheless, in late 1672, Leibniz argued that these varied definitions, so useful in discovery, were still supposed to lead one to, or to result in, clear and distinct ideas. In the discussion about the use of symbols in this letter, Leibniz used the language from his call for producing compendia appropriate to human reasoning. Many considerations simply demand using symbols, without "consideration of the ideas themselves," in what Leibniz called "blind thinking": "in this consists the art of thinking with symbols, so that there are more inclusive compendia through these ideas, yet free of confusion." Blind thinking still ought ultimately to result in clear and distinct comprehension: "if resolved into its ultimate elements, which is to say, clearly and distinctly understood."[156] Both pragmatically and epistemically, human beings need to use symbolic reasoning in order to come to true knowledge when working on many geometrical and arithmetical problems. Just as, for Descartes, practicing geometry involves a temporary and potentially blind use of symbols, in the end one should produce a result that is known clearly and distinctly. Even if Leibniz allowed here the possibility that the final answer might not be in this form, intuitive knowledge—the direct possession of ideas themselves—remained the quarry, the goal, in late 1672.

Such direct knowledge of ideas themselves was not at issue in his quadrature of the circle, nor was it in his late-1675 considerations on human knowledge. In his summer 1673 account of the history of mathematics, Leibniz confessed that human beings simply cannot have, in many important cases, the constructive, or clear and distinct, knowledge long demanded of geometry. Mathematicians ought often to be content with symbolic knowledge and its power to aid discoveries.[157]

In his late-1675 philosophical reflections on mathematical knowledge, Leibniz insisted that, although humans cannot obtain intuitive knowledge of complex things, they can gain knowledge about them. Considering the circle, Leibniz argued that human beings can have some knowledge about the essence: "In the meanwhile, we know the essence of the circle by thinking of its requisites part by part." He continued to describe the techniques that could compensate for this lack of intuitive knowledge. Only external *material* techniques, what Leibniz often called a palpable thread of reasoning, could make partial knowledge of essences possible. These techniques could provide temporally located and finite representations of the qualities of some things, such as mathematical objects: "The defect of idea in us is filled out by a certain image of a sensible thing or a definition, that is, an aggregation of characters. . . . Always the place of the idea is filled out by a certain phantasm, all of which is sensed at once. Images excite the senses; characters excite thought; the former are more appropriate for action; the latter for reasoning."[158] A symbolic expression replaces the idea not just in the middle of a computation but potentially at the beginning and the end. To gain knowledge part by part one must *produce* an external phantasm that is composed of images and characters and is capable of being viewed, of being literally sensed all at once. Simultaneity came only in looking at and thinking using the written expression about something, not in grasping the idea of that thing all at once. By late 1675 Leibniz argued that human knowledge of many complex things would come only from viewing such written forms of representations and thinking in terms of them.[159] Just a few months later, Leibniz concluded that the circle itself is a fictive entity, one enormously useful for thinking and computing: "even though this ultimate polygon does not exist in the nature of things, one can still give an expression for it, for the sake of the abbreviation of enunciations [*compendiosarum enunciationum causa*]." Despite their potential sometimes to mislead us into believing them to be real, such fictitious entities, he emphasized, are "excellent abbreviations of enunciations, and for this reason extremely useful."[160]

In the case of the circle we can generate a drawing of a circle or the infinite series expressed in a simple rule. In fact, we can easily create many different representations of the circle and its area, some far more useful than others. Producing varied expressions of formulas and geometric objects was central to Leibniz's mathematical practice and his reflections upon it—hence his praise of Pascal's call to transform enunciations to discover new truths. In considering sequences of numbers, for example, Leibniz stressed the utility of finding various "expressions" for the same series, such as expressing an arithmetical sum using its first and last terms exclusively. If some $d, a, b, c,$ and e are an arithmetical sequence, then[161]

$$d + a + b + c + e = \frac{(4 + 3 + 2 + 1)d + (1 + 2 + 3 + 4)e}{4}$$

Different forms have different uses: sometimes one might want a finite algebraic equation; in other cases, an equivalent infinite sum.[162] In working papers of the spring of 1676, he reflected upon the methods used in his quadrature and remarked upon the power of different algebraic expressions to lead to many new theorems, to reveal "wonderful identities," and to yield "beautiful consequences."[163]

Leibniz's work in mathematics powerfully illustrated how notational practices could enhance human ability. His hopes for potent compendia capable of extending human cognitive abilities had been dramatically fulfilled within mathematics, and he worked to extend the power of expression to other domains. Written expressions are available to us; not to use them is to deny some part of our postlapsarian capacity. Human physicality limits human knowledge, but it also points to means for perfecting knowledge: "It is only the obstacle of the body that prevents comprehending everything all at once, but even this makes it apparent that it is capable of a perfection infinitely greater, to the extent that corporeal movement can become faster to infinity."[164] Just as the limitations of human vision should encourage us to perfect telescopes, the limitations of reasoning should motivate us to perfect representational and written technologies, such as infinite series.

The philosophical and practical achievement of allowing representations to become legitimate objects of knowledge independent of the cognition of clear and distinct ideas rested on a remarkable confluence of Leibniz's practices in mathematics, optics, and philosophy. This confluence continued to provide Leibniz with powerful heuristics for extending his thought and practice. The term "expression"—which he came to use systematically by late 1674 to refer to his infinite series and algebraic representations—soon figured centrally in his philosophy.

His hermetic essay "What Is an Idea?" (probably 1677) offered an important discussion of "expression," which he now explained to be an extremely general category of relationship among any number of things: "Something is said to express another, in which there are relations that correspond to qualities in the expressed thing."[165] Leibniz stressed the variety he allowed under this definition. In writing his essay, he first focused on the circle and its equation: "The expressions of things are, however, varied, for example: the nature of the circle expressed by a certain algebraic equation."[166] He reworked this explanation using a wider variety of examples taken from diverse representational practices that had interested him from the early 1670s: "But these expressions are varied: for example, a schema of a machine expresses the machine itself, a perspectival drawing [scenographica] of a thing in a drawn plane expresses a solid, a speech expresses thoughts and truths, characters express numbers, algebraic equations express a circle or another figure."[167] Perspectival representation, algebra, series, words—the stuff of his call for compendia and his practical effort—all became prime examples of expression. Expression often reduces something with infinite complexity into something finite and graspable or, rather, into a representation that allows us to grasp some features of the complex thing. The two examples most central to my discussion, algebra and perspective, illustrate that an algebraic formula and a perspectival drawing accurately express true aspects of a circle and a solid but do not capture and represent their entire essences. To understand completely the essence of a building—its plan—one needs infinite perspectival drawings, each contributing some certain knowledge.[168] Leibniz in turn used such a metaphor of perspectival representational practices to help articulate and develop his account of the diversity in a harmonious, interconnected world. The variety and interconnection exemplified in the play of shadows, or in a series of sketches of the elevations of a building, helped him to explain the variety and interconnection of nature itself.

CITYSCAPES, MIRRORS, AND EXPRESSIONS

Throughout his mature philosophical and mathematical career, Leibniz turned to the analogy of perspectival views, particularly those of cities, to clarify the difference between divine and human knowledge and to explain the interrelatedness and harmony of all beings in the universe. In *Monadology* (1714), for example, he explained: "And so the same town, looked at from various sides, appears quite different and becomes, as it were, perspectivally numerous; it happens in the same way that, because of the infinite number of simple substances, it is as if there were so many different universes, which are nothing but perspectives of a single universe, according to the special view of each Monad."[169] While he was far from his later language of monads in the 1670s, Leibniz had developed many of his central metaphysical and epistemological doctrines and had already begun using similar analogies and metaphors.[170] An early example of the analogy used a more technical vocabulary of perspectival representations to express an analogous point: "multiple finite substances are nothing other than the diverse expressions of the same universe, following their diverse points of view and their own limitations. In the same way an ichnography has infinite scenographies."[171]

Techniques for creating such ichnographies (plans of buildings and other things) and scenographies (perspectival representations of them) fascinated Leibniz, much as they excited the curiosity of contemporary Jesuits and others interested in natural magic. In another early text, from 1678–79, he used a metaphor centered on another technology: "There are as many mirrors of the universe as there are minds; for every mind perceives the entire universe, but confusedly."[172] In his various analogies and metaphors, Leibniz moved freely between describing beings as scenographies—perspectival representations—and as mirrors.[173] This apparently strange conflation follows from the prevalent understanding of mirrors as optical devices important for producing and for thinking about perspective during this period.

As shown above, Leibniz was much taken with the sundry optical devices beloved of his contemporaries and important in their mathematical and astronomical practice—such as the pantometrum of Kircher, the pantograph of Christoph Scheiner, and the "polemoscope" of the great astronomer Hevelius.[174] Among his many other machines, Harsdörffer included one useful for learning how to make perspectival views and for

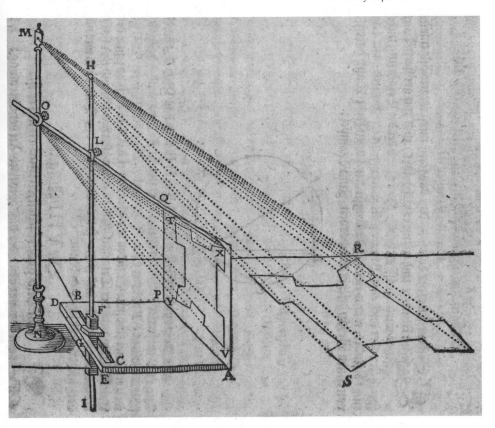

Figure 5.10. Instrument for making perspectival views. From Georg Philipp Harsdörffer and Daniel Schwenter, *Deliciae physico-mathematicae, oder Mathematische und philosophische Erquickstunden...*, vol. 2 (Nuremburg, 1651–53), p. 199. Courtesy of Rare Book and Manuscript Library, Columbia University.

"deforming" and reforming pictures (see fig. 5.10).[175] The great encyclopedist Johann Alsted defined "scenography" as nothing less than "the Optical-mechanical art of drawing well."[176] The manual for Kircher's pantometrum included detailed instructions on using it to make ichnographies and scenographies.[177]

In letters, his published "Notice" of 1671, and his private optical sketches, Leibniz praised both lenses and mirrors as magnifying and representational devices and made audacious claims about his own inventions using them.[178] Knowing that all actual lenses and mirrors create some confusion in focusing rays, he sought technical solutions to make them show images less confusedly with his *pandoche* lenses. To portray the confused sensations of every being in the universe, Leibniz drew upon his understanding of the aberrations and other problems attendant upon using mirrors and used his vocabulary of confusion to characterize them. He carefully and precisely used the metaphor of a mirror when he noted that all minds are *confused* reflections of the universe: they are representing things limited by their physical qualities.[179]

Even after his great mathematical discoveries, he still held mirrors to be necessary and powerful scientific instruments. In a 1677 consideration of the techniques for the proper analysis of bodies and the cause of things, he explained the superiority of mirrors as magnifying devices: "It is not so much the microscope that is to be used but also a perfectly polished concave mirror made from a large sphere. Lenses with smaller diameters enlarge more, whereas mirrors with larger diameters do so, and thus the diameter can be increased to infinity.... thus, the mirror appears more useful ... and from there an entire body will be accessible in a single glance in a single mirror, which does not happen in a microscope."[180] Leibniz thought that magnifying mirrors had great potential to bring together in a single optical field many things distinctly and all at once; they offered tremendous possibilities for increasing human perfection by allowing us to see much all at once and thereby to come empirically to better knowledge of the fundamental constitution of all things. Unfortunately for Leibniz, concave mirrors adequate for observation were far beyond the technology of his time.

Mirrors were also thought to be useful for scenography—for perspectival representation. Using mirrors and lenses to project images and aid in the drawing process were standard topics in treatises of natural magic. In the giant German compilation of natural magic, the *Physical-Mathematical Delights,* the German poet Harsdörffer commented: "The painter is a dead mirror: for a mirror can be called a living painter, as it represents not only the picture with its natural colors but also the movement of the same, which the brush can only begin to describe on the canvas."[181] Like the natural magicians, Leibniz devised portable devices using mirrors to facilitate representing scenes in perspective. He claimed to have developed the "best means of representing ... that is, of representing easily the entire field without a camera obscura.... This thus is to be done. A mirror is taken up and set immobile. The mirror is marked so the object becomes covered up, or in any event its extreme lines and the lines of all its parts [are marked on the mirror]. This delineation from the mirror is expressed on paper; the mirror is rubbed clean easily. The utility of a mirror is great" (Ger 105). Similar methods and goals figured prominently in the discussions of optics and mirrors in the texts of Harsdörffer, Kircher, Niceron, and others.[182] With his habitual humility, Leibniz deemed his technique superior and appropriate for all figures: "There is no better means of painting; furthermore, people will thus best be painted, as well as other moving things" (Ger 105). As with so many of Leibniz's projected devices, we can say little about whether this technique for representing ever came to fruition.

Leibniz developed his analogies and metaphors of scenographic representations of towns and of mirrors after he had envisioned practical, portable, and accurate means of representing viewpoints using mirrors, and after he had praised the particular potency of mirrors to improve human observational abilities. The casual interchange in his philosophical analogies between mirrors and perspectival representation is a telling historical artifact of his interest in concrete practices for making representations. He projected this interchange of analogies into his account of human knowledge and of variety in the world.

Just after describing mirrors as living painters, Harsdörffer praised God: "God himself was the first and best painter, who made and shaped human beings in his likeness."[183] Like Harsdörffer, Leibniz invoked God as a painter with profound expressive abilities. He used highly controlled analogies of mechanical and optical representational techniques in articulating his metaphysical and epistemological claims about God's creation of the world and human knowledge of it. His metaphors and analogies drew on his conviction about the necessity for representational technologies such as lenses, mirrors, and symbolic expressions as well as his awareness of their technical limitations, which could be minimized but never eliminated. By the mid-1670s, Leibniz had three major new heuristics to work through and to rearticulate his traditional, heavily Platonist metaphysical concerns about variety in the world, the harmony among that variety, and the production of that variety: first, optical technologies for representing viewpoints; second, the expression of mathematical areas, lengths, and other quantities though equations and infinite series; and third, the harmony among mathematical objects exemplified by the perspectival (projective) relationship among the conic sections. By following the changing significance of the town analogy before and after the early 1670s, we can see Leibniz drawing on his achievements in mathematics and his optical techniques as he developed and articulated his epistemological and metaphysical views.[184]

The town analogy mentioned above first appears in Leibniz's writings of the late 1660s, before his trip to Paris, his quadrature, and his optical efforts. The subtle changes in the analogy track well his changing account of the means of gaining knowledge and the place of perspective in his thinking.[185] In his earlier writing, he used the town analogy to clarify the difference between knowing the essence of something and merely observing its outer appearances. In a 1669 letter to Jacob Thomasius, he wrote: "Just so, the same city presents one aspect if you look down upon it from a tower placed in its midst; this is as if you intuit the essence itself. The city appears otherwise if you approach it from without, which is as if you perceive the qualities of a body. And just as the external aspect of a city varies as you approach it differently, . . . the qualities of the body vary with the variety of the sense organs."[186] The superior perspective from the tower stands for privileged epistemic access to essence. The horizontal perspectives stand for lesser forms of knowledge of external appearances. In this early use, the contrasting types of perspectival views distinguish certain knowledge of essences (*scientia*) from mere knowledge of accidents. In this perspectival analogy, all points of view are not equally valid.

Another early use of the analogy suggests that the external viewpoints, that is, external appearances, offer merely enjoyable views, suitable for delectation but not for genuine knowledge of essences: "As a plan of a city, looked down upon from the top of a great tower placed upright in its midst, differs from the almost infinite horizontal perspectives with which it delights the eyes of travelers who approach it from one direction or another. This analogy has always seemed to me excellently fitted for understanding the distinction between natures and accidents."[187] In the years after his development of the quadrature and his subsequent philosophical reflections about human knowledge, Leibniz ceased to distinguish in this way between grasping an idea or concept directly and merely grasping external appearances.[188] He transformed the town analogy to illustrate that much knowledge comes from producing located viewpoints. The qualitative difference between types of viewpoints disappears as the top of the tower becomes one viewpoint among many.

In his account of the mid-1670s, Leibniz maintained that God alone possessed all possible views simultaneously, which served as the key metaphor for having total knowledge of something. Leibniz related the production of physical representations of images or characters to the creative action of God. "So God, by the creation of many minds, willed to bring about with respect to the universe what is willed with respect to a large town by a painter, who wants to display delineations of its various aspects or projections. The painter does on canvas what God does in the mind."[189] Leibniz explained how God created variety in the world with this image of concrete representational practice. Where his earlier uses of the analogy feature a sightseeing traveler, the examples in the mid-1670s involve an active painter. God created each being based on one point of view of the essence of the universe he chose to create. Human beings can learn an important lesson from the way God creates. Human beings, beset by the loss of global knowledge, constrained to a point of view, should mimic God's production of views. The limited epistemic access of human beings ironically demanded mimicking God's *physical* production through writing, painting, or other forms of representation. Rather than the famous "I believe, that I might know," Leibniz suggested something like "I represent and express using appropriate tools, that I might know (some properties of the totality and some of their interconnections)."

Alongside the town analogies produced during his Paris stay, Leibniz compared God's active *production* of multiple perspectival views to the decomposition of numbers into various kinds of progressions. Variety in the world, he explained, is "the same essence related in various ways, as if you were to look at the same town from various places; or, if you relate the essence of the number 6 to the number 3, it will be 3×2 or $3 + 3$, but if you relate it to the number 4, it will be $6/4 = 3/2$ or $6 = 4 \times 3/2$."[190] The act of a subject considering objects produces expressions of their essences, which God alone cognizes directly and all at once. Numerical decompositions are at once the results of God's action in creating all aspects of essences (that is, 6 in all its possible decompositions) and the only means for our partial cognition of them. Leibniz here quickly and characteristically moved from considering arithmetical expressions as an epistemological tool to using arithmetical expressions produced from God's different viewpoints as a metaphysical description of the nature of essences and the production of variety. "It seems to me that the origin of things from God is of the same kind as the origin of properties from an essence; just as $6 = 1 \mid 1 \mid 1 + 1 + 1 + 1$, therefore $6 = 3 + 3, = 3 \times 2, = 4 + 2$, etc."[191] Relating an equation to some other term in order to produce another expression of the same equation was a central aspect of Leibniz's mathematical practice, an aspect he constantly underscored; solving an integral, for example, meant substituting one sum of differences with another sum of differences that was more tractable.[192] One powerful way to invent "elegant theorems and demonstrations" consisted in changing "one formula into various equivalent formulas" whose nature is better known.[193] Leibniz seized on one of his central mathematical practices—producing multiple symbolic expressions for the same equations or relations—to exemplify how God could create different instantiations of the same essence and to suggest that our knowledge of that essence will come through exploring diverse expressions of it. Leibniz translated a central mathematical practice into his account of creation.

In articulating his account of how God produces variety in the world, Leibniz crafted a brilliant analogy combining his optical and symbolic representational practices. He noted that this numerical analogy was the only one he had to explain this production: "How things result from forms I cannot explain except from the similitude of Numbers [resulting] from unities, but with this difference, that unities are all homogeneous but forms are different."[194] The thought seems to be this: one way for Leibniz to explain the production of the various forms in the world is by an analogy with numbers. All counting numbers are made of a sum of 1s; every 1 is like every other 1. In contrast, the "difference" is that real things and the "forms" produced from them are not all homogeneous as are numbers; and they cannot be reduced to homogeneous parts as numbers can be. The point is decidedly self-reflective. Leibniz in part needed to use numerical analogies to express his doctrine of expression—a doctrine that is rather difficult to discern. The very terms used in the explanation exemplify the human necessity for humble and apt expression in fathoming nature and coming to appreciate the ways of God.[195]

From the 1670s forward, expression remained central to Leibniz's mathematical, philosophical, and practical endeavors.[196] He went so far as to claim at times that all of his philosophy stemmed from the doctrine of expression.[197] This doctrine of expression, glimpsed in the papers just discussed, held, roughly, that all things are but different expressions of God's chosen essence of the creation. From this, it follows that all things express one another and are in harmony with one another.[198] As he has a character describe in a dialogue from 1679, "each mind is a new manner of expressing or representing the universe as God regards it, so to speak, from a certain side."[199] Despite the many studies that carefully consider just how rigorous a relation Leibniz intended by the term "expression," its fundamental rooting in Leibniz's technical and mathematical activity and in his justification of that activity remains insufficiently studied.[200]

In his important revisionist account of the history of perspective, James Elkins has stressed that Renaissance perspective was a collection of practices, not a unified philosophical doctrine or a shared worldview. He marks Leibniz as the crucial transitional figure in the transformation of the manifold, loosely grouped set of perspectival techniques into the grand philosophical notion of perspective as a totalizing worldview, most memorably developed by Erwin Panofsky.[201] Insofar as Leibniz took some heuristics from perspectival practices and then translated them into a set of philosophical views and practices quite aptly labeled perspectival, Elkins seems correct. The account of perspective that Leibniz used from the 1670s onward shares little with the totalizing, controlling gaze of Italian perspective, which sought to provide a single view presenting the essential qualities or quintessence of its subject. Both his envisioned practical tools and his metaphors partake in the more mimetic perspectives of northern Europe, particularly of the Netherlands, so well described by Svetlana Alpers and exemplified precisely by the carefully noted cityscape.[202] In his "best means of representing," for example, Leibniz sided squarely with contemporaries like Girard Desargues and his follower Abraham Bosse, who set mathematical and mechanical technique above ineffable artistic skill in painting. Whereas the French Academician Grégoire Huret, for example, argued that geometry and its techniques were "of no utility in the Portraiture of animals," Leibniz argued that people and animated things would "best be painted" with his solution using a mirror (Ger 105).[203] His rather brash claims about the utility of his inventions such as his "all-seeing" lenses for precisely measuring dimensions further illustrate his interest in the precise measurement and depiction of appearances. His proposed techniques for best representing scenes evidently aimed to produce representations of external appearances, which only then would be used in order to begin to gain knowledge of essences.

By producing knowledge about the expressions of something, human beings can gain partial knowledge of the essence of that thing itself.[204] As Leibniz became more convinced that human beings do not know truths directly, he came to argue that perfecting knowledge demands producing expressions of things using the best of available representational tools. While Leibniz long praised the power of optical tools to aid inquiry, and often used perspectival metaphors to characterize it, his mathematical works from late 1672 onward revealed ever more the representational potency, and often necessity, of written compendia. In 1676, he noted that improving human knowledge meant accepting that these representations would often be nothing like the things represented themselves: "Since things themselves cannot be painted or heard, we paint and listen to representations of them. Even if these representations are not similar to them, we see nonetheless certain sensible beautiful things in them that make us understand a theorem, that is, a property of the intelligible thing itself."[205] Just as algebra often leads us to beautiful truths that we do not and perhaps cannot grasp when examining geometric figures by themselves, nonmimetic representations often lead us to knowledge of beautiful truths we could not directly grasp from mimetic representations, such as the scenographic ones his "best means of representing" offered. Points of view, literal as well as figurative, often must be expressed with characters, as in Leibniz's examples using different arithmetical expressions of the same essence. Using representational tools—expressions—we can move from perceptions (and representations) that are infinitely complex, confused, impossible to grasp in a single mental act, to representations that permit us to cognize all at once with some certainty a distinct property (or some distinct properties) of an infinitely complex thing.[206] By producing multiple expressions we can cognize an ever-larger number of distinct properties of the thing. By attempting to relate varied expressions about the same thing, we can begin to discern the harmony underlying the properties those expressions reveal—just as we saw above in Leibniz's call for a "method of universality." Attaining knowledge of this harmony demands working to relate the various expressions—symbolic and otherwise—we have of things. As we have seen, producing and relating sundry expressions of things were central to Leibniz's mathematical practice and his normative conception of mathematics alike. Human beings must practically develop different expressions of mathematical equations and enunciations in order to discern the harmony among different mathematical objects and enunciations about them. Likewise, human beings need to develop expressive technologies to permit them to grasp the harmonies connecting sundry observations and points of view more generally.

With his perspectival, optical, and arithmetical metaphors and analogies, Leibniz articulated the perspectival quality of all human knowledge; at the same time he emphasized that representational tools can enable human beings to discern the harmony beneath the apparent diversity within their own experiences and those of others. The perspectival quality of human knowledge sets it apart from divine knowledge: "If some mind thinks nothing in particular, but thinks nevertheless, it will be God, that is, it will think all things."[207] Using again an analogy of producing representations, however, Leibniz stressed a crucial continuity between human and divine knowledge:

> There is no doubt that God understands how we perceive things; just as someone who wants to provide a perfect conception of a town will represent it in several ways. And this understanding of God, insofar as it understands our way of understanding, is very like our understanding. Indeed, our understanding results from it, from which we can say that God has an understanding that is in a way like ours. For God understands things as we do but with this difference: that he understands them at the same time in infinitely many ways, whereas we understand them in one way only.[208]

God's knowledge comprises all views held simultaneously, not a single view from nowhere. Human beings are limited to but one view at a time, but that view is not altogether different from *one* of God's. To put it another way, human beings, like all of creation, are but limited expressions of the divine essence, but they, above all created things, are the expressions containing limited forms of the divine epistemic faculties. It is incumbent upon them (and possible for them) to combine their views so as to gain ever more perfect, Godlike knowledge.

Although human beings can know only small portions of what God can know, fundamentally they know the same sorts of things. "One can say, concerning the perfection of the mind, that there is more difference between man and the other creatures, who lack reason, than there is between God and man. In sum, there is a certain society between God and men."[209] In sharp contrast to Pascal's account of human cognitive ability, Leibniz viewed humanity as having cognitive capacities proportionate to the world and, therefore, capable of knowing parts of it with certainty. Technical means of expressing viewpoints, of combining them, of making them viewable all at once, work to bring human beings closer to divine knowledge.

Leibniz demanded of perfected representations the qualities he claimed for his quadrature of the circle—written forms allowing many distinct things to be known all at once. In a list of "our duties," from the late 1670s, Leibniz included seeking "sciences useful to our perfection."[210] Such sciences meant techniques that would enable us to comprehend many things all at once: "for singular Minds contain a certain representation of the entire world. There is a more perfect manner of thinking, in which a single act of thinking is made to extend to many objects all at once, for there is more reality in that process of thinking."[211] In pursuing optical techniques and mathematical expressions that extended thinking, Leibniz was fulfilling the human duty to seek sciences useful for our perfection. Better means to discern the harmony of the world were necessary to make embodied human beings properly rejoice in its providential design and justice.

KNOWING WHAT CAN BE KNOWN SHARPLY

In a 1676 letter to the natural philosopher Edme Mariotte, Leibniz explained that mathematics and natural philosophy provided some of the best natural means for improving oneself. "In my opinion, the greatest thing that a man can do naturally for himself is perfect his mind, which unifies him with God insofar as is possible by natural forces." Developing perception meant relinquishing the hope for Godlike knowledge. "Yet the perfection of the mind consists in a perception sharp and just and, in consequence, in the knowledge of things that *can be known* sharply."[212] Perfecting the mind meant not underplaying human ability by rejecting the tools available to us but finding those things we can in fact know sharply and the tools necessary for that knowledge. A modified form of definition was one such important tool.

Perfecting the mind requires definitions of things. In his letter to Mariotte, Leibniz explained, "definitions are not principles of truths; they are principles of the *expression* of truths, which is to say that definitions are the principles of propositions."[213] Leibniz put forward algebra and its equations as an important analogy for defining. At the "end of the accounting," algebra offers only more characters, namely "the value of one letter expressed by other letters." This, however, "suffices for understanding the thing itself." Definitions do much the same, for "an equation is in fact only a species of definition."[214] In considering the epistemological consequences of his mathematical discoveries, Leibniz set forth a path toward a reform of epistemic practices and demands necessary to lead to human epistemic perfection. Rather than documenting the vagaries of Leibniz's varied accounts of definition in the 1670s, this section seeks to note the radical shift in his view of definition made possible by his account of expression and his tools for producing various expressions. As we have already seen, before the quadrature and the defense of perspectival expression, Leibniz had advocated a nearly Cartesian account of definition in which written tools could aid in obtaining ideas themselves. After the quadrature and his articulation and development of expression, he reconceived definitions as a form of expression that could substitute for traditional definitions. Leibniz thus had new powerful tools to fend off sundry assaults on the utility and significance of definitions, such as those of Thomas Hobbes, or the concerns of Pascal and Mariotte.[215]

Hobbes had taken, in Leibniz's interpretation, a "supernominalist" position that truth itself came from arbitrary names set by human convention. For Hobbes, such imposed names were necessary to secure the order of knowledge and of society.[216] While this position had long worried Leibniz, it became ever more troubling as Leibniz became more committed to the necessity of symbols for nearly all human reasoning.[217] Hobbes was right that human beings must impose systems of representation in order to speak of and know the world. *Some* form of expression, Leibniz stressed, is always "necessary for thinking." Immediately after the Paris period, in August 1677, Leibniz responded by arguing that, while names might be arbitrary, the relationships among them are not: "although characters are arbitrary, their use and connection have something that is not arbitrary, namely a definite proportion between characters and things, and the relations which different characters expressing the same things have to each other. This proportion or relation is the basis for truth. For the result is that, whether we apply one set of characters or another, the products will be the same or equivalent or correspond analogously."[218] A proportion exists between things and all legitimate compound written expressions concerning them. The root words may be arbitrary, but the operations on them and the connections among them are not. A relationship exists between any two valid expressions about the same thing; this relationship preserves some common descriptive content. Whether one expresses an arithmetical relationship with grains or with numbers, in binary or base ten, one choice of algebraic characters or another, the same relationships will be expressed. $6 = 3 \times 2 = 3 + 2 + 1 = 1 + 1 + 1 + 1 + 1 + 1$, all capture some aspects of the same essence and all potentially make evident different qualities of that essence. These expressions offer noncontradictory descriptive content, and their diversity is useful for human cognition and discovery. For Leibniz, the nonarbitrary relationship between expressions answered, to some extent, the threat that the arbitrary choice of language in created definitions would make all reasoning arbitrary as well. Discerning the strength and legitimacy of this response matters less than acknowledging Leibniz's belief that the necessary connections among every legitimate expression about the same thing warded off the threat of Hobbes's supernominalism.

For other contemporaries, the danger with definitions came less from their arbitrary nature than from a mistaken understanding of their place in the process of coming to know the world. In his *Logic* (1678), Edme Mariotte had outlined the defects of traditional logic, particularly its treatment of definitions, as a primary cause of the evil of disagreement and discord in natural and moral philosophy.[219] Just as Pascal had argued, to begin the reasoning process by seeking definitions providing the essences of things was to conflate the endpoint of natural inquiry with its proper beginnings. Wrongly aping geometry, many philosophers such as Aristotle had attempted and failed to offer statements giving essential definitions of various natural and supernatural things: "it appears that the design of these Philosophers was to be able to explain the nature and all the essence of a thing in a single proposition akin to the definitions of geometry."[220] Unlike geometry, however, the subject matter of natural philosophy does not rest on constructions in the imagination; in natural philosophy, we cannot define how the world is; we must discern how it is insofar as we can. Explaining the nature and essence of natural phenomena will happen only after the hard work of inquiry, if ever. Confusing philosophical definition with geometrical definition, Mariotte emphasized, was a central cause for the countless disputes in natural and moral philosophy. Rather than pursuing experimental inquiries to determine the nature of something, philosophers constantly became mired in disputing about their definitions of the essence of that thing. Chapter 3 discussed Pascal's similar attack on mistaking nominal definitions for explications of essence. For Pascal, demanding such essential definitions entailed a dangerous, pride-laden overestimation of human ability, led to mistaken claims of understanding the principles of the world, and provoked fruitless disputes.[221]

Drawing upon his claims of the power of written symbols, Leibniz offered a new account of nominal definition; as he told Mariotte, definitions are the expressions of the principles of truth, not truth itself. Not only useful for picking out the object to be reasoned about, nominal definitions could offer a partial description of the essence of the defined thing. In his first public attack on Descartes' account of clear and distinct ideas, Leibniz published an account of nominal definition in the journal *Acta eruditorum* in 1684.[222] Modifying Descartes' famous terminology of clarity and distinctness, Leibniz sketched the account of definitions he continued to hold until his death. According to Leibniz, Descartes' account of intuitive truths rested on ineffable standards for truth and falsity rather than clearly articulated and systematic criteria.[223] Such subjective standards held little hope for producing wide assent to knowledge and thereby ending disputes. Worse, Descartes' account poorly described real human knowledge of complex notions, as it falsely subsumed all knowledge under intuitive knowledge—evident and simultaneously grasped knowledge.

Drawing on Descartes' terminology, Leibniz set out a series of dichotomies for classifying human knowledge: clear versus obscure, distinct versus confused, adequate versus inadequate. If we have a confused notion, for example, we cannot enumerate "sufficient marks [*notas*] to distinguish one thing from another." Although we can distinguish colors and tastes, for example, we cannot give a sufficient enumeration of the marks that identify those qualities. The ability to distinguish among such qualities is like the "je ne sais quoi" of a skilled painter or connoisseur able to point out good art without really being able to explain why it is good.[224] If we have a distinct notion, in contrast, we can, through "marks and other techniques of examination [*examina*]," distinguish one thing from another, just as an assayer can systematically identify gold using a standardized set of techniques.

A nominal definition, Leibniz argued, is such an "enumeration of sufficient marks." For complex things, however, the number of these marks is so great that human beings simply cannot cognize them all and certainly cannot think about them all at once. "There are many other things, however, especially in a long analysis, whose nature we do not grasp [*intuemur*] in its totality all at once." Such grasp—Leibniz again used the vocabulary of Descartes—requires techniques proper to human nature, such as it is in our current state: "in place of things we use signs, the explication of which we are accustomed to pass over in any existing thought as a shortcut, knowing or believing that this explication is within our power." He offered his wonted examples of the necessity of symbols. In considering a thousand-sided geometric figure, for example, we use words like "side" or "one thousand," all the while believing that we have a full understanding of those terms. "I am accustomed to calling such thinking *blind* or *symbolic*, which we use in algebra and in arithmetic, indeed nearly everywhere."[225] Only through such symbolic and blind knowledge can human beings know complex things and propositions. While such complex things cannot be known intuitively, they can, if presented in appropriate compendious definitions, be systematically distinguished from other things with certainty; many of their qualities can be known with certainty or at least moral certainty.

Having such a nominal definition of something does not then imply enumerating all its essential characteristics. Nor does it require that all the distinguishing marks used in defining something have themselves a nominal definition. We might distinguish gold with the quality of being malleable, for example, without having a nominal definition for "malleable." "Adequate" definitions have only marks that are themselves known distinctly.[226] Adequate intuitive knowledge of something would be most perfect; alas, God alone has a simultaneous intuitive grasp of the complete analysis of things. Nevertheless, human beings can and should always work to make their definitions, and thus their knowledge, more distinct. In other writings, Leibniz explained that natural philosophy and metaphysical inquiry—both experimentation and abstract reasoning—involve providing distinct definitions of things in the world. Elsewhere, Leibniz made clear that further inquiry could make definitions ever more distinct.[227] With more distinct knowledge comes pleasure in understanding more fully the beauty of creation, its harmonies, and its underlying rules.[228]

With his account of nominal definition, Leibniz presented his clearest account of what we might call partial knowledge.[229] The knowledge offered by a nominal definition is partial because it includes neither all the distinguishing marks or predicates of some thing nor recursive definitions for all of its distinguishing marks or predicates. At the very least, nominal definitions offer morally certain knowledge of marks that are always sufficient to distinguish one thing from all others. Through forming and defending nominal definitions one comes to more distinct knowledge of the invariant qualities underlying phenomena and thereby can grasp and appreciate some part of the harmony behind apparent chaos.[230]

In Leibniz, Pascal had a great successor in trying to balance the wretchedness of humanity with its greatness. Nominal definitions allowed Leibniz to accept human limitation, to stress the need for inquiries to refine definitions, and to insist on the ability to have some certain, or at least morally certain, knowledge of parts of essences. In the 1680s, Leibniz clearly argued that "perfect demonstrations of the Truth do not require perfect conceptions of things." By "perfect conception," Leibniz meant an expression that contained a sufficient enumeration of the predicates of a subject. "All of our concepts of complete things are imperfect."[231] Nevertheless, we can be certain of what small number of predicates we have, without ever needing to have knowledge of them all. Our finite expressions can be certain even if they are neither totalizing nor complete. Producing such definitions and the techniques for expressing such definitions means working to obtain the knowledge of things we can know sharply.

Leibniz decidedly transmuted Descartes' epistemological vocabulary. Descartes displaced scholastic definition of essence by substituting a form of *sufficient* enumeration or description. Leibniz turned to the expressed enumeration of *sufficient* distinguishing marks, cognizable at once, as the finite, humanly graspable form of definition. Like an algebraic equation, such a definition offers an expression of the principles of truths, as he noted to Mariotte. Leibniz retained Descartes' stark emphasis on producing knowledge grasped all at once. Rather than machines providing the set of interconnections that allow a set of clear and distinct ideas to become a new higher-level intuition grasped simultaneously, Leibniz held out the power of writing to make definitions *literally* seeable all at once: proceeding with definitions fixes our thought and allows "the entire process of our thinking to be perspicuous in *one glance*."[232] The formal qualities of reasoning, not a subjective skill in recognizing the clear and distinct, would secure knowledge: "I even dare to say that the accounts of a receiving officer, as well as the calculation of an analyst, provide an argument *in form* . . . since the form or the disposition of the entire reasoning is the cause of the evidence. It is only the form that distinguishes a book of accounts made following the practice called Italian . . . from the confused journal of someone ignorant in matters of business. For this reason I maintain that it is necessary to maintain some constant formalism in order to reason with evidence in all things."[233] For our knowledge of all but the simplest things, the evidence demanded by Descartes with his clear, distinct, and simultaneous intuitions of the essences of things was to be replaced with a concrete mode of making things evident, such as their proper enumeration on an actual sheet of paper. With this form of writing, to be sure, there would be "less eloquence," but there would be "more certainty."[234] Such formal certainty alone could overcome the squabbles and quarrels plaguing philosophy and theology.

Late in his *Essay concerning Human Understanding* (1690), John Locke claimed that demonstrations retained some reduced amount of the natural light of their intuited presuppositions: "like a face reflected by several mirrors one to another, where as long as it retains the similitude and agreement with the object, [a demonstration] produces a knowledge; but it is still in every successive reflection with a lessening of that perfect clearness and distinctness, which is in the first; till at last, after many removes, it has a great mixture of dimness, and is not at first sight so knowable, especially to weak eyes. Thus it is with knowledge made out by a long train of proof."[235] For Locke, as for Descartes, demonstrations preserved the evidence of their intuitions even if the light of evidence fades in long proofs. In his response to Locke, Leibniz demurred: the quadrature of the circle offered a powerful illustration that much of human knowledge gained its evidence from the form of its expression alone (NE 4.3.6; A6,6:376–77). Perfecting human epistemic ability meant abandoning the hope for intuitive knowledge and settling for the evidence and certainty available through written expression. Mere human beings, Leibniz was at pains to show, could discover, know, and organize much of their world with the help of such expressions. Such discovery would require natural philosophers to draw on any number of technologies to regulate their practices, even the technologies of the lowly merchant and the painter, too long denigrated and ignored by earlier philosophers. In the next chapter I explore the explosion of uses of concrete forms of notation, in Leibniz's work after he left Paris. In trying to create his "universal characteristic," Leibniz sought to transfer the virtues of his quadrature and its philosophical developments to natural philosophy, theology, and statecraft alike.

With the help of perfected means of writing, we could take stock of the affairs around us and reduce them to something we might grasp at once: "This characteristic would deliver us . . . ; the most scrambled matters will be developed, much as the affairs of finance are put into good order by certain fashions or formulas of Calculators and merchants and by a book of accounts which represents receipts and expenses in the blink of an eye; thus it is that this characteristic will sort out the mind."[236] Much as Descartes had offered his variant of mathematical knowledge as a key stepping stone of his spiritual exercises, Leibniz made his mathematics of the expression of viewpoints into a centerpiece of his spiritual exercises for perfecting human ability and happiness, a fundamental duty in his version of the charitable life.

ABRIDGING AND REWORKING DESCARTES

When Leibniz pondered composing a treatise of moral instruction in 1676, he followed his own standard advice and first made a compendium. Under the Senecan title *De vita beata*, he wrote up a précis of René Descartes' philosophical way of life by excerpting the Latin editions of the *Discourse on Method*, the *Principles of Philosophy* with its French introduction, the *Passions of the Soul*, and the correspondence with Princess Elisabeth, Queen Christina, and Pierre Chanut. Seeking out the structuring *ratio* behind the whole, he organized these excerpts into a pithy summary of Descartes' ethics, his natural philosophy, and his exercises. Wisdom meant using reason to know as far as possible the good and the bad, the big and the small, the true and the false. Virtue meant proposing to oneself the pursuit of the commandments of reason. Happiness, in turn, was knowing that one pursued, insofar as human beings possibly can, what one judged to be right.[237] With all this, Leibniz annotated, Descartes offered the antechamber of the true morality.[238] For all his errors in geometry, metaphysics, and epistemology, Descartes had glimpsed the way forward for the philosophical life—one grounded in the proper estimation of human epistemic and moral capacities, the self-appraisal Descartes named "generosity" in the *Passions of the Soul*.[239] Descartes had rightly grounded this self-estimation in exercises involving mathematics and other arts of discerning order and perfecting the mind.

Summarizing his readings and offering critical comments on those readings were central to Leibniz's own method of philosophical reflection and his exercises of philosophical self-perfection. In a series of comments on the art of remembering, Leibniz noted that "it is good to make a written *inventory* of the knowledges that are most useful. . . . And it is necessary to pull from them, finally, a portable *manual* of what is most necessary and most ordinary."[240] Leibniz duly adapted and translated into German and French his summary of Descartes' way of life. Using these summaries as a suggestive starting point, he began sketching out a new version of his own account of the means for living the philosophical life.[241]

In his short compendium, Leibniz listed Descartes' observations about "correctly directing reason" and his standards of clarity, distinctness, and simultaneity. Quoting Descartes' advice for seeking techniques for learning to distinguish things, Leibniz underlined Descartes' stress on grasping all the singulars in a glance, to become certain that none is missing. Noting that "the advantage of these observations chiefly rests on practice," Leibniz emphasized Descartes' account of the utility of algebra as an exercise.[242] Perfecting the mind would include accustoming oneself to making enumerations, to recognizing analogies, to seeing strings of things all at once, in "a single blow of the mind."[243] In a series of lapidary annotations to his summary, he noted what the antechamber of the true morality lacked: "the true method puts forward the proper way of dividing up and a meditative thread."[244] A mathematics with concrete notational practices, geared to discover the *rationes* underlying complex phenomena, was required for a true method capable of leading one to live a proper philosophical life.

In a document entitled "The Utility of Geometry as a Medicine for the Mind," Leibniz argued that the "cultivation of the intelligence [*ingenium*] consists in the improvement of thinking, that is, of the faculty of judging and discovering, inasmuch as it does not depend on chance or fortune; geometry certainly furnishes this faculty with very beautiful examples."[245] As we saw at the beginning of this chapter, Leibniz claimed that his mathematical work in Paris had sharpened his mind.[246] Descartes, as Leibniz himself stressed, made a properly defined geometry and algebra a central exercise. As we saw in chapter 1, geometry offered certain experience with the clear and distinct. Leibniz concurred that mathematics trained the mind through examples of certain reasoning and systematic discovery. In mathematics, there is no "danger of deceiving oneself nor is it hard to undeceive oneself."[247] Noting that mathematics ought not totally to occupy the mind, he underscored that the "greatest usage that one can make of it is to learn to reason with exactitude."[248] Mathematics provides experience with the sort of clear, demonstrative thinking necessary to attain wisdom and repose: "The majority of people are accustomed to confused ideas, [and so] the most beautiful truths do not touch them; they do not realize that things known clearly are necessary for wisdom, and that wisdom alone can make us perfectly happy."[249]

Regulating everyday life with philosophy meant constantly meditating upon one's conduct, abilities, and gifts. The need for constant self-reflection appears in a suggested daily schedule, written in 1680:

> Discover [*inventer*] and Meditate in the
>   morning before getting up.
> Work and perform duties during the morning.
> Divert oneself and visit others after lunch.
> Put things into order after dinner.[250]

This process included the traditional Pythagorean psychagogic practice: "Every day one must examine at night what one has done, what to improve upon, and what to do in the future."[251] Leibniz used the language of business accounting procedures to explain a crucial first step in moral development, the taking stock of the self and of one's affairs: "To *meditate* is . . . to make a general confession of one's life to one's self, to calculate often the receipts and spending of our talents, and to imitate the wise merchant, who relates all the substance of all his journals into a secret book, to the end of seeing in the blink of an eye the entire state of his business."[252] Pursuing a philosophical life required the written techniques and philosophical practices capable of producing such an inventory.

In his *Principles of Philosophy*, Descartes had stressed that mathematics offered salutary practice for perfecting the judgment, useful before beginning to ascend the tree of knowledge toward a life regulated by philosophy. Leibniz likewise offered a seven-step program, beginning with mathematics and ending with tranquillity:

1. study mathematics;
2. apply mathematics to the "somewhat subtle" questions about the laws of movement and jurisprudence, insofar as they are "in our power";
3. pursue first philosophy: knowledge of God and the soul;
4. establish a "good morality";
5. divide time between "the duties of life, conversation, the pleasures of the senses, imaginative things, [and] abstract contemplations";
6. practice the rules one has set for oneself;
7. "pass the rest of life in a profound tranquillity."[253]

In his great theological dialogue of the early 1670s, *Philosopher's Confession,* Leibniz explained that people have not properly grasped the harmony and perfection of creation and the need for a moral life because they, quite simply, have not willed themselves down a serious enough path of reflection and attention. If one had but once considered seriously the statements "*say why you are here now* [*dic cur hic*], . . . , or *see what you do*," anyone "in the blink of an eye, in an instantaneous metamorphosis, would become infallible and prudent and blessed." In the margin, next to this, he wrote, "NB. NB. NB." (*nota bene*).[254]

Even more than Descartes, Leibniz underlined the power of mathematics to upset one's perspective on life. After his great mathematical achievements of the early 1670s, he renewed his interest in these questions of moral transformation. He connected his older language of instant metamorphosis to mathematics, which would provide one with the experience of rigorous and clear thought: "Someone who has understood some demonstrations will admire the force and clarity of truth. . . . He will see himself transformed in an instant, and he will himself remark on the difference between his past and present judgments."[255] Extensive practice could cement such a transformation in judgment: "Whoever will give a year [to geometry] will feel just how much light has sprung forth in him toward judging rightly."[256]

In a remarkable letter to Duke Johann Friedrich in 1671, Leibniz had stressed that real moral transformation came, not from the "temporary rapture" flowing from affectively potent speech, but from the implanting of ethical and religious demonstrations in the mind.[257] In a draft introduction to his unpublished treatise *On the Arithmetical Quadrature*, he explained that mathematics and the investigation of nature could effect such intellectual, not just affective, transformation: "And I consider the investigation of nature (which rests on the perpetual application of Geometry) to pertain also to the perpetual perfection of the mind, for when we understand the divine artifices thoroughly, by which the author of things exhibits certain wonderful effects, we are not so much struck down in the admiration of it and inflamed by love that pertains to regulating our will, but rather we learn the art of discovery from the highest teacher, and we increase the faculty of our understanding."[258]

Grasping the complexity of the effects of nature further perfects our ability to discover the perfections of nature and to devise techniques useful to perfecting human beings and human societies. Every discovery heightens our grasp of perfection and further hones our ability to discern yet more perfection in creation.[259] Leibniz cited the Platonic commonplace that mathematics served actively to raise the soul: "I judge the utility of geometry to be, not so much from the great benefits that human life receives and awaits, but rather that the soul is raised to higher and divine things, is thrown from material things, and is accustomed to exact reasoning."[260] Contemporaries, such as some chiliasts and Jesuits like Ignace-Gaston Pardies, took too static a lesson from reasoning about the infinite: mathematics offered not so much proofs of the divinity of the soul but rather a means for perfecting the mind as much as humanly possible.

In making his abridgement of Descartes' philosophical way of life, Leibniz was critical of far more than his mathematics. His philosophy offered too dour a form of consolation and too narrow an art of living. Descartes' morality was grounded in an account of God bereft of his central qualities of wisdom and goodness. Consequently, Descartes pictured nature bereft of its harmony, reason, and order.[261] Like the Stoics whom he followed, Descartes established "the greatness and the freedom of their Sage, praised for the power of mind that he has in resolving to let pass the things that do not depend on us and to tolerate them, when they come despite us." Such morality Leibniz called a mere "art of patience," and he judged it insufficient: "it seems to me that this art of patience that he makes into the art of living is not the whole of it. A patience without hope hardly lasts and hardly consoles." Descartes' philosophy, in Leibniz's estimation, can only console through "patience," for his God need not be just and wise. Descartes' philosophy included no assurance that God had made "everything for the good of Creatures insofar as possible."[262]

A real consolation appropriate for human beings, in contrast, would offer true intellectual pleasure, joy, and, above all, hope because it would reveal a God who is just and wise: "to satisfy the hope of the human species, it is necessary to prove that the God who governs everything is wise and just, and that he lets nothing go without compensation and without punishment. These are the great foundations of morality."[263] True tranquillity could only come from a certainty that all of creation will ultimately prove just, and that none will go without proper reward or punishment. Proving that such a God reigns requires a metaphysics, a mathematics, a natural philosophy, and a hermeneutic practice, each going far beyond those of Descartes. In all those disciplines, new techniques for allowing human beings to grasp much all at once were central tools in Leibniz's arsenal for producing happiness and genuine contentment. Only with such tools could human beings, on their own, achieve the happy consolation of realizing the order and ornament of this universe and be most strongly impelled to perfect themselves and others. For consolation and tranquillity involve, not passive pleasure or annealing oneself against suffering, but a constant movement toward self-perfection sustained by recognizing the harmonies of the world and working to expand them: "our happiness will never consist, and ought not to consist, in a *pleine jouissance*, in which there would no longer be anything to desire, and which renders our mind stupid, but in a perpetual progress toward new pleasures and new perfections" (GP VI:606).[264] True happiness consists in a never-ending rational desire to grasp perfection and to increase it.

Redoubling Descartes' stress on charity toward others, Leibniz moved the stress from one's own republic, society, and family to humanity more generally. In Greek he commented, "Not a lover of Rome, but a lover of mankind."[265] Loving mankind entailed helping it. "More perfect people are those from whom more of perfection overflows into others."[266] Fundamentally, true charity involved more than mere charitable actions. "More perfect people, however, are not those who have done many things but rather those who are most internally impelled to do many things."[267] Not content with mere charitable action, Leibniz insisted on a transformation of the self, the gaining of a habitual disposition to constant spontaneous charitable activity. The true morality, the inner (not mere ante-) chamber of human morality, would have more tangible methods of meditation and intellectual perfection, would better reveal the perfection in the universe, and consequently would better direct one toward habitually perfecting others. In the late 1670s, Leibniz attempted both to create more tangible means for meditation and knowledge—his universal characteristic—and to devise decorous means for persuading others to follow a path of perfection. Our duty, he wrote, includes seeking out new sciences of knowing and of persuading.[268]

CHAPTER SIX

Seeing All at Once

Sometime in the 1690s, Leibniz reacted to the pessimistic lessons a notable predecessor had found in the infinite. "What Mister Pascal says about the double infinity that surrounds us, increasing and diminishing, when in his *Pensées* he speaks of the general knowledge of man, is but an entrance into my system." Leibniz lamented that this brilliant stylist and mathematician had not seen further: "He would not have said it, with the force of eloquence he possessed . . . had he known that all matter is organic, and that any portion, no matter how many times one divides it, contains representatively, in virtue of the actual diminution to infinity within it, the actual augmentation to infinity that is outside it in the universe; this is to say, every little portion contains, in an infinity of ways, a living mirror expressing all of the infinite universe . . . , such that a sufficiently great mind, armed with a sufficiently penetrating view, could read in it everything everywhere."[1] Had Pascal glimpsed the harmonious world of this metaphysics and natural philosophy, he need not have preached terror before the infinite or lamented the terrible disproportion between human beings and a mute universe. Infinity ought to console, for everything everywhere bore the handiwork of a divine creator. Rather than bespeaking a terrifying universe devoid of evident purpose, every single thing, properly investigated, testified to the finality and organization of all things, to the justice and harmony built into creation.[2] Recognizing these qualities in the universe demanded new techniques for expanding human knowledge.

Leibniz claimed that no fundamental inability in human nature precluded the discovery of rational means for overcoming despair before an apparently meaningless universe: "The greatest part of our misery arises and happiness is lacking, not because of some defect in human beings, but from either a defect of knowledge or of good will." In contrast to Pascal's pessimism, Leibniz argued, "certainly many truths useful to life are within human power, and could be drawn from given and known things by a certain and constant method, if only human beings were willing to use the powers and knowledge given to them by God by following a method of right reason."[3] Drawing on his successful mathematical practice, Leibniz attempted to produce this method of right reason. Called a "general science," this method was essential both for achieving happiness in recognizing the justice of creation and for perfecting others.

Even the limited knowledge available to human beings, almost always mediated through spoken language and written signs, bespoke less human shame and infirmity after the Fall than the working out of an exquisite plan. Demanding to grasp this order at this moment in human history, Leibniz wrote a few years later, "is like wanting to take a novel by the tail and pretending to decipher the plot from the first book; instead of which, the beauty of a novel is great to the degree that it finally produces more order from a greater apparent confusion." Revealing the plot too soon would be a fault in the work, not a virtue, for it would eliminate the salutary work of recognition: "the beauty and justice of the divine government have been partly concealed from our eyes, not only because it could not be otherwise . . . but also because it is proper in order that there be more exercise of free virtue."[4] Such exercise of virtue, Leibniz repeatedly stressed, ought to involve inquiry into nature and history; it demanded the further development of scientific instruments and intellectual tools so that we might come to raise our eyes to see the harmony of things ever more perfectly.

In studying human ability to cognize the infinite, Pascal and Leibniz found indications of human limitations and human capacities; both drew fundamental lessons from the knowledge human beings can have of the infinite. In Leibniz's view, Pascal drastically underestimated human ability to recognize order and purpose in nature because he lacked a sufficient metaphysics and the linguistic tools necessary for gaining knowledge of the world. As exemplified by his unwillingness to develop appropriate algebraic notations, Pascal failed to develop enough new tools—in mathematics, physics, and metaphysics—to discern such order. He failed to see the hand of a just and wise God in every minute particle, each a confused mirror of all of creation. He failed to see the finality built into all things and the universal law common to God and man. Where Pascal saw only terrifying possibilities that one could not rationally dismiss in considering the natural world, Leibniz saw an order whose evidence one could not rationally dismiss.

In late 1676 Leibniz reluctantly left Paris to take up a new position as librarian and counselor in Hanover to Duke Johann Friedrich of Brunswick-Lüneburg. For Leibniz, the horror of the recent wars in Europe had a root cause: theological misunderstandings and conceptual sloppiness that drove the Catholic and Protestant churches apart. As far as he was concerned, only a superior metaphysics and natural philosophy could begin to resolve these theological problems and thereby begin to bring forth a new harmonious Christian Europe. With his numerous philosophical and practical projects, Leibniz worked to create new natural and metaphysical grounds to secure and to improve the political, theological, and ecclesiastical order. No matter how unhappy Leibniz often was at Hanover, no matter how underappreciated he felt, no matter how unsatisfactory his rank and salary, his positions there at the very least offered Leibniz some space and resources to explore the implications of his Parisian discoveries and philosophical work.[5]

In the years following his stay in Paris, Leibniz worked to produce practical tools for nearly every domain of knowledge and action. Drawing heavily on his successes with his calculus and other mathematical subjects, he pursued his old dream of an art of systematic discovery and judgment: his famous universal characteristic. In seeking this art, he produced numerous concrete tools for discerning the hidden order and harmony in mathematics, nature, and human society. With these tools, human beings might work toward an ever-more-Godlike comprehension of the universe and toward a greater ordering of themselves and human society. Leibniz reckoned that early-modern Europe desperately needed knowledge of harmony in the universe and the technical means to help engender social, religious, and political harmony.

In this chapter I highlight Leibniz's philosophy as a practice. First, I emphasize his insistence on the materiality and palpability of his efforts to reform human knowledge. Second, I illustrate Leibniz's own philosophical activity: the inductive mathematics, experimental techniques, the excerpting of authors, and, especially, the writing up of manifold projects, many springing from a common set of heuristics and tools. This chapter combines the insights of recent studies on early-modern scholarly practices of reading and writing with concerns from the history of science about the place of writing and inscription in scientific work.[6] Leibniz's philosophical practices were central means for cultivating himself. His vast written corpus was in part the product of this exercise and self-cultivation.[7]

Many of the practices Leibniz advocated in the 1670s and 1680s involved using new linguistic tools to make things viewable all at once, so that their underlying order might be grasped more easily. As shown in the previous chapter, Leibniz was able to prove that the area of a circle was equal to a sum of infinitely many terms, so that

$$\frac{\pi}{4} = \frac{1}{1} - \frac{1}{3} + \frac{1}{5} - \frac{1}{7} + \cdots$$

In the late 1670s Leibniz struggled to transfer his practices for discovering and judging mathematical truths through expressions graspable in a blink of the eye to a surprising range of domains, from natural philosophy to the running of a state. He supplemented his old Platonist commitment to knowing many things all at once with technologies to facilitate such knowledge. Despite his well-known reticence to publish, he envisioned sundry ways to disseminate such techniques, to spread both the skill of discerning order and the habitual impulse to extend such order. Only then might the harmony of the universe be harnessed to secure a new harmony on Earth.

THE MATERIALITY OF WRITING: TABLES, INDUCTION, AND SEEING ALL AT ONCE

In a programmatic note from 1677, Leibniz explained that perfecting humanity required improving reason: "the religion that I follow exactly assures me that the love of God consists in an ardent desire to procure the general good, and reason teaches me that there is nothing that contributes more to the general good of all men than that which perfects it [reason]."[8] Appreciating God and the soul demanded "science"—knowledge of the causes and order of the universe. To love God meant producing ideas, procedures, and tools to perfect embodied human reason insofar as possible. Only with the knowledge generated by such perfected tools might human beings recognize the justice of creation, achieve steadfastness before everyday adversity, and pursue just and charitable lives helping to improve others.[9]

Attaining these goals demanded a "general science," which Leibniz described as "a way that teaches all the other sciences how to discover and demonstrate *from sufficient givens.*"[10] Combining an encyclopedia with new methods, this general science was to unify human knowledge deductively, insofar as possible, along with making judgments as certain as possible in subjects knowable only through experience, such as much of medicine, physics, biology, history, and revealed religion. While Leibniz defined the general science in various ways, we can safely say that it included techniques for

1. concisely displaying what is known and what is claimed to be known,
2. conspicuously showing the grounds for that knowledge, whether those grounds be empirical, revealed, or derived from first principles,

3. adjudicating among claims to knowledge from the "givens" so displayed, and
4. making new discoveries.[11]

All these tasks demanded new written techniques—what he called "palpable marks." With such a science, human beings would be able more quickly to promote "human happiness" than they could with "the labor of many centuries."[12]

The most potent tool for all parts of the general science, the tool that could best lead humanity to greater happiness and greater perfection, was the "universal characteristic"—a prospective system of writing, reasoning, discovering, and judging applicable to all domains. Just as reasoning with novel signs had transformed mathematics, reasoning with appropriate signs could help advance every domain of human inquiry. Extending the work of George Dalgarno and John Wilkins, Leibniz sought to produce more powerful systems of signs for communicating, organizing, and extending knowledge. These authors, he remarked, had thought up a "certain language or Universal Characteristic" for organizing "notions and all things beautifully" and for allowing diverse nations to communicate; "no one, however, has sought a language or characteristic that contains the arts of discovering and judging simultaneously." The characters of arithmetic and algebra, two sciences God "bestowed on the human race," should serve as a model for producing characters useful for discovering and judging more generally.[13] Like an air pump or a telescope, the characteristic was to be a tool, one tailored for human beings, who lack the epistemic abilities of angels and God.[14] Closely bound to Leibniz's profound work in symbolic logic, the universal characteristic is a distinct project that concerns inductive as much as deductive reasoning.[15] Inspired by his mathematical practice, he sought to generalize its power: "When I had the pleasure to improve considerably the art of discovery, that is, the analysis of the mathematicians, I began to have certain new views for reducing all human reasoning to a sort of calculus that would be of use in discovering truth insofar as it is possible *ex datis*, that is, from what is given or known."[16]

A model for deductive certainty, mathematics equally served Leibniz as a model for inductive discovery from what he called "sufficient givens."[17] The characteristic would serve the many domains of knowledge that did not allow certain deduction: "It is not that probable arguments could be changed into demonstrative ones, where sufficient givens are lacking, but that one could be able in such a case to estimate the degrees of probability and put the given advantages and disadvantages into a line of an account and, at least, to reason surely from given things [*ex datis*]."[18]

In his sketch of a general science, Leibniz called for techniques capable of providing certain, or nearly certain, demonstrations as well as techniques for discovering truths from things already known, whether through the intellect or the senses.[19] The first he described as "the form of argumentation by which, through means of calculation, one could suppress demonstratively every controversy and either determine the absolute truth or at least demonstrate, when the givens are not sufficient, the greatest probability that can be had from given things." The second he described as "the art of invention, that is, a palpable thread for directing investigation and its species: combinatorics and analytics."[20] Both parts of the general science concerned truths of fact as well as truths of reason, the empirical as well as the axiomatic, the uncertain and the indubitable.[21] Both required new forms of writing to aid reasoning.

Throughout his writings Leibniz emphasized the materiality and palpability of his prospective universal characteristic and general science. In terms he used pervasively, he argued that this new form of writing would conduct "one's hand and mind" much like arithmetic; it would offer a meditative thread (*filum meditandi*), "a sensible and crude method, but assured of finding truths and resolving questions from given things."[22] In stressing palpability and sensibility, Leibniz underscored that this thread would involve savvy, usefully visible, forms of notation and be subject to rules directing the reasoning process as if by external guides.

The need for such sensibility stemmed from human embodiment: "For whoever would diligently consider the nature of our mind that is in this body would easily see that, although we have within us ideas of things remote from the senses, we nevertheless cannot pin down or devote attention to the knowledge of these things unless certain sensible marks approach them, such as names, characters, representations, analogies, models, logical connections, consequences."[23] Human beings can have no sustained knowledge of supersensible things without material notations and techniques for manipulating them. Contemporary eclectic German scholastics such as Johann Heinrich Alsted had similarly stressed that human knowledge after the Fall required logic and mnemonic arts.[24] Descartes and Pascal had underlined how easily complex phenomena and logical proofs could distract rather than instruct or convince. In his manuscript copy of Descartes' *Rules for the Direction of the Natural Intelligence*, Leibniz wrote "NB" (*nota bene*) in the margin next to Descartes' complaints that making any part of proof dependent on memory undermined its evidence, certainty, and epistemic power.[25] In long chains of reasoning, Leibniz elsewhere wrote, the "human natural intelligence [*ingenium*]" loses its way "in the middle of the journey" unless regulated "through apt characters and sensible marks, which can be touched as if by the hand."[26] Much more than mechanical means for reasoning, proper written language and representative techniques could regain and hold the attention. His characteristic would generate the *phantasm*, "all of which is grasped at once."[27] A simultaneous grasp of marks on a page would allow human beings to focus their attention and thereby aid in discovering and adjudicating complex truths.

Gaining access to knowledge, even a priori knowledge, required perfecting material notations to be visually examined, to be sensed.[28] Even the truth of mathematics could be made visible: "thus, every mathematical truth can be transferred with numbers into an experience by the eyes."[29] Ending the sectarian and philosophical disputes of the day required a means for making truths visually evident: "of the greatest urgency is finding an empirical and palpable criterion by which every true demonstration can be exhibited irresistibly and the true discerned from the false."[30] Certain demonstrations could not really end disputes unless their truthfulness could be made *sensible*—literally apparent. Overcoming the spiritual and intellectual defects of embodied humanity required superior methods of practical representation and writing. Solving abstract problems in metaphysics and resolving religious controversy demanded practical notational systems.

Like Descartes and Pascal, Leibniz sought appropriate tools not for disembodied minds, but for embodied ones. Convinced about the need for material expression, Leibniz worked to improve formal systems of reasoning and his universal characteristic for discovering and judging. After his mathematical and philosophical work of the early 1670s, he no longer sought a language to complement an already-worked-out network of clear and distinct ideas, as Descartes had deemed necessary.[31] Having abandoned the pursuit of cognizing ideas in themselves, Leibniz maintained that written expressions could constitute knowledge; in fact, as we saw in chapter 5, he claimed that proper forms of written expression could alone produce almost all complex knowledge humans beings could actually have without divine intervention.[32] The evidence appropriate to human reasoning takes the form of written expressions in good notation, not a cognitive grasp that is clear, distinct, and simultaneous. "This characteristic art...would contain the true organon of a general science of everything that is subject to human reasoning, clothed in the uninterrupted demonstrations of an evident calculus."[33] Proper written expression would also provide the best means for humanity to augment its knowledge and improve itself physically and morally.[34]

From his early work on combinations in the mid-1660s onward, Leibniz pursued new ways of displaying characters in different orders and sequences.[35] His new mathematical techniques offered models for creating combinations. He worked to produce new techniques for representing many things all at once as a central aid for discovering new truths and judging arguments. From his mathematical work in numerical series and summations, to his logical writings, to experimental natural philosophy, to positive jurisprudence, to biblical hermeneutics, Leibniz stressed that perfected characters ought to be put into tables useful for discovery and judgment: "just as accurately marking out a circle requires an instrument by which the hand is regulated . . . , we are in need of certain sensible Instruments for thinking correctly, of which I refer to the two great sources, *Characters* and *Tables*. . . . But *tables* are nothing other than inventories of things through systems of characters."[36] The general category "tables" includes sundry ways of organizing information. Working notes of 1678 remark on "the many ways of making up tables, so that the same things appear in many different ways." He gave the tables of Petrus Ramus and Theodor Zwinger as examples. He discussed various sorts of tables: columns, meaning "the exhibition of series simply"; inventories, meaning "varying coordinates or indices of the same things."[37] With such tables, the rules and order underlying displayed phenomena are to be found more easily.[38] Displaying the same information in different ways allows one to discern different underlying harmonies and orders.

Beyond drawing upon the famous tables of Bacon, Ramus, Zwinger, and others, Leibniz modeled his general modes of discovery on his remarkably fecund mathematical practice. Every new publication of Leibniz's mathematical papers reveals the range of his inductive labor with infinite series, harmonic progressions, linear equations, and other mathematical subjects.[39] In his work on numerical series, algebra, and combinatorics, he drafted and redrafted hundreds of tables in his quest to discern principles of order underlying apparent chaos, the harmony underlying the diversity.[40] In his mathematical writings, he sometimes followed his working tables and calculations with instructive guides about using tables to recognize the relationships he had just discovered.[41] He wove these practices into his methodological papers on the characteristic and the general science.

Leibniz repeatedly drew upon examples of inductions from series of numbers as the key heuristic for recognizing the hidden orders underlying sets of given things. In documents meant to convince patrons, he invoked a simple example involving squares of integers:

Numbers	0	1	2	3	4	5	6	7	8 . . .		
Squares	0	1	4	9	16	25	36	49	64	81	100 . . .
Differences	1	3	5	7	9	11	13	15	17	19 . . .	

From this simple table one can easily discern a rule about the sequence of the squares of the natural numbers: the sequence of differences of the square numbers is the sequence of the odd numbers: $4 - 1 = 3, 9 - 4 = 5$, $16 - 9 = 7$. Once the relationship is found inductively, a rigorous demonstrative proof is trivial. Let a_n be the square numbers, and b_n the differences, then $b_n = a_{n+1} - a_n = (n + 1)^2 - n^2 = n^2 + 2n + 1 - n^2 = 2n + 1$. Calculation or mere chance might eventually have produced this relationship, but not so easily or quickly.[42]

In his "Example of a New Analysis through Which Errors Are Avoided and the Mind Is Guided as if by the Hand, and Progressions Are Easily Found," dated June 1678, Leibniz illustrated the potential of new techniques of writing and expression using numerical notations to solve algebraic problems.[43] In this example, Leibniz offered calculational means nearly equivalent to Cramer's rule for solving simultaneous linear equations in four variables, albeit with some errors.[44] He showed how proper modes of notation in tables could enable one easily to find the underlying order—the *rationes*, or reasons—behind the apparently complicated algebraic formulas.

Finding the value of one unknown in a system of equations demands considerable computational work. In modern notation, using subscripted variables such as a_{12} and x_{12}, an example might be

$$a_{12}x_2 + a_{13}x_3 + a_{14}x_4 + a_{15}x_5 - A_1 = 0$$
$$a_{22}x_2 + a_{23}x_3 + a_{24}x_4 + a_{25}x_5 - A_2 = 0$$
$$a_{32}x_2 + a_{33}x_3 + a_{34}x_4 + a_{35}x_5 - A_3 = 0$$
$$a_{42}x_2 + a_{43}x_3 + a_{44}x_4 + a_{45}x_5 - A_4 = 0$$

In his 1678 paper, Leibniz argued for the utility of a notation using "artificial" or "fictional" numbers—numbers added to aid calculation, such as the subscripts above. Consider writing out such equations without any numerical indices (I have created this example to capture his complaint):

$$ax + by + cw + dz - A = 0$$
$$ex + fy + gw + hz - B = 0$$
$$ix + jy + kw + lz - C = 0$$
$$mx + ny + ow + pz - D = 0$$

Writing with such "letters indistinctly," Leibniz said, will create "horrible confusion" and demand "immense labor."[45] Although all the coefficients and variables have distinct letters, nothing in the notation serves to remind us about their initial relationships to one another. Once we begin calculating, we have to struggle to recall, for example, that the coefficients a, e, i, m began in the first column; they are hard to keep distinct. The characters $a_{12}, a_{22}, a_{32}, a_{42}$, in contrast, all include numbers indicating their initial position and relationships; they are trivial to keep distinct. Basing a notation on such artificial or fictional numbers will "express exactly the varied orders and relations among quantities and characters, so that immediately at first glance it appears that certain known letters pertain to an unknown quantity."[46] Leibniz expressed the four simultaneous equations as

$$12, 2 + 13, 3 + 14, 4 + 15, 5 - A = 0$$
$$22, 2 + 23, 3 + 24, 4 + 25, 5 - B = 0$$
$$32, 2 + 33, 3 + 34, 4 + 35, 5 - C = 0$$
$$42, 2 + 43, 3 + 44, 4 + 45, 5 - D = 0$$

He designated both the coefficients and unknown qualities with their indices alone, so that the modern coefficient a_{12} is simply written 12, and the unknown variable x_5 as 5. The relations, he claimed, follow "exactly" from notations involving numbers such as a_{34} or 34, but not at all from mere letters, such as k. A savvy notation permits the quick visual inspection of a set of equations. It makes the various parts appear distinct to the eye, which can then seek out the unifying order behind the distinct parts. However combined, 12 retains its distinct identity and recalls its position in the original problem, no matter how complex the algebraic expression in which it may appear. Using his new indices, Leibniz began to explore the value of the unknown 5 in terms of the coefficients 12, 23, 44, and so forth. Finding the value of the denominator of the value of 5 using his notational system reveals an apparently complicated equation. Its denominator is

12, 23, 34, 45		12, 23, 35, 44		12, 24, 33, 45		12, 24, 35, 43		12, 25, 33, 44		12, 25, 34, 43
13, 22, 34, 45		13, 22, 35, 44		13, 24, 32, 45		13, 24, 35, 42		13, 25, 32, 44		13, 25, 34, 42
$-$ 14, 22, 33, 45	$+$	14, 22, 35, 43	$-$	14, 23, 32, 45	$+$	14, 23, 35, 42	$-$	14, 25, 32, 43	$+$	14, 25, 33, 42
15, 22, 33, 44		15, 22, 34, 43		15, 23, 32, 44		15, 23, 34, 42		15, 24, 32, 43		15, 24, 33, 42

"By the power of numbers, we write this easily." Inspecting the coefficients reveals a beautiful order among them, one difficult, if not impossible, ever to see had they been written indistinctly.[47] Look now at the *first* characters of each set of multiplied coefficients in every row and every column, for example: 12, 23, 34, 45, then 12, 23, 35, 44, etc. In every case, they are 1.2.3.4. Now abstract the second character in each coefficient and write it out. For the first and second rows, one gets

$$2345 \quad 2354 \quad 2435 \quad 2453 \quad 2534 \quad 2543$$
$$3245 \quad 3254 \quad 3425 \quad 3452 \quad 3524 \quad 3542$$

Each set of numbers along the first row evidently consists only in 2 along with all six permutations of 3, 4, 5. The second line merely substitutes 3 for 2 and 2 for 3. The third and fourth follow the pattern. Each line is a systematic transposition of the previous. Leibniz has here, in his language of permutations, essentially the formula for finding the determinant of the matrix of coefficients.[48] He thus has a general method for solving linear equations in multiple unknowns.

In this fine example of a palpable notation at work to help discovery, he laid out the "most important rule" of the characteristic: "characters express all that is hidden in the designated thing, which numbers do best by virtue of their copiousness and their facility for calculating."[49] In this example, appropriate indices, forms of expression involving numbers, allow one systematically and easily to write out the answers to such algebraic problems. They provide an ordered inventory that permits the easy, nonalgorithmic, nonmechanical discovery of the rules underlying phenomena and the distinct cognition of all the parts at all times. They permit human beings to focus their attention and thus discern the rule underlying the equations, rather than to attempt simply to memorize them or to plod through laborious, blind calculation. The example illustrated nearly everything he claimed in his so-called philosophical papers that the characteristic and tables would do for discovery more generally. Leibniz himself marked these procedures as examples from his mathematical practice to be transferred to other domains. He readily transferred them.

Based on such mathematical examples, he suggested that the practical tools of writing things in series and tables should be used more broadly: "In investigating natural things, it is useful to seek for them in series. And, if it is possible to find [*reperiri*] the same thing in many series and [to find], as it were, the knot or intersection of diverse series, it will be better known thereby." He considered the example of finding the quadratures of a series of curves to different powers. Using such a method initially involves more labor, he conceded; "however, the progression is revealed more easily; and since this discovery is good to infinity, it compensates for the labor copiously."[50] Putting series into tables allows us to see them as produced within a series of series—the great strength of Leibniz's harmonic triangle and Pascal's arithmetical one.

Whether discovered by chance, method, or wit, if a set of givens "are reduced into a *system* and this system is ordered and assembled together so that everything is visible at once [*in conspectu*] and can be easily compared, we will have Tables."[51] A table, such as the solution to the equations above, would let one grasp a systematized set of information, of givens, all in once glance. Leibniz here fused two parts of his mathematics: his commitment to finding techniques for formally displaying arguments and his actual inductive practice using series and tables. These palpable modes of reasoning would help fulfill his long-standing project of grasping the unity and harmony of the diverse things in the world. In his mathematics, he had well illustrated the power of these tables. He worked, less successfully, to create them for other domains of inquiry.

In discussing discovery using the characteristic and tables, Leibniz did not sharply distinguish the revealed phenomena of scripture, traditional law, and the experimental world from logical, natural-philosophical, and mathematical truths.[52] The characteristic "will have more or less the same effect in matters of movement, of natural philosophy [*physique*], of morality, and of jurisprudence as do characters in arithmetic and analysis."[53] The books of nature and of scripture needed similar inductive practices. Mathematics, natural history, exegesis, and experimental natural philosophy alike needed inductive tables written in their proper characteristics. Legal interpretation likewise demanded seeking out reasons underlying different laws.[54]

Whether for discovering the laws of motion or the intentions of authors, Leibniz praised the use of tables to discover the "rule," "ratio," "law," or "analogy" underlying a diverse set of givens, to discern the order underlying the entire series of things abridged, organized, inventoried, and displayed before the eyes. In considering experimental natural philosophy, for example, he drew upon a shopkeeper metaphor to complain that "to amass experiences without putting them into an accounting is like having a well-stocked, great store without an inventory." Only with a form of inventory proper to natural-philosophical phenomena such as motion could one hope to "pull from the givens" everything human beings can know.[55]

Making inventories, he claimed, is to a researcher as keeping books is to a financier—it is with such an inventory "that it will be necessary always to begin."[56] Such inventories, he claimed, were central to his practice.[57] Organized and visible all at once, inventories admit systematic and formal means of revealing errors and omissions in them. His inventories "will be composed only of lists, enumerations, tables, or progressions, which will serve for having always in view, in any meditation or deliberation whatsoever, the catalog of facts, circumstances, and the most important suppositions and maxims that should serve as the basis for reasoning." Above all, "the exact review of what we have already acquired will marvelously facilitate new acquisitions." Tables make gaps apparent and show where additional work—mathematical, experimental, or philological—is needed. These tables reveal, "in a single glance, a region of the mind, already peopled," and with them one "would notice soon the areas still neglected and empty of people." Seeing such empty land, "one will send colonies to make

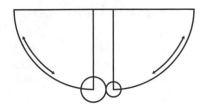

Figure 6.1. Schematic of Leibniz's experiment on the collision of bodies. After Leibniz 1994b, p. 131.

new plantations in the least well known part of the Encyclopedia."[58] While tables could never quite eliminate the chance involved in discovering the reasons underlying progressions and series, Leibniz claimed that they would significantly augment the probability and speed of grasping such relationships, just as they had in his work in algebraic and harmonic sequences. Using such characters arrayed in tables did not guarantee discovery in a mechanical or algorithmic way; it just greatly increased the likelihood of discovery.

The practice of using tables for aiding discovery was borne out in his experimental practice of the late 1670s, which is still too little known, despite fine recent scholarship. He drew heavily on his tabular methods in his most central physical innovation of the period. In January 1678 Leibniz undertook an intense examination of the rules of motion, particularly the rules of collision of bodies, all written up in a series of papers he entitled "On the Collision of Bodies."[59] He had long worried, on metaphysical, physical, and theological grounds, about the insufficiency of laws of motion grounded exclusively in physical extension and local motion, which could be represented as geometric (spatial) magnitudes.[60] Drawing on the rules of motions articulated by Descartes, Christiaan Huygens, Christopher Wren, Edme Mariotte, and John Wallis, Leibniz devised numerous candidate laws and then performed a set of measurements involving pendula with spherical wooden balls of various masses hitting one another from different heights[61] (see fig. 6.1). His goal in his mixture of theoretical and empirical investigations was to discern the rules, the *rationes*, behind the phenomena of impact.

Putting his results into tables revealed before long a startling gulf between his calculations based on theory and the results of his experiment. This gulf vitiated the best current theories: "By these experiences, the systems of Huygens, Wren, Wallis, and Mariotte are therefore overthrown."[62] The pages of manuscript following, which combine tables, detailed calculation, and metaphysical speculation, show Leibniz struggling with the aberrant results of his operations.[63] He concluded that the fundamental error was an assumption going back to Descartes, and held by nearly all mechanical philosophers, concerning the conservation of "force" in the universe: that the quantity of motion—quantity of matter times speed—is the "force" conserved.[64]

Combining calculation, experimentation, and metaphysical concerns led Leibniz through numerous conjectures until he proposed nothing less than what he called a "reformation" of mechanics.[65] Rather than a different measure of grace, this reformation required a different measure of force: the quantity of effect that, in the case of a falling body under the gravity of this world, equals quantity of matter (mass) times speed squared, what he would later call *vis viva*—living force. Having come to this conclusion, he worked through his manuscripts highlighting his errors. He duly wrote out anew the key table at issue, squaring the appropriate speeds, to show the match between his old experiments and his new theory.[66] With his new rule of the conservation of quantity of effect underlying all motion, Leibniz could insist that natural philosophy, based on the observed phenomena of motion as well as his metaphysical reasoning, required the reintroduction of something akin to substantial forms, eschewed and belittled by so many contemporaries such as Descartes and Pascal. Experiment confirmed his long-held metaphysical concerns about the insufficiency of the mechanical philosophy.[67] Just as his tables did not guarantee or cause his mathematical discoveries, tables in no way mechanically produced this key result in Leibniz's physics. They exemplify the heuristic aid that he claimed tables and varied enunciations provided: they helped to spur the creative discernment of rules that revealed the order behind well-exhibited phenomena, viewable all at once.

Far more interested in empirical methods and inductive practices than legend allows, Leibniz was no modest empiricist.[68] Any science aiming simply to generalize from empirical phenomena would never satisfy him in his quest for the sufficient reasons for the phenomena of the world.[69] However useful tables were as means for acquiring knowledge, such inductive knowledge did not satisfy Leibniz: "Concerning the progression of Series, whenever we have found a progression a posteriori by the power of a Table, we have produced something useful and splendid; nevertheless, we have not proceeded perfectly, for we have been able to discover the law of the progression a priori, when we can demonstrate it independently from the table." He added in the margin: "Rare is a discovery free of every chance."[70] The inductive work made efficient and easier through tables should later guide the creation of a priori deductions in those domains permitting such deductions.[71] Despite his concern with a priori demonstrations, he sometimes worried that concentrating too much on rigor would detract from pursuing new discoveries useful for improving human life.[72]

For all his success with formal logic and his evident mathematical innovations, inductive as well as deductive, Leibniz's art of discovery, his

dreams of characters and tables apt for writing and seeing, appears ultimately too vague and too ambitious. His inductive, tabular projects were
nevertheless as rooted in his mathematical practice as were his justly cel-
ebrated efforts in the more formal aspects of deductive logic. His belief in
the power of perfected language to allow overviews, the combination of
viewpoints, and the discovery of "reasons" through material aids such as
tables unquestionably served as a fundamental productive heuristic in his
work from 1676 to the late 1680s. He used this heuristic both to articulate
his long-held views more clearly and more powerfully and to transform
other domains of philosophy and practical activity.

In a much-quoted letter to his sometime mathematical collaborator
Tschirnhaus, Leibniz outlined the virtues of his characteristic: "there can
be no doubt that the general art of combinations or characteristics contains
much greater things than algebra has given, for by its use all our thoughts
can be pictured and as it were, fixed, abridged, and ordered; pictured to
others in teaching them; *fixed* for ourselves in order to remember them;
abridged so that they may be reduced to a few; *ordered* so that all of them can
be present in our thinking."[73] Picturing actual marks would help compensate for the deficiencies in human epistemic capacity and help us to teach,
to discover, and, quite simply, to know. These technologies for discerning
the *rationes* underlying everything in creation would serve mightily to improve bodies and souls alike: "I do not doubt that, from these experiences
and knowledge that we have already investigated . . . , we could draw out
many and great things through consequences that could not only perfect
the human mind and amend morals but could also make this life happier
and put an end to many ills that trouble our bodies."[74]

Wisdom, Leibniz wrote in 1679, "is nothing other than the science of
happiness." The most important benefit of knowledge is to understand the
perfection of God and the nature of the human mind, "so that we can,
even amid the greatest calamity, uphold an interior happiness" and in this
way become content with the past and the present: "if we were to inspect
the series of things, we would come to realize that nothing could be made
better."[75] Only then would we possess a certain hope about the future,
essential to achieve a "durable joy." The general science and its auxiliary,
the universal characteristic, matter, above all, in producing the "durable
joy"—the sustained happiness that alone can bring personal tranquillity
and that rests upon our effort to recognize perfection and accompanies our
effort to further perfection in this world.

FROM RECOGNIZING TO PERFECTING ORDER

The characteristic occupied Leibniz while he served as librarian and counselor (*Hofrat*) to Duke Johann Friedrich and then to Duke Ernst August

of Brunswick-Lüneburg (later electoral prince of Hanover). While Ernst August proved more interested in politics and dynastic history than philosophy and religious reconciliation, he continued the efforts of Johann Friedrich to set the administration of Brunswick on a more centralized and organized footing and established administrative norms lasting well into the eighteenth century.[76] With little success, Leibniz encouraged a yet more radical transformation of Hanoverian governance. He tried to show that perfecting the state and its subjects demanded the skills of the duke's librarian and counselor.[77] As an integral part of his characteristic, Leibniz devised myriad schemes to improve governance, the economy, and statecraft. Leibniz worried about more than the theoretical foundations of the state: like early Cameralists such as Veit von Seckendorff and J. J. Becher, whose works he knew well, he concerned himself with the everyday routines of bureaucratic and administrative practice appropriate to a state seeking to perfect subjects and resources, to increase its wealth, and to secure its domestic order.[78] Like the Cameralists, he sought to combine the details of quotidian governance with the lofty goals of human moral and material improvement; he sought to do so through administrative versions of his notational systems and improved understandings of human cognitive and affective nature. Just as mathematics and experimental natural philosophy needed new methods for grasping manifold things in a single view through concrete notational practices, the state needed an accounting viewable all at once and produced through practices necessary to perfect the economic, social, and political order. Such procedures required new kinds of administrators, people like Leibniz, who were capable of combining innovative managerial technique with sufficient knowledge of metaphysics, natural law, and new notational practices.[79]

In Hanover in the late 1670s and 1680s, Leibniz wrote a series of memos outlining new procedures and bureaus for systematizing and "harmonizing" knowledge about a state and its people, resources, and laws.[80] Before he went to Paris, Leibniz had called for new ways to discern the unities underlying the law.[81] Now a councilor seeking promotion and preferment, Leibniz made such demands more concrete in describing new administrative procedures and new institutions for those procedures. Just as theologians have "harmonies of confessions," jurists have accounts of the "differences" of sundry laws, and so should the government: "it would be useful to have such a harmony and collation in governmental matters [*Regirungssachen*]." All this effort should happen within a new sort of recording bureau (*Registratur-Amt*). The aim of such a bureau would be to allow princes and administrators "to find easily" everything necessary for the goal of ruling well.[82] Improving governance, no less than improving mathematics, required lessening the burdens on the mind through conspicuous notation.

The solutions to the problem of justice and order would come from technical solutions to problems of knowledge—the knowledge of the prince: "I call a State-Table a written, short account of the kernel of all the news concerning the ruling of the country, . . . arranged with a great advantage: the Sovereign, seeing it all at once, can easily find everything that he needs to take into consideration in every event." Since he can use it himself, it is one of the "most convenient instruments for a praiseworthy self-rule." A well-organized and productive state required improved knowledge of people, resources, and places. Such knowledge would permit leaders to grasp quickly what order and potential for order existed and to perfect that order. With such an inventory, the details of the entire country can be "ordered into a harmony together, extracted, and concentrated."[83]

Leibniz reworked concerns common to many Cameralists with his own concepts of order and harmony and with his own techniques for allowing fallible human beings to achieve them. The state-table and the recording bureau would in principle unify all of what we would call the statistical work done on the kingdom. This work would require "perspectives" on all aspects of the kingdom.[84] In memo after memo, Leibniz stressed the need to make the entire situation in question visible all at once, perhaps in new scripts.[85] The "philosophical godfather of Prussian official statistics," Leibniz envisioned that producing these abridgements would involve a whole range of actuarial, mechanical, and mathematical techniques, as well as new institutions: academies, libraries, and surveys, all extracting, indexing, and harmonizing the diverse information collected.[86] Gathered in new institutions, advisors and officials would compile, organize, and index this information. Better bureaucrats with superior skills would make governance more efficient and less administratively intensive: "To administer everything with as few ministers [*bediente*] as possible: rather Justice, Police, and other such things. Many secretaries, few administrators [*Rathe*]: one must have universal men. With that, one can connect all things, more than with ten [others]."[87] Superior knowledge, presented in forms tailored to the limits of human attention and cognition, would lessen the administrative burdens of the state.

Compilations viewable all at once were equally needed for preparing information at levels of governance below the sovereign and his advisors. A unitary glance was necessary, for example, to optimize the productivity of a mine. Leibniz's years working on the mines of the Harz Mountains produced more than new windmill and pump designs, as he hoped to integrate his engineering projects within a new regime of information and control.

He proposed a digest with a perspectival viewpoint: "Hence it is necessary to make a scenographical or perspectival digest [*abriß*] of the mine, as if one's eye were floating in the air and the mountain were transparent."[88] Such a viewpoint would allow one to recognize resources and to optimize their use, including the labor of extraction.

Beyond knowledge of resources, proper rule calls for knowing the true principles of justice and morality, whatever skeptical naysayers such as Hobbes or Pascal maintained. Working from first principles to discover the metaphysical and theological basis for moral and political right was necessary, but human beings should also pursue alternate routes toward moral and political knowledge, particularly in practical decision making. In describing some means to attain political knowledge, Leibniz turned to his account of expressing points of view and discerning the reason underlying them. "The place of the other is the true point of perspective in politics as well as in morality."[89] If we want to know whether our actions are just, we must imagine ourselves to be an "advisor and minister of an enemy or suspect Prince." Only then could we know the "views" that "our neighbor *can* have against us." Leibniz immediately noted that he was merely stating that Jesus Christ's fundamental moral principle—putting oneself in the place of the other—applied equally to politics. He expressed this so-called universal moral principle in an idiosyncratic idiom of perspectival viewpoints. Finding such political knowledge required human beings to use perspectival fictions that provoke them to seek deeper connecting principles: "This fiction *excites* our thoughts." Much as Leibniz used the language of fictions to explain mathematical notations useful to discern general mathematical relationships, he used the language of fictions to motivate the discovery of universal moral principles and the discerning of political motives.[90] Such fictions were more than thought experiments useful for achieving purely theoretical knowledge. In a set of exercises for the judgment, Leibniz called for imagining "that you are another and especially an enemy ... and to see what you would do from there and have to say, as if you were your own sworn enemy and a bought-off opponent."[91] Leibniz moved consideration of the point of view of another—a moral precept about sociability—into a fundamental epistemic principle and called for practice in using it for everyday action.[92]

In some later notes Leibniz advised: "put us in the place of others and others in our place; the exchange of places in thought."[93] Religion, morals, and natural philosophy would advance only if we constantly determined and considered the located views of others. For Leibniz, recognizing the perspectival quality of all human knowledge implied no relativism. Insisting on a moral, as well as epistemological, need to consider and to weigh carefully the view of others, Leibniz called for technical means to recognize what aspects of the truth were to be found in these views. His perspectivism would structure his approach in his later works, for example, to Chinese morality and religion.[94]

Determining and compiling the views of others would be greatly simplified with Leibniz's palpable means of reasoning. His oft-quoted call "let us calculate!" as a means for resolving disputes involved a generalization of the inductive, and not just the deductive, aspects of his mathematical practice for judging and discovering the rules behind apparently complex phenomena.[95] In reporting a conversation with Duke Johann Friedrich, Leibniz described his technique for ending religious controversies: "I dare to say that the method, which I claim to draw on, cuts away all these confusions with a view of the eye." With his technique, "one will see a representation so true to the claims [*raisons*] of one side and the other that every reader will have need only of good sense to judge." His method "will abridge disputes insofar as it is possible, so that one can see all the economy of them."[96]

A fragmentary "Method of Disputing to Come to the Exhaustion of the Subject" illustrates the continuity in notational practice for producing such an abridgment from mathematics to natural philosophy to political dispute. Just as Leibniz illustrated the power of using indices, or "fictional" numbers, in mathematics, he insisted on using them for representing the entire economy of an argument in a visually clarifying manner:

$$
1\begin{cases}
1(1)\begin{cases}
1(1)((1))\begin{cases}1(1)((1))(((1)))\\1(1)((1))(((2)))\end{cases}\\
1(1)((2))\begin{cases}1(1)((2))(((1)))\\1(1)((2))(((2)))\end{cases}
\end{cases}\\
1(2)\begin{cases}
1(2)((1))\begin{cases}1(2)((1))(((1)))\\1(2)((1))(((2)))\end{cases}\\
1(2)((2))\begin{cases}1(2)((2))(((1)))\\1(2)((2))(((2)))\end{cases}
\end{cases}
\end{cases}
$$

Leibniz here married the dichotomies of the Ramist tables that figured so prominently in German philosophy of his day with his concrete notational innovations. He left little doubt as to the universality of the procedure: "The same technique is useful in every deliberation." And he underscored its use "in the ending of philosophical and theological controversy."[97] He elsewhere compared this form of resolving disputes to two merchants meeting and settling their accounts.[98] The legend that Leibniz envisioned two mathematicians meeting and performing an extremely long string of calculations and then announcing a final result, as if they were modern digital computers, mistakes how he envisioned that his characteristic could help in formally solving disputes. This legend mistakes what he meant by "calculate."[99] In making his famous exhortation "let us calculate," Leibniz did not call for reducing all judgments about the soundness of reasoning to an examination of deductions from first principles. Deductions from first principles and reasoning involving a posteriori knowledge alike were to be examined in his calculus. To operate with characters meant both to deduce from axioms and to examine a range of phenomena expressed in characters that helped one to discern patterns. Some patterns, notably in mathematics, are capable of being proven by human beings from first principles. But many patterns remain outside the human ability to prove with certainty; they require probabilistic reasoning, aided by the same variety of written characters useful in deduction. Inventorying and abridging multiple arguments, so as to bring out their evidence and gaps, their paralogisms and solid arguments, would quickly help to discern the true and the just underlying the various positions. Just as good forms of mathematical notation helped one recognize the underlying *ratio* of an infinite series, good forms of notation could make apparent the *rationes* behind two opposing arguments.

To create knowledge, the philosopher ought to imitate God's creative activity by producing physical expressions of viewpoints and then by attempting to integrate those viewpoints. In creating the just state, the ruler ought to imitate God's stewardship by having knowledge of the state made available to him. For Leibniz, the virtue and intelligence of a single monarch were crucial for constituting the just and productive state: "Those to whom God has given at once reason and power in a high degree are heroes created by God to be the promoters of his will, as principal instruments." Statesmen imitate God's creative power in producing moral harmony. "Governors of the public welfare," they ought to "strive not only to discover the brilliance of the beauty of God in nature" but equally to further its perfection through imitative works.[100] Just as the philosopher must create the means of expression making possible the greatest amount of human knowledge, rulers must have philosophers and advisors create means of expression so that they might know the state, its resources, and its people as much as possible. Power needs knowledge of principles and resources to be justified and successful. Only through such properly expressed and displayed knowledge might the ruler pursue policies to perfect, discipline, and integrate the state and its citizens, toward greater peace, creativity, and productivity. Seeing the order and harmony in nature should drive the ruler to create order and harmony in his realm.

In an important consideration on the goals of the state written around 1680, Leibniz called for the state to help all its subjects develop a startling array of qualities.[101] All subjects need to internalize something akin to the self-discipline of a spiritual exercise: the people ought to have tranquil and content minds and be moderate and prudent; they ought to love and honor the ruler and be pious, be friends among themselves, be knowledgeable in many domains, be beautiful, and be exercised in every mental and physical virtue.[102] The state must also provide the material conditions that allow such qualities to be cultivated—to eliminate many vagaries of moral luck, we might say, insofar as possible given a natural hierarchy of abilities, of perfections.[103] Briefly, "care must be taken so that the people are prudent, are endowed with virtue, possess faculties in abundance, so that they know, want, and can accomplish the best things."[104] The state must provide physical necessities, intellectual training, and affective support to help its subjects habitually desire and produce the most prosperous and virtuous society. In another document of 1680, Leibniz emphasized a "key political idea": that "oaths, laws, punishments [*straffen*] do not have enough power" to keep the people from sin and evil. Leaders must educate them, from youth, to develop a "second nature" to make "doing well . . . easy and natural." Such a habitual second nature is nothing less than "the natural antidote to original sin and the precursor and attendant of grace."[105]

How could people achieve such a habitual virtuous life, a unified transformation of their material means, knowledge, and desires? In answering this question, Leibniz considered what human perfection entailed. To perfect something meant augmenting its "power," either by eliminating obstacles or by increasing its liberty. To perfect human beings meant eliminating the obstacles to their thought and action and increasing their liberty to think and act. To find means of perfecting human beings thus required a proper understanding of thinking itself.[106]

His definition of thinking in this political document comes as no surprise: "Thought is some active representation of many things all at once achieved in a thing by itself."[107] By the late 1670s Leibniz had redefined thought itself as plural representation or expression in his technical sense. No longer simply metaphorically clarifying human epistemic activity as in the Paris papers, such representation now comprised the activity of knowing itself. "I understand by representation every expression of a thing by another."[108] His wonted examples follow: numbers by characters, lines by letters, solids by figures on planes, things by words. The marks, which Leibniz insisted upon in logic, mathematics, and accounting as means for improving and accelerating the thinking process, were but a rarified, finely tuned species of all thinking itself. Not to perfect means of palpable and concrete expression would be to deny the perfecting of human epistemic abilities necessary for the good life and the common good.

Leibniz here defined the mental states of higher-order beings such as human beings in terms of expressing multiple things all at once. Able not simply to perceive passively, human beings can actively combine perceptions and become conscious of such combinations. Personal identity itself should be understood as expression: "It is necessary that many things be represented all at once in a thing all by itself. This is finally called 'Me,' and I am assuredly something all by itself, and in this unity are contained all these representations of a thinking thing."[109] As representing things, human beings differ from paintings, mirrors, and other passive things because they constantly work to integrate expressions and to produce higher-level expressions capturing and integrating multiple experiences.[110] Leibniz mapped his exhortations concerning practical procedures to improve knowledge onto his account of human cognition and identity, all the better to understand how to improve society and its members.

An apparent digression, this abstract discussion about thinking shows Leibniz working to discern the proper techniques for perfecting the state by improving the thinking processes of all its subjects. Given his definition of thought, he outlined ways to perfect it in rulers and in the people. He offered three standards of superior representations in thought: "that the representation be exact, or, what is the same thing, that the thought be distinct," that it be "strongly active, which is to say, that the will required by proper reason be firm," and "finally, that the representation be spaciously spread out, which is to say that we comprehend many things all at once in a single blow of the mind."[111] The best policy for the state—and the duty of its ruler—involves nothing less than perfecting the minds of its subjects with tools such as Leibniz's various means for making perceptions distinct and their overall economy visible and evident. Habitually knowing and doing the good would come only through knowing properly, knowing many things distinctly and all at once. Such a move to perfect the knowledge of the subjects of the prince needed equally to transform their affective lives: just as their knowledge and productive activity needed perfecting and directing, so did their desires and wills.

Leibniz did not envision that such changes in knowledge merely involve what David Hume later dismissed as "the cool assent of the understanding."[112] A few months before writing his notes on the instructional duty of the state, Leibniz had pondered the emotional effects of different sorts of thinking in some working notes he entitled "On the Affects" ("De affectibus"). He characteristically worked by first epitomizing and then reworking Descartes' *Passions of the Soul* with his own philosophical concepts and tools, notably the representation of multiple things all at once.[113] After many pages of restatements and revision, he declared that affect is "a determination of the mind to the thinking of some series," just as impulse in motion is "the determination of a body to run through a certain line." He underscored that "a series is a multitude ordered by a rule."[114]

What makes a soul follow a determined path, to think a given series of thoughts and not another? Most basically, the mind follows a series that is "already in it" when "nothing has turned the mind away from it." "[A]mong many series," one will think "the one that appears to be the most perfect," unless some impediment distracts the attention. Descartes had argued that controlling the passions required breaking the hold of astonishment, the undue attention to the first impression of something, which hinders inquiry. Leibniz rephrased Descartes' argument as the un-due power of "singular" thoughts to turn human attention from the most perfect series.[115] Following proposals such as those in his "Funny Thought" (1675), which outlines ways to educate people by surprising and then illu-minating them, Leibniz sought means to turn people and monarchs alike from the deadening grasp of such diverting singular thoughts. To lead peo-ple to the true happiness that ensues from developing "the habit of acting following reason," he wrote later, "one must try to remove the obstacles that prevent them from finding the truth and from pursuing true goods."[116]

Absent singular thoughts, a more perfect series will determine thoughts. What qualities, then, make a series more perfect? "The determination to think a series is strongest in that series whose rule involves more reality." To make the point more emphatically to himself, he drew a box around a re-statement, this time noting that the best series involves the greatest number of things understood distinctly under a rule all at once. Much like a good optical device, a powerful notational system, or a beautiful mathematical theorem, the stronger thought exhibits more material "distinctly."[117]

Leading people to act well demands techniques to break their undue attention to their current series of thoughts and to turn their attention toward habitually considering and pursuing superior series of thoughts, series of thoughts that combine a greater number of things subject to a rule. One must also get the people to consider series that are subject to rules that contain the greatest reality, those series containing the most things all at once. Anyone whose attention focuses on such series will gain a determination to proper action.[118] Leibniz reworked the scholastic claim that the will inclines to whatever is understood to be the good in terms of a hierarchy of more or less superior series of thinking.[119] Much as Descartes had argued, our mistakes about the good and our tenacity in holding on to those mistakes often prevent us from pursuing the actual good. To spread ethical living rested upon the propagation of the rules, the *rationes*, that contain the greatest number of things, for such rules can most powerfully produce determinations to the good and nothing less than love of the most perfect things.[120] As Leibniz often wrote, from his early works to his late ones, "To love is to find pleasure in the perfection of another."[121]

A better logic could lead to love—to loving perfection in others and habitually acting to perfect others. Although formal reasoning might well be capable of discovering the true and the virtuous, Renaissance humanists had long complained that such reasoning was hardly capable of moving anyone to the good life, of effecting real moral and emotional change. For Leibniz, the pathologies attributed to scholastic dialectic ought not to be used to dismiss all forms of symbolic and formal reasoning. A proper art of discovering and judging, one capable of grasping many things all at once, was essential for moving rulers and their subjects alike to moral thinking and feeling as well as to habitual charity and productive activity.

In his sketches about reforming the state, Leibniz reworked the perfecting of the human mind in terms of the qualities evinced in his example of solving linear equations with superior notation. Perfecting the characteristic to allow one to see all at once was nothing less than perfecting, policing, and disciplining human beings as intellectual, volitional, and affective beings. Such perfecting was at once the duty of a monarch and of Leibniz himself. Such perfecting required superior means for disseminating the art of finding better series for thinking.

CHARITY, DECORUM, AND SELF-PERFECTION

The iterated themes of the two previous sections mirror Leibniz's collected works: filled with derivation after derivation, list of works after list of works, tables, charts, constant redrafts, detailed excerpts, critical notes, translations, all replete with painstaking modifications, emendations, elaborations, queries, and self-critiques. Pierre Hadot has stressed that Marcus Aurelius's *Meditations* should be read less as a finished philosophical discourse than as the written products created in exercising himself.[122] The sundry philosophical, mathematical, and political papers of Leibniz ought to be similarly seen, at least in part, as the products of his philosophical project of perfecting himself. His manifold accounts for making things visible all at once in epistemology, mathematics, and statecraft offered means for trying to recognize the order in creation and to increase the order of human society. His constant redrafts, translations of documents from one language to another, from one genre to another, form the written products of his exercise.[123]

Recognizing the harmonies of the universe was only part of the good life. Truly perfecting the self meant becoming impelled to help others:

> He who loves God loves everything. . . .
> Whoever is wise loves everything.
> Every wise person endeavors to be useful to all things.
> Every wise person is useful to many.[124]

Loving God meant truly knowing and, therefore, loving his harmonious creation; such true knowledge led to acting habitually from love of the individuals of creation to perfect them, insofar as possible. As Patrick Riley has stressed, Leibniz insisted upon a charity of the wise—a love of all things and people combined with wisdom, to the end of acting habitually to perfect all beings insofar as possible. Leibniz harshly criticized radical Christians, such as Chiliasts and Millenarians, who severed themselves from the corrupt world and did nothing to improve it. They little understood the world and lacked true Christian charity. Likewise, Leibniz cruelly mocked the anti-intellectual meditations of the Quietists. To reach their quietude or inaction would require "taking opium or drinking a good *rausch*," not thinking and acting charitably.[125]

Perfecting himself entailed attempting to perfect others, even the paternalistic attempt to change nearly everyone. Perfecting himself meant developing means for transferring the benefits of mathematics, of expression, of seeing many viewpoints all at once, into the domains of the *honnête homme*, the counselor of state, the monarch, and the people. Far from just some optional bonus or a shameful polluting, and thus epiphenomenal, attempt to secure patronage, making his techniques public was integral to perfecting himself in his theory and actual practice.[126] Leibniz in fact transferred his practices for unifying knowledge to many domains as we saw above; by virtue of his own understanding of the habitual charitable life, he simply *had* to do so.[127] Such an effort, Leibniz contended in his *Discourse on Metaphysics* (1686), was contained in Christ's message and its means of expression. Just as Pascal argued that "God speaks well of God" (S334), Leibniz contended that Christ "divinely well expressed [truths], in a very clear and very familiar manner, so that the most base minds understood them."[128] The whimsical "Funny Thought," discussed in chapter 5, offered just one of many proposals for perfecting people. There, Leibniz advocated strategies to deceive the members of the public into improving their intellect and morals. As the Jesuits had argued, teaching truth need not clash with its rhetorical accommodation; indeed, a commitment to spreading truth requires a savvy decorum. With such displays and writings, Leibniz sought to prompt others to perfect themselves, to become more harmonious by recognizing order and harmony. The more they did so, the more their society would become unified, harmonious, and productive.

Doing good and perfecting oneself and others required the esteem of the powerful. Although hardly a model courtier, in 1679 Leibniz composed a set of "rules of life," a prudential and rhetorical counterpart to his philosophical rules for self-perfection. These rules and suggestions sketched out Leibniz's personal book of the courtier or manual of *honnêteté*.[129] They offered means for balancing "being," that is, sincerity and being true to oneself, with the management of outward "appearance," necessary to gain favor and trust among the powerful: "One must seek two things: Being and Appearing."[130] To fulfill this demand, Leibniz offered a mix of rhetorical instruction taken from Cicero and Quintilian; tips from courtesy manuals, French, English, and German; Ficino's thoughts on a scholar's health; Jesuit casuistry; and the praise of the utility of natural curiosities. The rules for living suggested guidelines for producing the appearances—the rhetoric, the outward manner of being—capable of making princes and the powerful more philosophical and thereby helping to produce more just governance. The first rule recommended wonder as a key tool for the philosophical gentleman. "At all times, one must have something with him with which he can provoke astonishment. For example, the little machine that pulls so much weight."[131] Although using such natural-magic wonders to gain attention and approval at court had long been cliché, it fit well within Leibniz's understanding of the necessity of using the inclinations of courtly society to move it toward higher ends. The cultivated, decorous individual ought to make himself into a surprising and instructing display in microcosm, someone capable of obtaining at once esteem, financial support, and the power to effect epistemic and moral change in others.

The power of play and amusement to aid instruction and moral habituation had long fascinated Leibniz. In his 1667 work on new ways to teach jurisprudence, he argued, "Teaching is to the soul as medicine is to the body of an animal. Just as the physician aims to heal (1) carefully, (2) swiftly, and (3) pleasantly, so the same things are required in the care of the soul; teaching should be (1) sound, (2) swift, and (3) pleasant."[132] In revising this early work some years later, he greatly amplified his examples of playful learning with examples from Harsdörffer and others.[133] In a 1680 proposal for reforming the Hanoverian state, he called for the "easy and playful teaching of the sciences" among things central to a proper reform of the education of children.[134] In another remarkable proposal, Leibniz described a spectacular education for a young prince along these lines. He described with glee the profound subterfuge at work: "it is good that everything *be regulated under an apparent liberty . . . so that he learns all the while believing that he's amusing himself.*" Sundry machines, techniques, tables, exercises, and plays would make up this only apparently free education. Machines could profitably teach the young prince the fundamental principles of nature: "one can make *machines that respond to our questions* and offer reason [*raison*] at the same moment from their response . . . when these reasons are palpable and sensible, the satisfaction from them is doubled."[135] Such machines would playfully initiate the prince into the eternal laws of nature, the stuff of eternal law and justice that ought to organize his future reign.

Not just monarchs and courtiers needed decorous, often deceptive, pedagogy to perfect them. Leibniz wanted to institutionalize such prudential techniques for reforming others.[136] In a series of proposals, he described an "order of charity" or "society of the love of God" modeled in part on the Jesuits. This society would use "God's works worthy of astonishment" as well as medical cures and chemical mysteries to attract people, fight atheism, and lead them toward the love of God.[137] Members of this society were to speak "without scholastic style" and to strive "for the manner of common speech"—to pursue decorous, if sometimes temporarily deceptive, means to bring as many people into the fold of proper belief and Christian action as possible.[138] Just as the philosopher at court needed arts of being and appearing, of logic and of rhetoric, a society of charity required logic, natural philosophy, and metaphysics along with decorous ways of speaking, writing, and presenting.

CONVERTING A JADED HONNÊTE HOMME

Much taken with Platonism in form as well as content, Leibniz praised the decorous quality and instructive power of dialogues: "not only are souls imbued with truth through familiar conversation, but one can see the order of meditation itself."[139] His dialogues illustrate the charity involved in making philosophy public. In his "Conversation of the Marquis de Pianese and of Father Emery Eremite," written around 1679–81, Leibniz dramatized the making public of his philosophical practices and illustrated how they were to change one's way of life. A philosopher-hermit, highly learned in philosophy, the sciences, and practical arts, leads a semidissolute marquis, gentleman, and statesman, from a skeptical detachment toward a philosophical life of knowledge, love, and charity.[140]

Toward the beginning of the conversation, the marquis explains that he achieved a skeptical way of life by reading the modish skeptics Michel de Montaigne and François Le Mothe Le Vayer. He has achieved tranquillity and inner peace from following custom and the rites of his religion. These authors, he says, have "cured me" of the sickness of the troubled soul and mind produced by the dreams and fantasies of philosophers and mystics seeking more profound knowledge. "[I]nfected with the normal skepticism of men with an air of nobility," the marquis is a type needing a cure, an awakening to the sheer depravity and irrationality of his chosen course (2246).

The dialogue is set in the hermit's academy. Much as Leibniz claimed his projected "society of the love of God" might do, this hermitage "made a great noise in the world" (2246). Tales of magical cures, even rumors of the philosopher's stone itself, bring many who seek the hermit's spiritual and medicinal succor. Coolly disbelieving such frippery, the marquis comes nevertheless, seeking genial diversion from paradoxes and clever reasoning in the "spirit of the world, which seeks only novelties" (2246). Like Pascal's *demi-habiles*, the marquis holds himself to have a superior, enlightened attitude about such novelties. No savant has shown him anything capable of making him believe in a "knowledge of God and of nature above that of the people [*vulgaire*]." While savants claiming to have great knowledge amuse him, they have always failed to convince him to abandon his comforting skepticism and fideist beliefs. In the works of natural philosophers, he claims, there is "often much of fashion and of parade; some little curiosities or some austerities that dazzle the people." Such displays reveal no great wisdom or genuine knowledge: "We are all equally ignorant at heart when it is a question of anything of importance" (2247). Savants and the people alike abuse and pain themselves with dreams of higher understanding; few grasp, the marquis maintains, the tranquillity arising from the skeptical following of custom.

Few things confirm the marquis in his skepticism more than the proliferation of competing philosophical accounts of the world among those who style themselves learned. To break the marquis away from his skeptical path in life, the hermit must explain the cause of uncertainty amid the profusion of contradictory views; he must provide an etiology of error. He begins with the chiaroscuro of nature and life, the mix of dark and light Leibniz often used to capture natural and moral ambiguity: "There are conveniences and inconveniences, good things and bad things in all things in the sacred and profane world. This is what troubles men; this is what gives birth to the diversity of opinions: each envisions objects from a certain side" (2250). Leibniz here echoed Pascal's attempt to understand human error. To make someone recognize his errors, Pascal argued, "it is necessary to observe from which side he envisions the thing, for it is ordinarily true from that side, and to confess to him that truth but to reveal to him the side from which it is false. He will be happy with that, for he will see that he did not deceive himself; he simply lacked seeing from every side" (S579). Self-reflexivity is essential. Encouraging people to reflect on their own position can make them grasp the conditions under which they ought to have thought what they previously had thought and to understand the conditions framing their current beliefs.[141] They come to recognize how their previous position and interests helped to dictate their previous beliefs. Like Pascal, Leibniz drew on the figurative language of sides and viewpoints to explain the causes of error and to suggest a cure. In Leibniz's opinion, too few ever tried to overcome their single points of view. Few people had attempted to see with care from every side; this failure even to attempt to consider the views of others encouraged the false conclusion that the current multiplicity of viewpoints implied an eternal human inability to discern the truth and to reach consensus on it: "There are but few who have the patience to tour around a thing until they put themselves onto the side of their adversary, that is to say, whoever wants with equal application, and with the mind of a disinterested judge, to examine both the case for and the case against, in order to see which side tips the balance" (2250).

In this dialogue, Leibniz turned the heart of his practical knowledge-producing practice, of his picture of perfecting human cognition itself, into the hermit's explanation of error. Knowledge remains fractured and uncertain not because different people have diverse viewpoints about ambiguous things but rather because so few seek to consider, to weigh, and to reconcile these multiple viewpoints. A lack of attentiveness, a moral failure, leads necessarily to epistemic laziness and failure. The so-called educated do not follow Christ's call to take the viewpoint of another, in morals, natural philosophy, or religion. Little wonder that they fail to come to any solid knowledge. Leibniz's mathematical notations, his universal characteristic of tables and characters, his techniques of mediating controversies, all facilitated the metaphorical tour about an entire thing, all aided in weighing all views of it and in making them visible all at once—all helped one to follow Leibniz's rewriting of Christ's injunction.[142] Most philosophers, far from developing such techniques, did not even see the need.

An improper attitude toward philosophy compounds this epistemic failure. Seeing philosophy primarily as a source of diversion, as do too many of his contemporaries, the marquis has never plumbed its depth and found the superior way of life to be found within (2250). Popular among the learned, this attitude causes opinions to multiply rather than encouraging their reconciliation. True knowledge, true inner peace, and true interpersonal concord might come only through earnest seeking to combine viewpoints, observations, and arguments, not simply from encouraging their proliferation.

Descartes and his followers illustrated the worst qualities of philosophers. While admiring Descartes in many ways, Leibniz castigated him for seeking to appear novel by ever deprecating the viewpoints of others, past and present, even those from whom he had brazenly stolen. His success had turned him into the worst sort of example, one that had stoked ever more dispute: "He gave birth to jealousies and to contestations with the loss of precious time and rest necessary for discoveries of consequence."[143] The "certain spirit of contradiction" evident in Descartes and other philosophers amplified the disastrous epistemic effects of the failure to examine different opinions in depth. In seeking social distinction, "we study everything to oppose ourselves in appearance to what ordinary men are accustomed to judge and to wish for. Thus, we render everything problematic" (2250). Much as Pascal described the disdain of the *demi-habiles* for customary opinions, Leibniz's hermit explains that the desire "to speak well" and to stand out drives the proliferation of unreconciled viewpoints. The conventions of most forms of disputation can yield only skeptical withholding of belief when offered a plurality of strongly articulated views.[144] The standard practices of social distinction, even in philosophy and the sciences, reward clever contradiction without encouraging anyone to reconcile apparently contradictory viewpoints. Montaigne argued, "We learn to dispute only to contradict; each contradicting and being contradicted, it follows that the fruit of disputing is to lose and annihilate the truth."[145] Echoing such sentiments, Leibniz targeted a key pathology of witty conversation and disputational culture: "When we have found some skillful or ingenious retort that can refute or disconcert someone who has put forth some proposition to us, regardless of whether it is perhaps useful or well founded, we enjoy this victory, and we move on to other matters, without examining if the proposition was fundamentally correct" (2251). At root, these errors stem from treating thought as mere diversion, parade, or academic question, at most something to gain social distinction, and never as something useful for the "the practice of our lives" (2251). For philosophy to become autonomous, an academic matter or mere game, distant from everyday thinking and living, is to guarantee its fall into diversion and base sophistry. Not only undermining its purpose of finding the true and the good, such ludic autonomy renders it useless for securing religious and political order. Escaping such pathologies requires practices for weighing and contrasting opinions carefully, not simply producing or dismissing them.

The hermit's strategy in the conversation is that of Leibniz's prospective educational casino and academy of games, of his charitable societies, of his ludic forms of education: to make an elixir from the poison of curiosity. The hermit aims to use the marquis' desire for novelties to push him toward spiritual elevation: "I will profit from your penchant, to make you more attentive" (2250). As we saw in chapter 4, Pascal attempted in his *Pensées* to force readers to confront their contradictions and their ignorance of the real motivations of their way of life; likewise, the hermit strikes at the contradictions of the marquis' skeptical way of life. The marquis seeks to live a tranquil life according to custom, yet delights in paradoxes. The interlocutors in Pascal and Leibniz alike delude themselves in believing that they are superior by producing doubts about common beliefs and revealing their contradictions. Such "singular opinions give us an imaginary elevation above others; we would be annoyed to speak like the people, although we follow the general corruption" (2250–51). The lust of the *honnête homme* for paradox and novelty illustrates that his attitudes and ways of acting are as vulgar as the people's, but less honest. Like Pascal's *demi-habile*, Leibniz's skeptic is truly no better than the untutored, vulgar people he disdains; in fact, he is worse, for he preens his false superiority. Like Pascal's *demi-habile*, the marquis has intelligence and desires—his love of paradox and wonder—that present a great opportunity for provoking him to transform his way of life.

The hermit leads the marquis through a series of exercises dedicated to recognizing the order behind the mix of dark and light in nature and in human affairs. The hermit first praises the use of arguments in form. Reflecting the widespread early-modern distrust of such reasoning, articulated by Pascal for example, the marquis replies that the schools use arguments in form "without fruit." Reasoning in form should not, the hermit avers, be confused with the "boring *quicunque, atqui, ergo*" of the schools; it need not follow their "order and manner." Rather, "a chain [of reasons] or a sorites, a dilemma or enumeration of all the cases, finally every rigorous mathematical demonstration, even those of an algebraic calculus or an arithmetical operation, are arguments in form, just as are common syllogisms with three terms" (2260–61). In a fairly weak dramatic move, the marquis abruptly metamorphoses into an enthusiastic advocate of Leibniz's general science with its mix of the a priori and empirical: "If the principal axioms were arranged and demonstrated in the manner of mathematicians . . . and if experiments were well ordered and linked with axioms, I believe that one could form from these the marvelous elements of human knowledge, and distinguish the true, the probable, and doubtful. . . . I see very well that in matters where it is not possible to go beyond the probable, it will suffice to demonstrate the degree of probability and to show to which side the balance of experiments will tilt" (2262–63). All this certainly is "to be wished for," the hermit announces to his heated interlocutor, but is not necessary for their present purpose: discerning the sort of life the marquis should pursue. In the interim, the tools of this formal reasoning are still useful: "since you recognize that there is a means for assuring oneself about what one must decide [*juger*] about things from their appearances, let us content ourselves with making use of that rigor to consider the question of wretchedness versus supreme happiness" (2263). He advises the study of progressions discerned in nature to enable spiritual and mental progress, to grasp just how perfectly nature has been constructed. While never offering the certain happiness that a demonstrative metaphysics could offer, such study forms a potent exercise for perfecting oneself. By actively considering nature, the hermit contends, one quickly comes to appreciate that the universe could not have come about by chance or without design.

The hermit offers a set of practices, including prayer and observation of nature. He offers the marquis this advice: "Accustom yourself above all to notice that there are orders, liaisons, and beautiful progressions in all things" (2273). Developing a proper attitude and disposition toward creation means learning habitually to recognize its greater unities, the *rationes* underlying all phenomena. The central practices for this ethical life are those Leibniz sought to augment with his method of invention: expressive technologies aiding the discernment of order amid the apparent confusion of everyday experience.

Learning to seek out such interconnection and unity in all things prepares one to wonder properly at the spectacle of creation:

> Since God exercises our faith in the apparent muddles that he knows how to put into harmony with a happy future, we will do well in the meantime to excite and strengthen ourselves sometimes with the sensible experiences of the greatness and wisdom of God, which are found in the marvelous harmonies of mathematics and in the inimitable machines of God's invention that appear to our eyes in nature. . . . Physical marvels are a food proper to sustain without interruption this divine fire that excites happy souls, for there one sees God through the senses. . . . I have often remarked that those who are not touched by these beauties are hardly sensible to what one ought to call truthfully the love of God. (2273–74)

Not emotionally sterile, such understanding allows a glimpse of the harmony, the perfection, of the natural world. It permits one a foretaste of God's love and justice in that harmony. Just as Leibniz explained in other documents, the harmony and order evident in nature *excite* the soul— recognizing them offers real affective transformation, not the dead knowledge or logical contradiction of sophistry.[146] Wonder for Pascal involved acknowledging that creation was far beyond human ken. Salutary wonder for Leibniz involved achieving a knowledge of details that are beyond human ken ever to construct. They are readily observed in created things, should one look attentively with the proper tools. Given knowledge of the craftsmanship of creation, exemplified everywhere in nature, should one take the time to consider it seriously, mental and emotional anguish will fade. Far superior to skeptical detachment or the "patience" of the Stoics and Descartes, understanding the economy of creation will make us ever more certain about divine justice and goodness. Grasping the creation leads beyond oneself, beyond the narcissism of many spiritual exercises, toward a love of all things in creation. Understanding the harmonies of the universe impels one to promote order on Earth; the love that accompanies that understanding provokes a constant perfecting charity to fellow human beings, not philosophical retreat.

A genuine weakness remained at the heart of all Leibniz's writings that attempted to convey the orderly nature of the universe. The means for discovering order amid the perplexing and confusing diversity of things were hardly as simple as the hermit maintains. The method of inventing, so extremely powerful in mathematics, never quite came to fruition within experimental natural philosophy, hermeneutics, or statecraft. Leibniz often happily brought forth examples of natural and created order as synecdoches for creation as a whole. To make the ubiquitous orders truly obvious, however, he needed nothing less than the perfected tabular reasoning of his universal characteristic. That he never quite completed.

CODA: DEMONSTRATING THE FORMS OF CHARITY

A blizzard in February 1686 offered Leibniz a break from supervising the construction of his new windmills for the silver mines in the Harz Mountains. Not prone to wasting time, he wrote a short introduction to his metaphysics, *Discourse on Metaphysics*, that focused on the rehabilitation of substantial forms and finality, the complementarity of such explanation with the mechanical philosophy, and the universal law behind all of creation, which ought to regulate human communities.[147] This endeavor returns us to the beginning of this chapter and his response to writers such as Pascal who saw no consolation or guidance in the natural world accessible to human reason.

In a letter to the Catholic nobleman Landgrave Ernst of Hesse-Rheinfeld urging him to examine the *Discourse*, Leibniz stressed that learning true natural philosophy and metaphysics would transform one's intellectual, volitional, and emotional life.[148] Like the Jesuits skilled in conveying the truths of scholastic metaphysics in a nonscholastic style, he aimed to present the "veins of gold" present amid the sterile rocks of the scholastics, who argued only through "disputes, distinctions, and word games." Proper thinking would yield real change: "what makes us more capable of thinking about the most perfect objects, and in a more perfect manner, perfects us naturally." However useful for everyday living, the confused thought of, say, artisans fails utterly to perfect us; in contrast, the distinct knowledge of a Galileo or Archimedes offers "knowledge of reasons [*raisons*] in themselves." Reasons that are "most comprehensive and that have the greatest relationship with the sovereign being can perfect us."[149] This letter offers

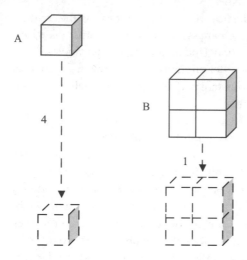

Figure 6.2. Leibniz's 1686 conservation of force argument. After A6,4:1557, 2028.

a good guide to understanding at least in part the strategy and purpose of Leibniz's *Discourse*.

At the heart of the *Discourse* is a straightforward natural-philosophical thought experiment to convince readers that natural philosophy demands something like the substantial forms of the Aristotelians.[150] Having set forth some of his metaphysical doctrines, which he admitted might seem a bit bizarre at first, Leibniz offered in the middle of the *Discourse* a "most simple" demonstration from natural philosophy to make the reader recognize the need for his apparently outlandish metaphysical system.[151] This demonstration stemmed from his discovery of the new conservation principle in 1678 discussed above.

The demonstration began with a point widely accepted among the new philosophers: that some "force" is conserved in nature. But what "force" and how to measure it? Descartes and his followers maintained that God conserves a constant "quantity of motion" in the universe, where quantity of motion is like quantity of matter (mass or bulk) times speed, something like $m|v|$.[152] Other qualities of force were widely consented to, including

1. that the force gained by something in falling a given distance equals the force required to raise that thing to a height of that distance,
2. that the force required to raise an n-pound weight one foot equals the force required to raise a one-pound weight n feet, and so on.

Consider two objects, one made up of one unit of matter, the other of four (see fig. 6.2). By the second supposition, the same "force" will raise the first object to a height of four units and the second to a height of one unit. What happens if they are dropped from those heights?

Galileo had shown that the speed of a falling object is proportional to the square root of the distance fallen: $|v| \propto \sqrt{d}$. So when the two objects fall, the first falls four units and reaches a speed of 2; the second falls one unit and reaches a speed of 1. What quantity is conserved? Quantity of motion, $m|v|$, as the Cartesians maintained? For the first object we get $mv = 1 \cdot 2 = 2$, whereas for the second, $mv = 4 \cdot 1 = 4$. No, the Cartesian claim that quantity of motion is conserved is trivially false. Force must be something more, something more subtle.

What then is the force conserved? Leibniz focused on the first supposition, measuring force using the effort required to raise something. Force "must be estimated by the quantity of effect it can produce, for example, by the height to which a heavy body of a certain size and type can be raised, altogether different from the speed that this situation can give it" (§17, 1558).[153] The world conserves the ability to cause *future* effects, not any particular amount of the *actual* motion of quantities of matter. Leibniz's quantity had a contentious career as an estimation of value, of work, and later of potential energy.[154]

For mechanical philosophers like Descartes and Pascal, only things fully describable in terms of their physical shape and motion made up the world, along with a small number of incorporeal things (such as the minds of human beings and angels). Quantity of motion included only quantities dependent on the physical disposition and motions of objects: their quantities of matter (mass or "bulk") and their speeds. These could be mechanically conceived at every moment. Leibniz's measurement of force rested on something less mechanical, for potential effect is not a quantity that can be measured merely by examining the physical disposition of bodies and their speeds. One had to cause the thing to act, to bring forth its potential, and then one could measure it. Such potentiality literally has no place in a mechanical world.

From the otherworldliness of this force, its existence as something hidden in matter, stemmed its persuasive and philosophical virtue for Leibniz: "everything that is conceived in a body does not uniquely consist in extension and its modifications, as our moderns persuade themselves" (§18, 1559). Leibniz's simple thought experiment, grounded in his real experimentation, revealed reality to be more than mechanical, a world not unlike the scholastic world replete with substantial forms, long decried by philosophers such as Descartes, Hobbes, and Pascal.[155] In both natural and political philosophy in the seventeenth century, the question of the existence of forms and finality was an empirical question about the way the world is, not a silly question about names. The world could have been produced with or without forms and final causes. For scholastics as well as thinkers such as Hugo Grotius, it was metaphysically possible but nevertheless absurd and evidently untrue that the world worked only through efficient causes—simply consider the horror of the vacuum.[156] For Pascal, Descartes, and numerous others, it was absurd to attribute finality to brute matter. The world simply did not include matter that had attributes such as minds, love, or ends.

His rather simple demonstration illustrated, Leibniz claimed, that a merely mechanical world could hardly account for such simple phenomena as falling blocks or swinging pendula: "If the rules of mechanics rested only upon geometry without metaphysics, phenomena would be completely otherwise" (§21, 1563). Acknowledging this proof led to awareness of "something more real," a world more fundamental than the material. The new philosophers were right to ridicule and banish the improper use of forms and explanations involving forms. They were wrong to banish forms completely. The laws of nature simply require them, appropriately used: "we are yet obliged to reestablish some beings or forms that they have banished" (§18, 1559). The rehabilitation of substantial forms permitted the return of arguments of finality, such as least-action principles, like those useful in optics. Descartes, Pascal, Arnauld, and other mechanists had failed to penetrate deeply enough into physics to recognize the evidence for finality in the most basic laws of motion. Recognizing the existence of substantial forms and therefore finality in the world "purges the mechanical philosophy of profanity" (§23, 1566). With his demonstration, Leibniz confronted a dangerous implication of the mechanical philosophy: that God had created a world without evident purpose and ends (§19, 1560). With such proof of divine purpose, Pascal need not have been so terrified of those infinite spaces and infinitely detailed mechanisms after all. The universe was ordered and purposeful—far more so than even the scholastics ever recognized.

In *Discourse on Metaphysics*, Leibniz illustrated his role as a decorous mediator between the scholastic and mechanical philosophies; he showed that each required the benefits of the other.[157] This was no minor point. Elsewhere, he described the significance of reconciling the two forms of explanation: "For by this understanding will the internecine war of philosophers cease, for of late not only schools and academies have been thrown into confusion, but thereupon have the church and commonwealth as well."[158] Philosophical reconciliation was essential for reconciling Catholics, Lutherans, and Calvinists—for healing the rifts of Europe. Like his English contemporaries Robert Boyle, Isaac Newton, and Robert Hooke, Leibniz worked to reestablish experimentally and observationally the existence of principles in the world beyond the merely mechanical. He sought to secure the role for God in the world and to counter the threat that a purely materialist philosophy posed for proper divinity, political order, and moral organization.[159]

Considering the nonmechanical reality evident in physical phenomena would raise the "minds of our philosophers from merely material considerations alone to more noble meditations"—the Platonist commonplace Leibniz often invoked (§23, 1566).[160] As he noted in a draft of an introductory text on natural philosophy: "Physical reasoning about the causes and ends of things is of the greatest use for the perfection of the soul and of devotion to God."[161] Leibniz's simple proof about the nature of the conservation of force belongs among his material procedures for recognizing order amid disparate phenomena. Its inclusion in the *Discourse* shows that it belongs among Leibniz's decorous ways of publicizing knowledge of that natural order, so important to religious and political order, so central for regulating thoughts and emotions to follow the right path toward harmony.

For Stoics and Epicureans, as well as for Descartes and many Jesuits, knowledge of the natural world brought consolation. For Pascal, it brought despair. To resist Pascal's despair required proof that nature involved more than just brute matter lacking any certain trace of divine justice and wisdom. With his reintroduction of substantial form, Leibniz claimed to offer the road to a truer consolation, a reasoned satisfaction with the chiaroscuro of the created world. Not only found in "the mechanical structure of some particular bodies," the wisdom of God "must be demonstrated in the general economy of the world and in the constitution of the laws of nature" (§21, 1563). Understanding such economy demands the means to seek out and find the *rationes* behind the world. To accept something as true need not require knowledge of all its mechanisms, which may very well be infinitely complex, as Pascal had argued. To recognize the overriding economy of all things does not require understanding all the necessary compromises inherent in making the best possible world, but simply grasping the evident existence of such a harmonious economy of choices. "To explain always the admirable economy of this choice—this will not happen while we are but travelers in this world: it is enough to know it, without understanding it" (§30, 1577). Human beings, mere *viatores* in this world, cannot know or understand all of God's choices and his reasons for them. They can know with confidence that these choices all conform to a divine economy of the best choices. In infinity one ought to see not terror but divine bounty, an infinite number of forms, each signaling the divine economy of the best possible combination of choices and of a just universe. "In fact," Leibniz argued, "those who are not satisfied with what he has wrought seem to be similar to discontented subjects whose intention is little different from that of rebels" (§4, 1535). To complain about God's economy of choices was both ignorant and impolitic.

Given knowledge of the underlying order of the world and its future, "one must be neither a quietist nor wait ridiculously with crossed arms." Rather, "one must act according to the presumptive will of God, insofar as we can judge, trying with all our ability to contribute to the general good, and particularly to the ornament and perfection of those things which affect us, or what is next to us." (§4, 1535–36). Understanding the just economy of nature spurs on the charitable action of recognizing natural and social order in the world and working to increase it. More than just knowledge, or the removal of terror, proper knowledge of the comprehensive economy of things, its reasons, moves one to improve the world, to discipline it, to bring it into greater harmony.

In 1671 Leibniz wrote, "Everyone would love everyone if we were to grasp, if we were to lift our eyes to recognize, universal harmony."[162] Getting others to lift their eyes required him to work on techniques to help cognize this harmony as well as techniques to motivate others to attempt to grasp harmony. Throughout the 1670s, Leibniz worked hard at both. Although Pascal was right to encourage his contemporaries to reflect upon nature, he misjudged how nature should affect them. Serious reflection upon the cosmos—a beautiful, just, ornate creation—should impel everyone to strive to make the small place of human beings ever more beautiful, just, and ornate.

Epilogue

Well into his *Critique of the Power of Judgment* (1790), Immanuel Kant confronted the dangers the sciences posed for morals: "There is no denying the preponderance of the evil showered upon us by the refinement of taste to the point of its idealization, and even by indulgence in the sciences as nourishment for vanity, because of the insatiable host of inclinations that are thereby aroused." Much as Jean-Jacques Rousseau had complained about the products of human civilization, his diligent reader Kant argued that the sciences, replete with subtle problems and tricky solutions, could easily lead one away from the good life. And yet, Kant maintained, the sciences were essential for collectively learning to live well: "Beautiful arts and sciences, which by means of a universally communicable pleasure and an elegance and refinement make human beings, if not morally better, at least better mannered for society, very much reduce the tyranny of sensible tendencies, and prepare humans for a sovereignty in which reason alone shall have power."[1] Although they failed to make human beings moral, the sciences (and arts) disciplined and civilized. Rather than revealing human beings to be naturally inclined to society, the sciences helped make human beings into sociable beings, capable of ordering themselves and their society. The sciences helped human beings to become autonomous, subject only to an ethics grounded in reason itself, not to the tyrannical dictates of natural desires or external coercion.

For Kant, the early-modern revolution in modern science rested upon escape from the tyrannical direction of nature. In the second edition of his *Critique of Pure Reason,* he explained that this revolution occurred when people ceased approaching nature "in the capacity of a pupil who lets the teacher tell him whatever the teacher wants" and chose instead to approach nature "in the capacity of an appointed judge who compels the witnesses to answer the questions that he puts to them." So long as science rested on merely observing what nature offered, anyone who attempted to know nature could only grope about blindly. Figures such as Galileo Galilei, Evangelista Torricelli, and Georg Stahl understood, in contrast, "that reason has insight only into what it itself produces according to its own plan; and that reason must not allow nature by itself to keep it in leading strings as it were, but reason must . . . proceed according to constant laws and compel nature to answer reason's own questions."[2]

Science did not emerge because human beings learned to become passive mirrors, capable of mimicking nature with no human artifice and without the use of subjective qualities. Science had been obstructed so long as human beings refused to impose their demands and rational schemes upon it. The *absence* of human intervention, not its presence, precluded science. Science could discover true knowledge precisely because it rested on human artifice and disciplined subjective capacities. Hobbes was right: "It is as *Hobbes* maintains," Kant remarked, "the state of nature is a state of injustice and violence, and one must necessarily abandon it and subject oneself to the constraint of law." Only human law brought knowledge and peace.[3] Genuinely human action and thought came in breaking away from the guidance of nature. Science and ethics alike rested on a civilizing process that permitted human beings to detach themselves from the deceptive mimicry of nature and allowed them to regulate themselves and their societies with the artificial products of human reason.

Kant captured something central to the transformations of natural knowledge in the seventeenth century. He mistook what early-modern western Europeans imposed upon nature in order to know and control it. Seventeenth-century natural philosophers and mathematicians in France and Germany did critique and then apply a disciplined reason to nature and mathematics. Descartes, Pascal, and Leibniz did not merely impose a disembodied pure reason onto nature. Drawing on extant resources for civilizing and educating, they attempted to apply the full range of human epistemic, physical, and social faculties to know nature. They imposed disciplined human capacities, with what they understood to be real human powers and limitations: powers of reasoning and intuiting, inclinations and problematic sensory powers, linguistic competencies, rhetorical techniques and physical skills, forms of organization and means for producing legitimate assent. They sought to discern human subjective capacities and incapacities, not to disregard them but rather to perfect them insofar as possible. Descartes' mathematical exercises, Pascal's arithmetical triangle, and Leibniz's instructive displays and notational innovations all testified that perfecting human ability would require disciplined work on oneself and on others. Such labor was a precondition for better knowledge and the good life.

My argument is not that these seventeenth-century figures came to understand human capacities correctly and, therefore, could produce real science by refining those capacities. Rather, in so trying to understand and perfect the faculties, they developed sundry innovations useful in scientific and mathematical work. Within their dour Christian conceptions of the fallen qualities of humankind, they worked to produce and to distribute intellectual, material, and social techniques appropriate to their wrongly universalized understanding of human greatness and wretchedness. They brought these techniques to the study of the natural world and mathematics. Instead of thinking humans divine, capable of knowing and changing the world easily, they sought knowledge by humiliating humanity, by critiquing and bracketing human ability. They sought knowledge of the world in knowing themselves. Fulfilling Socrates' old injunction required no small mathematical and natural-philosophical labor.

Since antiquity, the desire for self-cultivation had offered powerful incentives to know oneself. In an early-modern Europe wracked with dissent and war, this desire spurred the search for new means to know nature. The real difficulties of knowing nature in turn dramatically illustrated the strengths and weaknesses of early-modern Europeans for coming to know the natural world and for creating consensus about it. Self-cultivation involving natural and mathematical work fueled no small part of the dynamism in natural knowledge making in the seventeenth century.

How natural philosophy and mathematics later lost their perceived power to cultivate the moral person is a story yet to be told. Yearning for a new naturalization of ethics, sundry recent policy makers, philosophers, and scientists would make ethics once again follow nature. The stark separation of the moral from the natural nonetheless largely continues to structure our considered ethical life; we rightly worry about the proper ethics to guide science. It should please us that we must work hard to imagine a historical period when ethics needed science, rather than science needing ethics. We should not let the need for such imagination slip away.

Notes

INTRODUCTION

1. Written by the moralist Saint-Evremond, "Jugement sur les sciences où peut s'appliquer un honneste homme" was originally printed anonymously in Boileau 1666 (quotations on pp. 26, 28, 29); see Saint-Evremond 1965, pp. 6, 11, 12. For a fine discussion of this essay, see Bensoussan 2000, esp. pp. 183–85.

2. Pierre Nicole, preface to Arnauld 1781, p. 5.

3. See now Smith 2004 for a survey of artisanal knowledge in the Scientific Revolution and its historical effacement. See citations in chapter 1 for different sorts of mathematical practitioners.

4. While I present a claim about some important longer-term continuities, I do not offer a new master narrative for the "Scientific Revolution," nor do I attempt to resuscitate that periodization. Many of the developments in natural and mathematical knowledge in the seventeenth century had no connection to the technical considerations or the concerns about self-cultivation central to the subjects of this book.

5. This study surveys neither seventeenth-century philosophies understood as spiritual exercises nor such spiritual exercises, nor does it extensively document the major ethical systems prominent in early-modern France and Germany.

6. Seneca to Lucilius, 108.36; Seneca 1965, vol. 2, p. 109.

7. Most recent scholarship, following Oskar Kristeller, does not ascribe a robust common philosophical program or outlook to humanists. See Rummel 1995, pp. 30–34. For the Christian background, see Trinkaus 1970, vol. 1, pp. 46–50.

8. For entry into the literature, see Rummel 1995, esp. pp. 182–89. For a recent introduction to the relationship of philosophy with philology, see Kraye 1996. Recent scholarship has stressed the interplay of scholastic philosophy and Renaissance humanism in both the fifteenth and the sixteenth century. For ethics, see Kraye 1988 and now Lines 2002.

9. For skepticism about humanist education in practice, see Grafton and Jardine 1986.

10. Scholarship in the Renaissance clearly pursued both an ethical path and one of technical or "scientific" classical scholarship, such as philology and numismatics, which

often had little or no ethical drive. For this split, see Grafton 1991b, pp. 25–26. For the continuing unity of scholarship and virtuous living for key late humanists, see Miller 2000.

11. See now Moreau 1999, 2001; Miller and Inwood 2003.

12. Cottingham 1998, p. 69.

13. Seneca to Lucilius, 16.3; Seneca 1965, vol. 1, p. 42; trans. modified from Seneca 1969, pp. 63–64.

14. Hadot 1998, pp. 211–12. Whether or not Hadot's formulations are adequate for all or any ancient philosophies need not detain us here, as the elements he stresses were certainly among those that most keenly interested early moderns. See also Hadot 1981, 1995, 1996, 2002; Voelke 1993; Foucault 1984, esp. pp. 51–85; Foucault 2001; Davidson 1994; Hadot, Carlier, and Davidson 2001.

15. The dominant forms of "spiritual exercises" in early-modern Europe were religious and often mystical forms of exercise, prayer, and meditation, all of which fall outside the scope of this study.

16. Levi 1964, p. 8.

17. Bury 1996b, p. 205.

18. Pierre Gassendi to Louis de Valois, 6.2.1642, in Gassendi [1658] 1964, vol. 6, p. 131; cf. Gassendi 2004, vol. 1, p. 239.

19. For intimations on early-modern philosophy as spiritual exercises or arts of living, see variously Spink 1960, esp. ch. 8; Cottingham 1998, ch. 3; James 1997, 1998. For women philosophers later in the century, see Conley 2002. For a major study of German models for self-cultivation, see Hunter 2001. For the uneasy coexistence of the ideals of pagan and Christian antiquities in the period, see Zuber 1981. For the connection between a life of scholarship and the arts of living, see Miller 2000, pp. 14–15 and passim. Sarasohn (1996) has emphasized how Gassendi's physical thought figured within his Epicurean ethic of self-cultivation; for philosophy as an exercise in Gassendi, see Taussig 2003, esp. pp. 220–22. For seventeenth-century Stoicism, see Lagrée 2004, esp. pp. 154–58 on exercises and self-cultivation. Cf., for Germany, Smith 1994, pp. 41–44; Hotson 2000. For England, Gaukroger 2001; Solomon 1998, pp. 37–43; Tully 1993. For the humanist tradition of history reading as moral instruction, and the collapse of this tradition in the late Renaissance, see Hampton 1990. For differing accounts of the pedagogic role of logical treatises, see Risse 1964, vol. 1, ch. 6.

20. See, e.g., Sorel 1671, mostly on literary examples.

21. In the seventeenth century, the term "spiritual exercises" usually referred to religious spiritual exercises; I apply it to a wider range of texts and philosophies. In calling something a spiritual exercise, I mean the following:

1. It comprises a set of practices, often including logic and mathematics, intended ultimately to lead one's self or soul toward some goal of self-cultivation. These practices necessarily involve the development of various faculties, including *ingenium, ingenio, esprit,* the wit, memory, and so forth.
2. In its more philosophical guises it might include some or all of the following:
    2.1. a metaphysical account of the world,
    2.2. a natural-philosophical account of the world,
    2.3. an account of the faculties to be improved,

2.4. the appropriateness or inappropriateness of those faculties for gaining access to knowledge of the fundamental truths of the world, facts about nature, and ethical truths, and

2.5. the ethical code to be followed (such a code may or may not be naturalistic, often depending on 2.1 and 2.2)

3. a specification of the social field or group at which it is aimed.

Ignatius himself stressed the plasticity of the term: "The term 'spiritual exercises' denotes every way of examining one's conscience, of meditating, contemplating, praying vocally and mentally, and other spiritual activities.... For just as strolling, walking and running are exercises for the body, so 'spiritual exercises' is the name given to every way of preparing and disposing one's soul to rid herself of all disordered attachments" (§1, Gueydan 1986, p. 65, trans. in Ignatius 1996, p. 283).

22. For example, in 1634, the famous correspondent Marin Mersenne weighed the respective values of mathematics versus natural philosophy as forms of self-cultivation. See Mersenne [1634] 1972a, esp. pp. 74–79. Compare Gassendi's views on the proper role of science, discussed in Taussig 2003, pp. 132–33.

23. For mathematics as *paideia* in the earlier Renaissance, see Rose 1975, ch. 1. For early-seventeenth-century accounts of the utility of mathematics, see Jungius 1929; Clavius [1611] 1999, p. 6.

24. Adam 1986, p. 195.

25. See especially "L'homme, qui veut connoistre toutes choses ne se connoist pas luy-mesme," in Saint-Evremond 1965, pp. 116–33; see also Bensoussan 2000.

26. Complementing recent work showing the importance of Renaissance scholarship for invigorating the new sciences, this book underscores the ethical, as well as the philological, significance of Renaissance humanism for the development of some central examples of early-modern scientific practice. See Rose 1975; Grafton 1991a, 1996, 2001, ch. 5; see also Blair 1992, 1997.

27. For "popularization" and its dangers, see the sources cited in chapter 4, n. 8.

28. See Hacking 2002a, 2002b; Davidson 2001; Daston 1994. Both Hacking and Davidson have been much inspired by Foucault 1971, see especially the English preface. In contrast with those projects, my reconstructions all deal with explicitly articulated accounts of truth and falsity, not the underlying "positivity" or "style of reasoning" that permits some concepts and excludes others. Nevertheless, I am interested in getting at the historical standards and forms of reasoning used by my subjects, without attempting to reduce them to some other standards of reasoning such as propositional logic, modern real analysis, or forms of scientific experimentation.

29. The study of theoretical practice in the modern exact sciences has recently focused on exercise and training within distinct pedagogical traditions. See especially Warwick 2003; Kaiser 2005.

30. For an important reminder about the dangers of collapsing the accounts of scientific actors of their method with their actual scientific practice, see especially Schuster 1984, 1993.

31. For postlapsarian anthropology and early-modern knowledge, see Harrison 2002. For the importance of theological considerations about human cognitive capacities to the

sciences during the seventeenth century, see, e.g., Funkenstein 1986; Osler 1994; Wojcik 1997; see also Force and Popkin 1990.

32. The transition from the late Middle Ages to the Renaissance and the seventeenth century has often been characterized as centrally including a new certainty about human power to know and to transform the world. The term "humanism" is often loosely used to mean precisely such a new belief. To take just the study most familiar to historians of science, in her great and highly problematic work on Giordano Bruno, Frances Yates (1964) set forth Marsilio Ficino, Pico della Mirandola, and Agrippa von Nettesheim as great renewers of confidence in human ability. Several generations of scholarship on the Renaissance have replaced this beguiling picture of a teleological and mostly secular "Renaissance philosophy of man" with a panorama of diverse Renaissance, Reformation, and Catholic Reformation understandings of the cognitive, affective, and spiritual capacities of humanity. Crucial here is Trinkaus 1970. For France, see, for indications, Gouhier 1987; Levi 1963, 1964; Hatfield 1998; James 1998; Keohane 1980.

33. Recent history of science has emphatically emphasized, however, the intense labor of "Aristotelian" experimentalists. See Heilbron 1982; Dear 1995a.

34. In addition to the classic study of the self, Taylor 1989, esp. chs. 8–10, see Seigel 2005, chs. 1–3.

35. Among the many studies, see Shapin 1994; Moran 1991a, 1991b; Daston 1995; Biagioli 1996; Licoppe 1996; Findlen 1994; Smith 1994; Schuster and Taylor 1997. For courtly and noncourtly models among Jesuits, see Gorman 2002. For conversation, see the literature cited in chapter 3. For legal culture and doubts about the importance of gentlemanly culture, see B. Shapiro 2000, ch. 6.

36. Cf. Shapin 1994, esp. at pp. 164–65; Shapin 2000; Dear 1998.

CHAPTER ONE

1. Cicero 1971, 3.3.5, 6, pp. 228–29, 230–31.

2. For the theme of philosophy as a therapeutic for the soul in Hellenistic philosophy, see Voelke 1993.

3. Charron [1601/4] 1986, p. 369.

4. See the 1591 *Ratio studiorum,* "Regulae professoris humanitatis," Lukàcs 1965–, vol. 5, p. 303.

5. See the *"circinus ad angulum in quotlibet partes dividendum"* and other machines at AT X:240–42; and the claims for finding "medians" at AT X:229. For the early development of Descartes' mathematics, see Bos 2001, pp. 231–53; Sasaki 2003, pp. 109–32.

6. Cf. CSM I:35; see Sepper 1996, p. 135, and his comments on the expansion of *ingenium*'s power through exercise, at p. 140.

7. My analysis of the geometry rests on a number of specialized studies, above all the work of Henk J. M. Bos.

8. For metamathematical concerns in the seventeenth century, see Schüling 1969; Mancosu 1996.

9. Recent studies of seventeenth- and eighteenth-century mathematics and physics have carefully detailed the specific mathematics of different thinkers. For examples, see Mahoney 1993; Bertoloni Meli 1993; Blay 1992; Goldstein 2001; Jesseph 1999; Stedall 2002; among others.

10. See, e.g., Biagioli 1989, 1993; Johnston 1991; Westman 1980; Bennett 1986; Dear 1995a; Mahoney 1994, pp. 1–14, 20–25; Schneider 1997; Gorman 2002.

11. See the more formal definition in the introduction along with the citations to Hadot, Foucault, and Davidson.

12. I know of one major exception: "To imitate Descartes' example," David Lachtermann rightly notes, "one will need to practice and apply it, not memorize or passively receive it" (1989, p. 134). In a penetrating study of Descartes' *Meditationes*, Foucault insisted that recognizing the intelligibility of Descartes' choices rested on a "double reading" of his text as both a *system* and an *exercise* (1998, p. 406). Compare the considerably more historically grounded Hatfield 1986; for a negative assessment of Hatfield, see Rubidge 1990; and for positive ripostes, Dear 1995b; Sepper 2000.

13. Descartes made this claim even though Jesuit mathematicians such as Christoph Clavius had echoed Plato's claims about the cultivating effects of mathematics in arguing for the place of mathematics in Jesuit pedagogy. Following Plato and his interpreter Proclus, Clavius stressed that mathematics can help the mind ascend from the material to the eternal. For Plato, see *Republic*, 7.522c–527c; Clavius [1611] 1999, p. 6; Crombie 1977; Sasaki 2003, p. 59. For Jesuit mathematics pedagogy in France and at La Flèche, see the traditional source, Rochemonteix 1889, and now Romano 1999; Rodis-Lewis 1987; Sasaki 2003, pp. 13–30, 45–49, 84–93.

14. See Guez de Balzac 1995b, pp. 82–83. In defending Guez de Balzac's rhetoric, Descartes offered a historical account of the loss of true rhetoric and the production of rules and *sophismata* to replace it. See AT I:9 and the discussion in chapters 2 and 3.

15. See Bury 1996b.

16. On *honnêteté* and Descartes' work in natural philosophy, see Dear 1998, esp. pp. 62–63.

17. For more on *honnêteté*, see chapters 3 and 4; Zuber 1993; Bury 1996a; Magendie 1925; Keohane 1980, pp. 283–88.

18. AT X:376; CSM I:18; cf. *Discours*, AT VI:7–8; Marion 1975, p. 151.

19. Cicero, *De inventione*, 2.51.156, in Cicero 1949, p. 324; cf. 2.55.166, p. 332, and pseudo-Cicero 1954, 3.2.3, pp. 160–62. See also Quintilian 1920–22, 3.8.22, vol. 1, p. 490.

20. See Guez de Balzac to Boisrobert, 28.9.1623, quoted in Beugnot 1999, p. 542. For Descartes' views, see chapter 2.

21. See Zuber 1997.

22. AT III:333 (my italics), as reported by Schooten to Christiaan Huygens. See also Milhaud 1921, p. 160.

23. Compare the similar attack on Descartes's friend Jean-Louis Guez de Balzac as a "Gascon"; discussed in Jehasse 1977, p. 117.

24. Descartes to Mersenne, 12 or 1.1637–38(?), AT I:490.

25. See Bos 1984, p. 363. See also Gaukroger 1992, pp. 106–8.

26. Nicely noted in Kline 1972, p. 308.

27. For example, Girolamo Cardano, Lodovico Ferrari, and Niccolò Tartaglia participated in such mathematical duels. See Ore 1953, pp. 53–107; Mahoney 1994, pp. 6–7; the review by Keller 1976.

28. For a thorough introduction to early-modern problem solving, see Bos 2001, ch. 4; Bos 1990, pp. 352–56.

29. Bos 1981, 1990; see now the more detailed account in Bos 2001, pp. 293–301. Descartes defined multiplication as follows. Let AB be unity. If one wants to multiply BD by BC, then one only needs to join C to A and draw the parallel from D to E. Then BE is the desired product. Since $AB:BD::BC:BE$ and AB is unity, $AB \cdot BE = 1 \cdot BE = BC \cdot BD$. Similarly, he defined the square root geometrically.

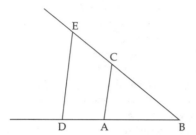

Multiplication. After René Descartes, *Discours de la méthode pour bien conduire sa raison* ... (Leiden, 1637), p. 298.

30. Not all commentators agree on this point, in large part because Descartes was well aware that his algebra could produce nongeometric solutions; at times he accepted and developed these nongeometric solutions. Nevertheless, his philosophical account of mathematics excluded such solutions in principle if not in practice, and many of the most important and odd features of his mathematics stem from his demands about constructability. For important accounts focused on algebra, see Schouls 2000, ch. 5; Sasaki 2003.

31. The Pappus problem: Let there be n lines L_i, n angles φ_i and a segment a, and a proportion α/β. From a point P, draw lines d_i meeting each L_i with angle φ_i, find the locus of points P such that the distances of the lines d_i maintain a set of proportions: for $n > 2$, $2n - 1$ lines, $(d_1, \ldots, d_n) : (d_{n+1}, \ldots, d_{2n-1}a) :: \alpha : \beta$, and for $2n$ lines, $(d_1, \ldots, d_n) : (d_{n+1}, \ldots, d_{2n}) :: \alpha : \beta$. I take this description from Bos 1981, p. 299; Bos 2001, p. 314.

32. The Pappus problem itself acted as a sort of machine that produced an extended family of smaller problems, each with its family of orderly solution curves. On this, see especially Grosholz 1991, ch. 2.

33. See also Gaukroger 1995, pp. 153–54.

34. This latter point appears most prominently in the fourth meditation. See Garber 2001, esp. pp. 283–88, for a fine study of Descartes' account of the cultivation of the intellect; see also Dear 1995b.

35. In *L'homme*, Descartes explained how focusing the attention on one sense often hinders the reception of the impressions from another. See AT XI:185–86.

36. Trans. in Sepper 1996, pp. 76–77.

37. Descartes was no kinder about commonplaces in his later writings. See his bitter denunciation of the use of commonplaces and his call for the repeated rereading of entire good works in "Epistola ad G. Voetium," AT VIII/2:40–41. For humanist commonplace methodology, see, e.g., Blair 1992; Blair 2004, esp. p. 427, on Descartes.

38. For the concept of monsters as beings lacking some fundamental unity, see my discussion in chapter 4 and the literature cited there.

39. Descartes to Hogelande, 8.2.1640, AT III:722–23. Cf. *Regulae*, AT X:367.

40. Robinet 1996, pp. 191–96.

41. Cf. CSM 1:36, and see Marion's annotations in Descartes 1977, pp. 217–18.

42. For this notebook and the vicissitudes of its transmission, see Gouhier 1958, pp. 11–18; Rodis-Lewis 1991.

43. See AT X:230, discussed in Sepper 1996, pp. 76–77. See also AT X:94, 204, and the discussion in Sepper 1996, pp. 44–46.

44. Trans. in Sepper 1996, p. 77.

45. See Bos 2001, pp. 239–45. I skip completely over the difficult question of Descartes' encounters with mathematicians while in Germany. For a persuasive recent view, based on careful analysis of algebraic procedures, see Manders 1995. See also I. Schneider 1997; and for the wider context, see now Mehl 2001.

46. The equation $b = a^3$ encodes both a curve and the progression $1:a::a:a^2::a^2:a^3$. This algebraic progression encapsulates the constructive process of his compass. See Lachtermann 1989, pp. 165–66. Timothy Lenoir argues that, for Descartes, algebra "served as a device for the easy storage and quick retrieval of information regarding geometrical constructions" (1979, p. 363).

47. The best account of deduction in Descartes' work is Recker 1993; see also Van De Pitte 1988a; Gaukroger 1989; Belaval 1960; Clarke 1991.

48. Most famously in Leibniz, "Meditationes de cognitione, veritate et ideis" 1684, A6,4:590–91. See above pp. 219–21.

49. Descartes' ambivalence toward algebra has divided commentators, with some seeing the algebra as heralding modernity and algebraic liberation, and others stressing how tied the algebra remains to the geometry. For the former, see Schouls 2000; Sasaki 2003; and to some extent Mahoney 1980. For the latter, see Lenoir 1979; Grosholz 1991. For Descartes' refusal to pursue the potential for formal reasoning latent in his algebra, see Gaukroger 1989, pp. 72–88.

50. For the centrality of habit in *Passions de l'âme*, see §50, AT XI:368–70, and the more detailed discussion below.

51. *Dictionaire de l'Académie française*, 1st ed. (1694), p. 551, s.v. "Habitude." Cf. Dupleix 1984, p. 137; Paulo 1640, *Ethica*, p. 28; Du Moulin 1638, vol. 2, pp. 72–73.

52. See *Nicomachean Ethics*, 1105a27–1105b4.

53. See, e.g., AT X:232–33, on a machine for making conic sections.

54. For the demand of sufficiency, see AT X:384–87; Serfati 1993, pp. 213–14.

55. I owe this simile to Michael Gordin.

56. This claim does not mean that Descartes always effaced them in practice; he did not.

57. For another way to define simplicity, see Newton's alternative standards for *geometric* simplicity: Newton 1967–81, vol. 4, p. 345.

58. For detailed study of the parabola, see Bos 1992.

59. AT X:438–39, on the movement from talk of "knowns" and "unknowns" to mechanical explanations.

60. Dupleix 1994, pp. 174–75.

61. Cf. CSM I:18. The quotations from this section draw on the similarities between Rule 4A and 4B and do not highlight their differences. For the debates about the dating of the parts of the *Regulae*, see, e.g., Schuster 1977; Van De Pitte 1991.

62. Cf. Gaukroger 1992, pp. 102–4.

63. See, for example, the defense of the status of mechanics in Laird 2000 and the citations above on the social and epistemic status of mathematical practitioners. For the elevation of painting to an intellectual art in early-modern France, see Heinich 1993, esp. chs. 4–5.

64. Fréart de Chambray, *Idée de la perfection de la peinture* (1662), quoted in Heinich 1993, pp. 148–49.

65. Gaukroger 1989, pp. 72–88.

66. AT VI:389–90. Molland (1976, esp. pp. 35–37) shows how Descartes misrepresented the ancients better to make his case.

67. AT VI:389–90 (my italics). Henk Bos has nicely outlined the twists of Descartes' account, with its multiple, apparently disparate accounts of acceptable curves. I largely follow Bos's account here (Bos 2001, esp. ch. 24; see also Bos 1981, 1990, 1992).

68. Noted as a "parallel" in Bos 1981, p. 310.

69. Clavius considered the quadratrix at some length. See Bos 2001, pp. 160–65; Mancosu 1992, pp. 93–95; Sasaki 2003, pp. 47–48, 69–72.

70. See Molland 1976, pp. 26, 36; Bos 2001, p. 342; Bos 1981, p. 314; Hofmann 1974, pp. 101–3.

71. See Bos 1981, pp. 313–15; Bos 2001, pp. 341–46.

72. Elisabeth to Descartes, 8.1645, AT IV:280.

73. See Bos 2003; Bos argues that the exchange helped to sharpen and develop Descartes' account of the relationship between algebra and geometry (pp. 210–11).

74. The problem proved far more difficult than he had foreseen initially; Elisabeth's acuity in attempting to solve the problem convinced him of her mathematical abilities. See Descartes to Pollot, 21.10.43, AT IV:26–27; Bos 2003, pp. 206, 210.

75. Descartes to Elisabeth, 11.1643, AT IV:38.

76. Descartes' answer is deceptively straightforward. Let x be the radius of the circle to be found. Let $AD = a + x$, $BD = b + x$, $CD = c + x$, and $AE = d$, $BE = e$, $CE = f$. Introduce two new variables y and z such that $DF = GE = y$, $DG = FE = z$, $AF = d - z$, $FD = y$; since $AD^2 = AF^2 + FD^2$, then $(a + x)^2 = (d - z)^2 + y^2$; likewise, $(b + x)^2 = z^2 + (e - y)^2$, $(c + x)^2 = (f + z)^2 + y^2$. Given these three equations, one can eliminate y and z, giving an equation in x and x^2. Descartes performed only a step or two more and thus did not run into the great computational complexity involved in actually solving the problem algebraically. See Bos 2003, p. 205.

77. Descartes to Elisabeth, 11.1643, AT IV:42.

78. In his next letter to Elisabeth, he commented that he usually worked problems out only until he could tell what sort of construction—ruler and compass or some other constructing machine—would be necessary to solve them, without actually performing the construction itself. Producing the construction, "in hiding the Algebraic procedure, is only an amusement for little Geometers," he said; "it does not require much of the mind or of knowledge [*d'esprit ny de science*]" (Descartes to Elisabeth, 11.1643, AT IV:46–47).

79. In a letter to Elisabeth in 1643, Descartes elaborated various exercises necessary to understand souls, bodies, and their relation: "metaphysical thoughts, which exercise

the pure understanding, serve to make the notion of the soul familiar to us; and the study of mathematics, which exercises principally the imagination in the consideration of figures and movements, accustoms us to form quite distinct notions of bodies; and finally, it is only in making use of living and of ordinary conversation, and in abstaining from meditating and studying the things that exercise the imagination, that one learns to conceive the union of soul and body" (Descartes to Elisabeth, 28.6.1643, AT III:692).

80. For a helpful overview of Descartes' ethics, its relations to his natural philosophy, and its distance from Stoic ethics, see Cottingham 1998, ch. 3. Much recent Anglophone scholarship has substantially improved the understanding of his ethics in the English-reading world. See especially Marshall 1998; James 1997; Williston 2003, esp. pp. 301–15, along with the other essays in that volume. Levi 1964 remains essential.

81. Preface to *Principes*, AT IX/2:13–14.

82. Preface to *Principes*, AT IX/2:14.

83. Descartes to Elisabeth, 1.9.1645, AT IV:283–84.

84. *Passions de l'âme*, §73, AT XI:383. See the helpful discussion in Williston 2003, pp. 311–12.

85. *Passions de l'âme*, §76, AT XI:385.

86. Descartes to Elisabeth, 15.9.1645, AT IV:291.

87. Descartes to Elisabeth, 15.9.1645, AT IV:296.

88. Descartes to Elisabeth, 15.9.1645, AT IV:296.

89. See the remarks in Cottingham 1996, pp. 73–74.

90. Descartes to Christina, 20.11.1647, AT V:84.

91. Because Descartes stressed happiness so centrally as a result of the good life, many commentators, most importantly Martial Gueroult, have judged his ethics to be in the end little more than a variety of hedonism. Recent commentators have stressed that Descartes argued that pleasure is not the goal of the good life; nevertheless, pleasure accompanies it. See Marshall 1998, pp. 63–70, 88–90; Williston 2003, pp. 305–6; cf. Gueroult 1953, vol. 2, chs. 19–20.

92. Descartes to Christina, 20.11.1647, AT V:85.

93. In addition to the satisfaction supervening upon our self-knowledge of our habitual training and acting, we also become happy in controlling our desires: "In making us know the condition of our nature," the correct use of reason "so limits our desires that it is necessary to acknowledge that the greatest happiness of man rests on this proper use, and in consequence that the study that results in acquiring it is the most useful occupation that one can have, just as it is without doubt the most pleasant and sweetest." Descartes appears to have believed that this satisfaction is qualitatively greater than any pleasures to be desired. See Descartes to Elisabeth, 4.8.1645, AT IV:267.

94. He noted that his remedies could eliminate only the suffering of the soul, not of the body. See Descartes to Elisabeth, 5.1646, AT IV:411.

95. I take this point from James 1997, p. 107.

96. In this short discussion, I skip over the different sorts of intellectual and corporeal love that Descartes considered.

97. *Meditationes*, IV.

98. For coming to self-knowledge of the physical limits of human beings, see, e.g., *La description du corps humain*, AT XI:223–24, 227.

99. *Passions de l'âme*, §161, AT XI:453–54. For "generosity," see Cottingham 1998, pp. 100–101; Marshall 1998, ch. 9, esp. pp. 150–52; L. Shapiro 1999; Williston 2003, pp. 302–3.

100. *Passions de l'âme*, §153, AT XI:446.

101. What meditations might produce this virtue? Descartes counseled "occupying oneself" with thoughts of the "advantages that come to one" with a firm resolution and also with thoughts of "how vain and useless are all the cares that the ambitious fix upon." With these thoughts, "one can excite in oneself the Passion, and thence acquire the virtue of Generosity." Once made habitual, this virtue offers "a general remedy against all the rulings of the Passions," for it establishes firmly the suzerainty of the reason in judging our possible choice of actions. See *Passions de l'âme*, §161, AT XI:453–54.

102. *Passions de l'âme*, §156, AT XI:447–48.

103. Descartes to Elisabeth, 1.1646, AT IV:357.

104. Compare Hume's inclusion of the sociable virtues in Hume 1998, §8.

105. Descartes to Elisabeth, 15.9.1645, AT IV:293–94.

106. Descartes to Elisabeth, 1.1646, AT IV:355.

107. Descartes to Elisabeth, 15.9.1645, AT IV:292.

108. *Principes*, III, 2, AT IX/2:103–4. The second recommendation does not appear in the Latin.

109. See Des Chene 1996, pp. 391–98, esp. pp. 391–92 n. 1, for Descartes' other, more central reasons for banning such teleological thinking.

110. Descartes to Elisabeth, 15.9.1645, AT IV:292. Cf. AT IX/2:103–4; AT V:53–55, etc.

111. Descartes to Chanut, 6.6.1647, AT V:56.

112. Rousseau 1959, pp. 238, 232.

113. Lamy 1966, pp. 67–68.

114. Lamy 1966, p. 67.

115. Lamy 1966, p. 104. His textbook itself expands these points; see Lamy 1684, esp. fols. [à2r–v], [à5r–6r].

116. See Karpinski and Kokomoor 1928, p. 27; Kokomoor 1928, pp. 86, 94–97.

117. Nicole, preface to Arnauld 1781, p. 7.

118. Nicole, preface to Arnauld 1781, p. 7. Kokomoor underlines the influence of Arnauld's geometry text; see Kokomoor 1928, p. 86; Karpinski and Kokomoor 1928, p. 22.

119. Nicole, preface to Arnauld 1781, p. 9.

120. Nicole, preface to Arnauld 1781, p. 10.

121. Although Malebranche's *Recherche de la vérité* is rarely seen nowadays as a logic, contemporaries grouped it along with the work of Arnauld, Nicole, and Locke. See Schuurman 2001.

122. Malebranche 1991, p. 401.

123. Malebranche 1991, pp. 403, 401, 402.

124. Brockliss 1987, p. 382. Earlier Jesuit mathematics texts referred to Plato but with strikingly Cartesian-sounding terminology, such as "clarity" and "evidence." See Tacquet 1651, fol. a4r–v; the great Jesuit commentary on Descartes' geometry, Rabuel 1730, does not mention cultivating the intellect.

125. Descartes 1657, fol. ã3v. For the editorial work, see Van Damme 2002.

126. Descartes 1657, fol. ã3v.

127. For the mathematical reception history, see Bos 2001, pp. 416–22; Sasaki 2003, pp. 274–80.

128. On the reception history of Descartes, see now Van Damme 2002. For intended audiences, see Cavaillé 1994. For reception of *Discourse,* see Garber 1988; Méchoulan 1988. For "Cartesian women," see Harth 1992; O'Neill 1999.

129. Descartes to Mersenne, [12.1637?], AT I:478.

130. See especially Ribe 1997; Sutton 1995; Dear 1998; Shapin 2000.

131. *Passions de l'âme,* §153, AT XI:453–54.

CHAPTER TWO

1. Casaubon 1654, pp. 129–31 (italics in original). See also Heyd 1995, chs. 3–4; Heyd 1990.

2. Armogathe, Carraud, and Fernstra 1988, p. 126; trans. in Carr 1990, p. 1.

3. Descartes was not alone in declaring mathematics to be evident and certain, but calling a discipline or practice "evident" could be faint praise. One of the major critics of the scientific nature of mathematics, Benito Pereiro, for example, held mathematics to be evident precisely because it was close to quantities in the material world: "Mathematical demonstrations are the most certain, the most evident, because of their subject matter, i.e., quantity, because quantity is the most sensed, since it is reached by all the senses, and is the medium or principle of mathematical demonstrations" (1576, pp. 73–74; trans. and quoted in Homann 1983, p. 239; see also Crombie 1977, p. 68). While late-Renaissance debates about the certainty and scientific quality of the mathematical disciplines have been well treated in recent years, to my knowledge the "evidence" of mathematics and other sciences has not been systematically studied. "Evidence" does not appear to figure centrally in Jesuit defenses of mathematics by Clavius and others. For the debate over the certainty of mathematics, see Jardine 1988; Romano 1999, ch. 3, esp. pp. 153–62, with citations to the large literature.

4. For the history of "evidence" in early-modern philosophy, see Halbfass 1972; Mazzantini 1967. In Gilson's index, e.g., *"evidentia"* does not have its own entry in the main listing; looking it up in the auxiliary index leads one to *"notitia"* (Gilson 1963, nos. 312–13).

5. See Gaukroger 1997.

6. For Boyle, see Shapin 1994; Shapin and Schaffer 1985. For the more nuanced empiricism of Boyle, see Sargent 1995.

7. Two prominent examples are Ginzburg 1999 and Vickers 1988. The reemphasis on Renaissance rhetoric and its importance during the early-modern period has been proceeding for some time now. A smattering of examples might include Fumaroli 1994a, 1999; Skinner 1996; Eden 1997; Shuger 1988; Cave 1979; Goyet 1996.

8. *Olympiques,* AT X:184. For citations to the large literature on Descartes' dreams and their relations to Renaissance culture, see Vasoli 1999. I omit the large literature on Descartes and the Rosicrucians.

9. Manuscript fragment, cited by Poisson, in AT X:255, trans. modified from Rodis-Lewis 1998, p. 237 n. 35.

10. Trans. in Sepper 1996, p. 44; cf. Descartes 1990, p. 51.

11. Sepper 1996, p. 45.

12. For Renaissance music and control, see Reiss 1997, pp. 159–60; Walker 1985. For divine furor and the connection to Ficino, see Tomlinson 1993. For music and natural philosophy in the period, see also Cohen 1984.

13. Reiss 1997, p. 190.

14. The pervasiveness of Platonism is a theme throughout Levi 1964.

15. Quotations from Yates 1947, pp. 127, 79.

16. Clements 1942, pp. 30–33.

17. Tyard, *Discours philosophiques,* fol. 2r, quoted and trans. in Yates 1947, p. 78.

18. Yates 1947, p. 79.

19. On the centrality of sloughing off the body in spiritual exercises, see Hadot 1995, p. 103.

20. For hints of a relationship between the Pléiade and Descartes, see Rodis-Lewis 1998, pp. 41, 237 n. 36. A central connection is the famous corresponding monk Marin Mersenne, whose knowledge of the Pléiade was great, and his sympathy for them not insignificant. See Yates 1947, pp. 284–90 and passim; Descartes 1990, pp. 26–31, 139 n. 56.

21. Descartes would have also received at La Flèche a combination of a scholastic philosophical and Platonist mathematical education.

22. The French historian of rhetoric Marc Fumaroli has shown how this view drew on the mysticism of Ignatius's *Spiritual Exercises* and a Platonist account of poetical creation. See Fumaroli 1980a, pp. 1281–82, 1285–86.

23. Quotations from Campbell 1993, p. 65. See also Shuger 1988, pp. 87–89, 136–37; Caussin 1619, pp. 107–8, 154.

24. One technique Descartes praised as a means for "ascent" was the use of figurative description. See AT X:217; Sepper 1996, p. 47.

25. In his French translation of *Rules,* Jean-Luc Marion translates the term as *régarde,* which prevents one from too easily assimilating it to "intuition." But the English "gaze" carries different connotations. Douglas Sepper leaves the term untranslated. I use "intuition" because its vagueness leaves considerable room for different meanings.

26. Vinci 1998, p. 16; see also Van De Pitte 1988a.

27. Gaukroger (1995, p. 118) rightly stresses simultaneity and Descartes' rather hazy commitment to "the idea of the instant" both here in his epistemology and in his microphysics elsewhere.

28. See also CSM I:14; Descartes 1998, p. 79.

29. This essential point is from Van De Pitte 1988a, pp. 457–58.

30. Gaukroger 1997; Gaukroger 1995, pp. 119–24.

31. Gaukroger 1995, p. 122.

32. The two terms appear independently throughout, e.g., at AT X:407: "ut propositio clare et distincte . . . intelligatur." Compare, however, "donec assuescamus veritatem distincte et perspicue intueri" (AT X:399).

33. See Heffernan's comments in his translation, Descartes 1998, p. 18.

34. Quintilian 1920–22, 6.2.29–32, vol. 2, pp. 433–37. Cf. Skinner 1996, pp. 183–84.

35. Quintilian 1920–22, 6.2.32–33, vol. 2, pp. 436–37.

36. Quintilian 1920–22, 6.2.30, vol. 2, pp. 434–35; see also Vickers 1988, p. 321.

37. For a schematic account and references to other rhetoricians, see Lausberg 1998, pp. 117, 361.

38. No one study captures the plurality of uses of *enargeia* in the period. For literary uses in the Renaissance, see Cave 1976; 1979; Galand-Hallyn 1991. For literary uses in antiquity, see Eden 1986; Trimpi 1983. For historical uses, see Ginzburg 1989. For uses in Renaissance history, see Couzinet 1996, pp. 234, 244–45. For grand style rhetoric and Augustinian anthropology, see Shuger 1988; cf. Moss 2001.

39. For an emphasis on the similarities between Protestants and Catholics in this regard, see Shuger 1988, esp. ch. 5. For great detail on the various rhetorical movements in France and their internal differences, see Fumaroli 1994a, pp. 138–52 and passim. For Augustinianism as a key strand of late humanism, and the movement to use the problematic will as a crucial resource, see Bouwsma 1990; Mouchel 1999. See also O'Malley 1979.

40. See Shuger 1988, pp. 230–31.

41. Shuger 1988, pp. 199–200.

42. Ludovicus Carbo, quoted in Moss 2001, 389; see also Campbell 1993, p. 57; Caussin 1619, bk. 8, ch. 4, p. 313.

43. This emphasis on *moving* the soul had deep philosophical ramifications. For example, manuals on the soul—*de anima* treatises—were closely connected to rhetorical treatises, for they offered an account of the internal human faculties, including the passions, upon which rhetoric was to work. Indeed, Aristotle's *Rhetoric* was often read as an account of the passions of the soul. These treatises, modifying Thomism with Augustine, Ficino, and pseudo-Dionysius, offered the accounts of the passions upon which the Reformation's and Catholic Reformation's evangelical "grand style" of rhetoric gained its legitimacy. See Shuger 1988, pp. 132–36, 194, 209. For an exemplary rhetorical study of the passions, see Caussin 1619, bk. 6, "De affectibus"; on this, see also Campbell 1993.

44. Beyond the foundational studies, Gilson 1930 and 1963, see, e.g., Marion 1975; Van De Pitte 1988a, 1988b; Garber 1992; Ariew 1999; Des Chene 1996; Biard and Rashid 1997; among others.

45. See O'Malley 1993, pp. 257, 244; and also Blum 1985, pp. 104, 109.

46. Fonseca "semble faire fusionner argumentation dialectique et argumentation rhétorique, qu'Aristote distinguait avec soin" (Fumaroli 1994a, p. 145 n. 206). Fonseca's logic text, originally published in Coimbra, was reprinted at La Flèche in 1609 (Fonseca 1609). Not enough work has yet been done on how humanist these late scholastics were—in their means of editing texts, in their approach to the reconstruction of Aristotle's thought, and on their appropriation of the focus of humanist dialectic on commonplaces. See, however, comments in Lewalter 1935, pp. 21–28. For the distinctiveness of Spanish Scholasticism, see Sirven 1928, pp. 182–83 and passim; Mora 1953, esp. pp. 535–36; and the remarks on the effects of humanism in Coxito and Soares 2001, pp. 456–57, 472–73.

47. Dainville 1940, p. 87.

48. For a recent survey on the questions of when Descartes studied at La Flèche and with whom, see Rodis-Lewis 1998, 1987. For Jesuit mathematics education in France, see Romano 1999; Huppert 1984; Martin 1988; Scaglione 1986; and the still essential Rochemonteix 1889.

49. See the good description in Fumaroli 1994a, pp. 233–42.

50. Dainville 1940, pp. 78–79.

51. Paraphrased from Dainville 1940, pp. 188–89.

52. For the Jesuits' account of the suitability of theater, see Oliazola 1999; for the use of theater in schools, see pp. 388–91. For Seneca taught as theater among the Jesuits, see Garciá-Hernández 1997. See the examples in Rochemonteix 1889, vol. 3, pp. 217–353.

53. Soarez 1569, I, 1, fol. 1r. See also Bayley 1980, p. 24. Even if Descartes did not use Soarez's *De arte rhetorica libri tres*, which is highly unlikely, it was the basis for most subsequent Jesuit rhetorics. For Soarez's centrality in Jesuit pedagogy, see Moss 2001, pp. 383–84. On Soarez and La Flèche, see Flynn 1957, pp. 261–63; Rochemonteix 1889, vol. 3, pp. 28–31.

54. Soarez 1569, II, 27, fol. 38v, misnumbered as 34v. See also Quintilian 1920–22, 6.2.29–30, vol. 2, pp. 432–34; cf. 9.3.62. See Bayley 1980, p. 27.

55. For the centrality of the idea that one must move oneself before moving others in Renaissance and early-modern rhetorical theory, see Shuger 1988, pp. 227–32.

56. Pseudo-Cicero 1954, 4.55.68, pp. 405–7

57. See Gaukroger 1989; Recker 1993; Fumaroli 1980b; Clarke 1991.

58. Caussin 1619, p. 283.

59. See Toledo's three sorts of definition: essential definition, causal definition, and descriptive or accidental definition (1985b, bk. 1, ch. 4, pp. 6–7). Cf. Fonseca 1609, p. 116; Fonseca 1964b, vol. 1, p. 284. All of Fonseca's book 5 considers definitions. For the relationship of description and definition in Jesuit scholastics and rhetoricians, see Fumaroli 1980b. For considerations on definition in late Scholasticism more generally, see Kuhn 1997, esp. at p. 328.

60. Soarez 1569, I, 15, fol. 6r. Cf. Caussin 1619, p. 141; and a key source, Cicero, *Topica*, 5.27–7.32.

61. Fonseca 1609, bk. 5, ch. 3, p. 120; Fonseca 1964b, vol. 1, p. 292.

62. Soarez 1569, I, 15, fol. 6v. Cf. Cicero, *Topica*, 7.32.

63. Caussin 1619, p. 148. For the centrality of *concinnus* and composition in the Renaissance, see Witt 2000; Baxandall 1971.

64. Fonseca discussed enumeration of parts in great detail; where Caussin tended to cite Cicero's *Topics*, Fonseca cited Aristotle's. See, e.g., Fonseca 1964b, bk. 7, ch. 17.

65. Philosophers and rhetoricians alike agreed on the existence of such higher knowledge, although they did not agree on its availability to human beings in this life or about the means to acquire it.

66. Fumaroli 1980b, p. 44. For these *translationes*, see, e.g., Soarez 1569, I, 15, fol. 6v, quoted above.

67. Fumaroli 1980a, pp. 1281–82, 1285–86.

68. *Principia philosophiae*, I, §45, AT VIII/1:22. Compare the French: "I'appelle claire celle qui est presente & manifest à un esprit attentif: de mesme que nous disons voir clairement les objets, lors qu'estant presents ils agissent assez fort, & que nos yeux sont disposée à les régarder" (AT IX/2:44; cf. trans. in CSM I:207).

69. See, e.g., *Passions*, §20, AT XI:344; *Meditationes*, 4, AT VII:53.

70. *Passions*, §26, AT XI:349.

71. *Passions*, §§44–45, AT XI:361–63 (quotation at 362). Descartes strongly insisted on the particular power of *true* representations (§49). For a subtle consideration of the relationship between *Passions* and ancient rhetoric and its medicinal claims, see Struever 1993.

72. For examples, see *Spiritual Exercises*, §§47–50, 103–26, English trans. in Ganss 1991, pp. 136–37, 148–51.

73. See the collection of documents in Iparraguirre 1955; on these points, see pp. 394, 651–55.

74. For Jesuit writings as instantiations of Ignatian procedure, see Fumaroli 1994a, pp. 354–91, esp. 365 for Caussin.

75. For the difficulties of connecting Descartes closely with Ignatius's exercises, see Hermans and Klein 1996.

76. For a list of uses of *"ingenium"* in the *Regulae*, see Robinet 1996, pp. 201–4. In claiming that mathematics trained the *ingenium*, Descartes echoed Clavius's gloss on Plato: see Clavius [1611] 1999, p. 6. For a study of the variety of *ingenia*, see Caussin 1619, pp. 104–10; see also Kemp 1977.

77. Caussin 1619, p. 107.

78. Vives [1782] 1964, vol. 3, p. 374. For the centrality of *ingenium* in Vives, see Nero 1992, pp. 199–207; Hidalgo-Serna 1983. For a brief description of Vives and Melanchthon on these points, see Altman 1987, pp. 137–38, 154 n. 22.

79. Vives, *De tradendis disciplinis*, quoted in Nero 1992, p. 201.

80. Vives [1782] 1964, vol. 3, p. 374.

81. For a tour through the whole range of antischolastic logics from Agricola through Ramus to sundry less well known names, which made up the "immediate context" around Descartes, see Robinet 1996.

82. See, e.g., Goclenius 1964, p. 241. For classical and French Renaissance definitions, see Lausberg 1998, p. 502. A whole new set of dense meanings would be added by Juan Huarte, who produced a Galenic theory of different sorts of *ingenia*. See Huarte 1989; and for his reception in France, Pérouse 1970.

83. Caussin 1619, p. 108.

84. See Caussin's book 3, "De adminiculis eloquentiae ingenio, doctrina, et imitatione."

85. Quintilian 1920–22, 10.2.12, vol. 4, p. 81; 10.2.12, vol. 2, p. 81; see also 10.7.15; 12.10.6. For a list of citations to *ingenium* in Quintilian, see Descartes 1998, p. 69 n. 11.

86. For La Flèche, see Rochemonteix 1889, vol. 3, pp. 30ff. The relation between the Jesuit *Ratio studiorum* and Vives is complex and unclear. See Batllori 1986, esp. pp. 143–45 and the works cited there. For training the *ingenium* in Jesuit pedagogy, see, e.g., "De explicatione Orationis in Classe Rhetoricae," Lukàcs 1965–, vol. 5, pp. 339–40; see also Demoustier, Julia, and Compère 1997, pp. 251–52.

87. Trans. modified from Carr 1990, p. 14. On Descartes and Caussin here, see Fumaroli 1994d, esp. pp. 386–87.

88. This strain of argument offers Descartes' answer to the contention that rhetoric and the epistemic standards associated with it concern only the production of effects (affects, really) in listeners. Concerned primarily with emotional swaying in this view, rhetoric was and is opposed to truth production. In his defense of Guez de Balzac, Descartes held that the highest and true form of rhetoric necessarily carries with it a sincerity validating its truthfulness.

89. For the context and major arguments of this attack, see Verbeek 1992, pp. 19–29; and the introduction to Verbeek's crucial French translation of Schoock 1643 and associated documents, in Descartes and Schoock 1988.

90. Schoock 1643, pt. 2, ch. 10; trans. following Descartes and Schoock 1988, pp. 249, 251.

91. Schoock 1643, pt. 2, ch. 11; Descartes and Schoock 1988, pp. 253, 254.

92. Recker 1993, p. 236.

93. Early-modern scholastic philosophers saw causal explanations involving the essences of things—Aristotelian forms involving final and intermediate ends—as complemented by explanations in terms of their efficient causes, including their local motion. A draught with a dormitive property ought to be understood both in terms of its end (to cause sleep) and in terms of its means (the material and efficient causes through which it produces the physiological changes leading to sleep). True knowledge of the natural and moral world involved understanding both ends and means. For a thorough examination, see Des Chene 1996; Hutchison 1991.

94. See Maclean 2002, pp. 144–45; 2005, pp. 163–68. For a similar use of descriptions in legal texts, see Maclean 1992, pp. 98, 108–11. I thank an anonymous referee for these references. For the explosion of descriptions in sixteenth-century botany, see Ogilvie 2003.

95. Early in *Regulae*, Descartes temporarily seemed to allow deduction to comprise a series of intuitions in memory and time; in Rule 7 he clarified his position.

96. It is a good *philosophical* question whether Descartes' account of deduction has much validity; my concern here is more with its historical genesis and character.

97. See also AT X:407–8; CSM I:37.

98. A logical or algebraic deduction is legitimate, not in virtue of its form, but in virtue of being cognized in a continuous sweep (which may be possible only because of its form).

99. See the discussion and translation in Recker 1993, p. 240.

100. See Caussin 1619, p. 148.

101. Quintilian 1920–22, 7.10.16–17, vol. 3, p. 171–73. On rhetorical *oeconomia* and its importance in hermeneutics, see Eden 1997, pp. 27–32, 82–83.

102. To be sure, Descartes' account of a sufficient enumeration includes essential *invisible* mechanisms that rhetoricians would have ignored.

103. *L'homme*, AT XI:132; I draw here on the fine remarks of Des Chene 2001, p. 73.

104. Soarez 1569, II, 27, fol. 34v.

105. Among his many other important points, Des Chene (2001) elucidates the profound difficulties any mechanical philosophy has in showing that individual beings are unified wholes distinct in some meaningful way from the rest of the world.

106. Second set of replies to *Meditationes*, AT VII:159; cf. trans. in CSM II:113: "totum corpus Meditationum mearum intueri, & simul ipsarum singula membra dignoscere."

107. I refrain from a more detailed analysis of *Meditationes*; see the fine articles by Hatfield, Dear, Rubidge, and Sepper cited on p. 275 n. 12.

108. "Epistola ad D. Voetium," AT VIII/2:41 (my italics). See Verbeek 1992, p. 26.

109. For the definition, see Shuger 1988, p. 210; see also Marion's commentary, Descartes 1977, pp. 199–200.

110. Ockham 1967–88, vol. 1, pp. 5–6. The literature on Ockham's account of evidence is confusing and often contradictory. I draw this definition of evidence primarily from Karger 1999, p. 208.

111. See Alanen and Yrjönsuuri 1997, pp. 159–60.

112. See Nuchelmans 1983, pp. 47–48. See *Summa theologiae*, IIaIIae 2.1, among many others. "Evidence" is used by later scholastics to gloss Thomas's account.

113. Ockham 1967–88, vol. 1, p. 68, trans. McGrade 1988, p. 425.

114. See, e.g., *Principia philosophiae*, I, §§43–44, AT VIII/1:21.

115. Duns Scotus maintained that some theologians could attain tremendous scientific knowledge of God. Not intuitive cognition of God, reserved for the time after, this scientific knowledge comprised distinct abstractive cognition—the best knowledge of the divine in this lifetime possible without direct divine illumination. In fact, Scotus devised the category of abstractive cognition precisely to account for this sort of nonintuitive scientific knowledge of God. See Dumont 1989, pp. 590–93.

116. Kenny (1972) argues that Descartes radically shifted his views sometime after he composed *Regulae*. Kenny holds that the notion of the will making judgments does not appear anywhere in *Regulae* or early correspondence. In contrast, Caton (1975) forcefully contends that the doctrine remains more or less constant over Descartes' entire career, but that Descartes, by the time of *Meditationes*, had adopted a scholastic vocabulary to describe that doctrine. Caton shows that in *Regulae* Descartes uses the term *intellectus* in three senses: "the mind in all its powers; or the power of perception; or pure understanding, as distinguished from sensation and imagination" (p. 101). Caton points to AT X:395, 411, 415–16. *Cognitio* and *ingenium* are also used in these senses. What in *Meditationes* is described as the action of the will is described in *Regulae* as a "faculty" of the *intellectus* (AT X:420). A few pages later, Descartes talks of a judgment being made "à propriâ libertate" (AT X:424). Moreover, there are significant hints of the later use of the terms "will" and "liberty" at AT X:370–71, 424. See also Van De Pitte 1988a, p. 468 n. 69. In his fourth meditation, Descartes sharply distinguished free will from mere indifference. Free will demands only the possibility of choosing otherwise than one does; or put another way, that we do not feel that some external force determines our choice and confirms or denies that which the intellect proposes to us. To be free does not require that the intellect propose all possible choices as indifferent equals. Descartes regarded choosing among equal options a debased—"the lowest"—form of freedom, attributable to a defect in cognition or something similar. Freedom becomes ever more noble as one retains the ability to choose despite an ever increasingly apparent choice: "Neither divine grace nor natural cognition diminishes freedom; rather, they strengthen and augment it" (AT VII:58).

117. *Passions de l'âme*, §152, AT XI:445.

118. Consider, for example, the stress in *Meditationes* on the need for "attentâ & saepius iteratâ meditatione" by which "ita habitum quemdam non errandi acquiram." He furthermore calls this certain habit of not erring the "maxima & praecipua hominis perfectio" (AT VII:62). See Levi 1964, p. 296, discussing Descartes to Mesland, 2.5.1644, AT IV:117. For the will's inconstancy, not its corruption, as the source of error, and the need to train the intellect to produce constancy, see Descartes' "discussion" with Burman, 16.4.1648, AT V:158–59. In the *Principia*, Descartes contends that the state of being deceived often arises if someone attempts to make a judgment about truth or falsity when that person has not yet become able to recognize clarity and distinctness. See *Principia*, I, §42, AT VIII/1:20–21, and the amplified French version at AT IX/2:43.

119. Even if Descartes radically changed his account of the relationship of the will and intellect, as Kenny and others argue, he insisted on the exercises necessary to train the intellect to recognize clarity and distinctness in the early *and* the later texts.

120. See *Passions de l'âme*, §§149–52, AT XI:443–45.

121. Toledo 1985a, qu. 2, p. 8. Descartes wrote that he had read and forgotten Toledo; see Descartes to Mersenne, 30.9.1640, AT III:185.

122. Fonseca 1964a, bk. 4, ch. 1, qu. 1, §3, vol. 3, p. 9.

123. Paulo 1640, *Logica*, p. 136. For Descartes and Eustachius, see Van De Pitte 1988a, 1988b.

124. Roughly, distinct knowledge meant grasping all the necessary features of a nature: "Distincta est, qua totum omnibus, quae in eo continentur enucleaté inspectis, cognoscimus; veluti cum hominem sic intelligimus, ut explicita perceptione cuncta eius essentialia praedicata teneamus." Following Scotus, the Jesuit Coimbra commentators noted that the "progressus naturae est ab imperfecto ad perfectum per medium." Confused cognition offered a medium level of knowledge set "inter ignorationem & inter notitiam distinctam." See Coimbra 1984, bk. 1, ch. 1, qu. 2, a. 1, vol. 1, p. 57.

125. See, e.g., Coimbra 1984, bk. 3, ch. 8, qu. 4, a. 1, vol. 1, p. 393.

126. See the consideration of such concerns in Larmore 1984.

127. See Alanen 1999.

128. *Directoria conscripta iussu et auctoritate R. P. Cl. Acquaviva*, in Iparraguirre 1955, p. 701. This is the official guide for directing the *Spiritual Exercises* and was promulgated in 1599. See the discussion in Hermans and Klein 1996, p. 437 and n. 54.

129. See Certeau 1965.

130. Among a large literature, see, for Spain, Haliczer 2002; and, for France, Certeau 1982.

CHAPTER THREE

1. Pascal to Fermat, 10.8.1660, M IV:923. For Pascal's changing opinion on the worth of mathematics, especially in later life, see Mesnard 1953.

2. Pascal to Fermat, 10.8.1660, M IV:923.

3. Pascal to Fermat, 29.7.1654, M II:1142.

4. Compare Pierre Nicole's comments on poetry: "One can produce an ode or a sonnet without being a poet by profession, without making it one's occupation and employment. It can pass for the diversion of an *homme d'esprit*" (Nicole 1996, p. 176).

5. Méré made similar claims about how the *honnête homme* could perform the actions of a craft without becoming a craftsman. See Méré 1682, vol. 1, p. 100; vol. 2, p. 80; see also Tourneur 1933, p. 115.

6. François du Verdus to Thomas Hobbes, 4.8.1654, in Hobbes 1994, vol. 1, p. 187. For Fermat's communication of the porisms to his Parisian friends, see M II:1159–60.

7. The great collection of midcentury gossip recounts that Le Pailleur "sçavoit la musique, chantoit, dansoit, faisoit des vers pour rire . . . il fit la desbausche à Paris assez longs-temps," all on top of being a fine mathematician (Tallemant des Réaux 1961, vol. 2, p. 99). Other sources on Le Pailleur and his relations with the Pascal family include Pascal 1923, vol. 2, pp. 115–21; Strowski 1921, vol. 2, pp. 12–16; Mesnard 1963; Mesnard 1965, pp. 165–67; Pintard 1943, pp. 348–51; Mazauric 1997, ch. 2. For Le Pailleur's poetry, see Arsenal MS 4127, Receuil Conrart, vol. 22, esp. nos. 33, 35, 38, 43.

8. Mesnard 1965, pp. 164–67. For these groups in general, see Mazauric 1997, chs. 1–2; H. Brown 1934.

9. See Le Guern 2003.

10. Mesnard 1953; Descotes 2001, pp. 50–53.

11. *Lettre dédicatoire à Monseigneur le Chancelier,* 1645, in M II:335; see also Jones 2001.

12. On conversation and its epistemic, social, and rhetorical dimensions, see Waquet 2003, esp. pp. 296–99; Fumaroli, Salazar, and Bury 1996; Bury 1996a, ch. 3; Goldsmith 1988; Burke 1993, esp. pp. 102–8; Shapin 1994, pp. 114–22.

13. For humanist, same-sex conversational culture, see Miller 2000, ch. 2. For women philosophers and salon culture, see Conley 2002. The importance of polite spaces of conversation, often gendered female, for early-modern science has now been well documented; see, e.g., Sutton 1995; Harth 1992; Walters 1997; Terrall 2002. See also Goodman 1994.

14. Morhof 1708, bk. 1, ch. 15, "De conversatione erudita," p. 180. See Fumaroli 1994c, p. 75; Waquet 2003, pp. 296–99 (cf. pp. 260–63).

15. Méré 1687, no. 88, p. 38. Such examples are easily multiplied: see Boileau 1666, p. 29.

16. "Art de conferer," Montaigne 1652, p. 685; Montaigne 1992, III, 8, pp. 922, 923.

17. See Mesnard's introduction to the speech and Mesnard 1963 for his arguments about the intended audience.

18. Guazzo 1581, fol. 14v. See also Waquet 2003, p. 296. Guazzo viewed the pursuit of glory as a powerful contributing factor to the production of rigorous arguments. See also Anthony Grafton's portrayal of the competitive, humanist culture of emendation (2000, pp. 53–57).

19. Guazzo 1581, fol. 15v. See also the discussion in J252–53.

20. Montaigne 1652, p. 686; Montaigne 1992, III, 8, p. 924. Montaigne denounced the common art of merely contradicting as opposed to truth production: "chacun contredisant, & estant contredit, il en advient que le fruict du disputer, c'est perdit & aneantir la vérité" (1652, p. 687; 1992, p. 924).

21. Montaigne 1652, p. 686; Montaigne 1992, III, 8, p. 924.

22. Waquet (2003) rightly stresses the continuity between such conversation and the institutions of modern science dedicated to perfecting and developing works in progress.

23. Mesnard cites Ismaël Bouillau's *Exercitationes geometricae tres:* Fermat had sent his friends in Paris "propositiones quasdam subtilissimas et porismata" (M II:1159). For the porisms, see M II:1159–65. Pascal sent Fermat a copy of *Treatise on the Arithmetical Triangle,* one of the few copies we know he sent out before his experience of late November 1654 (see below); the rest were by and large distributed only after his death (M II:1153–54). Fermat's correspondence with Pascal on probability is one of the most important examples of mathematicians working together through correspondence.

24. Pascal to Fermat, 27.10.1654, M II:1158.

25. Shapin 1994; Shapin and Schaffer 1985. For the different forms of civility in seventeenth-century academies and their relationship to different possible sorts of the truth claims, see Biagioli 1996. For mathematics as problematic for civility in the Royal Society, see Shapin 1988. For different French models of civility and kinds of truth claims, see Licoppe 1996.

26. "De l'art de persuader," M III:423.

27. Montaigne 1652, p. 685; Montaigne 1992, III, 8, p. 923.

28. He immediately noted a troubling regress: "And one cannot make this choice if one has not already been formed and not spoiled. Those who escape this circle are fortunate" (S658).

29. See Mazauric 1997; H. Brown 1934, pp. 75–76. For the theory of exclusion, see Fumaroli 1994c. For the exclusion of women, see, e.g., Terrall 1995.

30. Mersenne [1634] 1972b, pp. 87–88.

31. See also Descotes 2001, p. 27.

32. See I. Bolliau to Christiaan Huygens, 6.12.1658, in Huygens 1888–1950, vol. 2, p. 287, discussed in Biagioli 1996, p. 198, and H. Brown 1934, pp. 87–88.

33. Gilles Filleau de Billettes to Leibniz, 27.10.1696, A1,13:307 (italics in the original), and the discussion in Mesnard 1965, p. 371.

34. This distinction is most clearly seen in aesthetic works. See, e.g., Nicole's account of the place of rules in poetry (Nicole 1970, pp. 144–45), the letters of Méré cited above, and the attack on Abraham Bosse in 1662, mentioned in chapter 1.

35. For Le Pailleur's works in manuscript, see the evidence in Mesnard 1963, pp. 5–6.

36. Cf. Shapin 1994, pp. 181–82.

37. For Méré and Mitton as mathematicians, see, e.g., Gilles Filleau de Billettes to Leibniz, 23.2.1697, A1,13:574; see Mesnard 1965 in general for the extent of Pascal's business, intellectual, and affective relations with these *honnêtes hommes.* For his concern with speaking to and convincing wider audiences, see Descotes 1993, pp. 444–45.

38. "Amateurs" in its modern English sense is a term of convenience, given that few of the savants were employed as mathematicians or natural philosophers.

39. "Celeberrimae matheseos academiae parisiensi" (1654), M II:1031–35. It is not clear whether he delivered this speech as written.

40. "Le Mémorial," 23.11.1654, M III:50.

41. The complicated composition and printing of the various parts of the *Traité* are untangled in M II:1166–75. For the uses of French and Latin in it, see Descotes 2001, pp. 59–60.

42. Mesnard prints both sets of texts in M II.

43. For an introduction and historical background, particularly strong on indicating first discoveries of different relations, see Edwards 2002 (chs. 6–7 focus on Pascal's work). A fine exposition can be found in Shea 2003, chs. 10–12; see also Descotes 2003, pp. 23–31.

44. In some cases, versions of the arithmetical triangle were used for finding combinatorial numbers; in others, for binomial numbers; for the history see Edwards 2002, 1–56, 140–43.

45. A rigorous proof by induction is a two-step process: first, prove the proposition for the first case; second, assuming the proposition to hold for some arbitrary case n, show that it holds for the next case $n + 1$.

46. Edwards 2002. Pascal may not have known of many of the earlier discoveries or may have simply omitted giving credit.

47. Although Pascal initially referred to both the figurate numbers and the "orders" of numbers interchangeably, in his French versions he dropped the use of the term "figurate."

48. Edwards 2002, p. 4.

49. Edwards focuses particularly on the use of the triangle to solve problems in probability. See Edwards 2002, ch. 7, and appendices 1–2, which reprint his important articles on the subject; Edwards 2003.

50. Although Pascal identified the numbers of the triangle with the combinatorial numbers and with the binomial coefficient, he did not explicitly note that the coefficients of the binomial expansion are the combinatorial numbers:

$$(a + b)^r = \binom{r}{0}a^r + \binom{r}{1}a^{r-1}b^1 + \binom{r}{2}a^{r-2}b^2 + \cdots + \binom{r}{r-1}a^1b^{r-1} + \binom{r}{r}b^r$$

Newton later expanded the binomial theorem to noninteger values of r; he was far more interested in such forms of algebraic expression. See, e.g., Newton to Oldenburg, 13.6.1676, in Newton 1959–77, vol. 2, pp. 20–21, 32–33. For Pascal the relationship with the triangle itself seems to have been of greater interest than such a particular connection to algebraic expression. Cf. Edwards 2002, p. 79.

51. For the theme of invention in Pascal's mathematics and thought more generally, see Bold 1996.

52. Once a set of relationships is discerned with the help of propositions from another sort of number, one can prove the results more directly. Pascal did so; he discovered many of his results independently of the triangle.

53. The model rests on the distinction between liberal arts and mere mechanical arts. See Pascal's discussion of these issues in his account of the development of his arithmetical machine, in M II:335.

54. Montaigne 1652, pp. 691, 692; Montaigne 1992, III, 8, p. 931.

55. Preface to *Oeuvres de poésie chrétiennes et diverses*, edition in Nicole 1996, p. 142.

56. For Pascal's account of the importance of discerning the economy of a work and its connection to the rhetorical tradition, see Force 2005, pp. 31–33. See also Descartes' insistence on discerning the economy unifying the works of great authors, including himself, discussed in chapter 2 above.

57. For the single principle, see Descotes 1993, p. 147.

58. Méré 1682, vol. 1, p. 111.

59. For the infinite number of possible truths, see Descartes to Mersenne, 31.12.1640, AT III:274.

60. See Jones 2001, upon which this section draws. The poems can be found in Dalibray 1653. A sense of these discussions can be found in Arsenal MS 4119, Receuil Conrart, vol. 14, which includes letters of Pascal and Noël, one from Descartes to Balzac, and much material on changing views on rhetoric.

61. Charles Vion Dalibray, "Au même sur le vide," in M II:693; Pascal 1923, vol. 2, p. 44 n. 1.

62. Pascal to Le Pailleur, 2.1648, M II:562–63.

63. Pascal to Le Pailleur, 2.1648, M II:563. He later noted that mathematicians alone may allow their knowledge to rest on mere possibility. In mathematics, "possibility is a sure mark of the truth in these sorts of knowledge, since it is a question of the essence of things" (M IV:1299).

64. Pascal had left himself open to misinterpretation on this point because he had proffered a very Cartesian-sounding account in a previous letter. See M II:519; discussed in Jones 2001, p. 170.

65. Pascal did not claim some clear and distinct knowledge of these entities here. He claimed that we know that time, space, movement, and unity exist, but they remain inexplicable. See *Entretien avec M. de Sacy,* in M III:142.

66. Noël to Pascal, early 11.1647, M II:538; Pascal to Le Pailleur, M II:564.

67. Pascal to Le Pailleur, M II:564.

68. The call for regulating tropes was central for rhetorical reformers. See, e.g., Lamy 1688, pp. 100–106; the theoretical claim in Nicole 1996, pp. 72–74, and his examples on pp. 76–93.

69. Etienne Pascal to Etienne Noël, 4.1648, M II:591.

70. Guez de Balzac 1995, p. 127. See also Beugnot 1999, p. 559.

71. Descartes to [?], 1628, AT I:8–9.

72. Fumaroli 1994a, p. 300; Fumaroli 1994b. The demand for such a "natural" habituation to good rhetoric took many forms; for example, it took a Jansenist guise with Saint-Cyran (Fumaroli 1994a, p. 639). See also Fumaroli 1979, pp. 363–64; Descotes 1993, pp. 23–25.

73. Le Moyne 1645.

74. Fumaroli 1994a, pp. 620–22, 677–84, quotation on p. 678; Fumaroli 1994e, esp. pp. 97–99. Compare the similar English "non-rhetorical rhetoric" of humility and probabilism in Wintroub 1997, pp. 192–95; Shapin 1994, pp. 222–23.

75. Montaigne 1652, p. 648; Montaigne 1992, III, 5, p. 874; noted in M III:428 n. 1.

76. Each number is different by some common whole number. I examine only the case of differences of unity, but the method is more general.

77. Pascal uses a concrete example involving the series 5, 8, 11, 14, and does not give a general proof like the one I supply here. I nevertheless follow his procedure to make apparent the intuition and show some of its generality. See Shea 2003, pp. 295–304; Edwards 2002, pp. 82–83; Descotes 2003, p. 30.

78. We would probably write this as an expression involving a sum of combinatorial numbers and sums.

79. This discussion and example are drawn from Gardies 1984, 62–63; and, with greater attention to Pascal's method, Loeffel 1987, pp. 101–2, and Edwards 2002, pp. 82–83. For the general approach Pascal used to find these quadratures, see Andersen 1986, pp. 22–23. For Roberval's results, see Andersen 1994, p. 301.

80. As *n* goes to infinity, terms with *n* in the denominators become negligible compared with the one-third term. Where we would probably use a limit argument, Pascal argued, "In the case of a continuous quantity, quantities of some order [*generis*], of whatever number, add nothing to quantities of a higher order" (M II:1271). Mathematically, this meant something like: "thus, points add nothing to lines, lines nothing to surfaces, surfaces nothing to volumes." See also "De l'esprit géométrique" (1654?), M III:408–9; and his more lengthy discussion in "Lettres de Dettonville" (1658), M IV:423–26.

81. Andersen 1994, p. 301.

82. M II:1272; see Gardies 1984, pp. 63–64.

83. Pascal's reworking of Euclidian geometry as the study of abstract space is one of his major innovations: "The object of pure geometry is space"; "Space is infinite in all directions"; "Introduction à la géométrie" (1655?), M III:435, 436. See also Mesnard's comments at M III:433.

84. Mancosu 1996, pp. 130–39; Descotes 1990.

85. See Descartes, *Principia philosophiae,* I, §§26–27, AT VIII/1:14–15; AT IX/2:36–37; see also AT V:167.

86. Mancosu 1996, p. 136.

87. Pardies 1710, fols. A7r–v, [A8r] (my italics). See also Mancosu 1996, pp. 142–43; K132–33.

88. Coimbra 1984, vol. 1, p. 51. Cf. the discussion in Mancosu 1996, pp. 142–43.

89. Coimbra 1984, vol. 1, p. 396, marginal comment.

90. Strictly speaking, the space is finite; the surface containing it is infinite. I follow Pascal in referring to the space as infinite.

91. See Pucelle 1963 for connections to Descartes; for longer-term connections to theology, see LG II:1187 n. 2.

92. In *Pensées,* such entities are the prime examples of the knowledge offered by the faculty Pascal called the heart (S142).

93. In *Meditationes,* IV, Descartes defended the limited range of the certain principles given by natural light, that is, those that can be known clearly and distinctly. Pascal accepted both the power of evident principles and the fact of limitation but maintained that the principles available to human beings were more limited than Descartes had claimed.

94. For Descartes on the evidence of deductions, see AT X:369, discussed in chapter 2; cf. the interesting intermediate position of Locke 1975, 4.2.6, p. 533. See also Schuurman 2001.

95. The break with Descartes is quite sharp, since Descartes grounds certainty in evidence, and not vice versa.

96. In more formal terms, Pascal accepted both the categorematic infinite (roughly, for the infinitely large case, there exists some number greater than all other numbers) *and* the syncategorematic infinite (roughly, that, for any number, there exists another that is larger); in fact, he held the latter to imply the former. In contrast, Descartes accepted only the syncategorematic infinite in mathematics and natural philosophy, which he called the "indefinite." Knowledge of the existence (but not nature) of the infinite that is God was crucial for Descartes, especially in *Meditationes,* III (see AT VII:44–45); he explained that "I never use the word 'infinite' to signify the mere lack of limits (which is something negative, for which I have used the term 'indefinite') but to signify a real thing, which is incomparably greater than all those which are in some way limited" (Descartes to Clerselier, 23.4.1649, AT V:355; CSM III:377). For Pascal's account of the infinite, see Gardies 1984, esp. pp. 114–23.

97. Again an important contrast: Descartes argued that "in seeing things in which . . . we do not discern any limit, we do not affirm that they are infinite, but we will consider them only as indefinite things" (AT VIII/1:15; AT IX/2:36).

98. Pascal's citation of *"bornes"* (limits) suggests a reading of Descartes' *Principia,* as cited above, perhaps guided by the letter to Elisabeth of 15.9.1645 (AT IV:292–93) on imagining a universe without limits, and perhaps his letter to Clerselier of 23.4.1649, the final letter in the first volume of Descartes' letters of 1657 (AT V:356; Descartes 1657, pp. 661–62).

99. For Descartes' view of wonder, see *Passions de l'âme,* §§70–77, AT XI:380–86; considered in Daston and Park 1998, pp. 316–17.

100. Caroline Bynum has recently contrasted early-modern and medieval accounts of wonder. Her stress on the cognitive aspects of wonder captures one part of Pascal's usage: "No medieval theorist reduced wonder to the physiological reaction of the wonderer. The amazement discussed by philosophers, chroniclers, and travelers had a strong cognitive component; you could wonder only where you knew you failed to understand. Thus wonder entailed a passionate desire for the *scientia* it lacked; it was a stimulus and incentive to investigation" (2001, p. 72).

101. For the house of the duc de Luynes as a center for Cartesian conversation, see the introduction to *Entretien avec M. Sacy,* M III:126; see also Mesnard 1953, p. 16.

102. For the collaborative nature of this venture, see Le Guern 2003, chs. 11–12, with a particular focus on Pascal's contributions to the more formal aspects of the *Logic.* For the ample bibliography on *Logic,* see Arnauld 2001.

103. Nicole made similar arguments about the utility of literature: insofar as morally improving, it was crucial; as an end in itself, it was useless. See Guion, introduction to Nicole 1996, p. 38.

104. J306, J307 (my italics). See also Gardies 1984.

105. Mancosu 1996, pp. 100–102.

106. Pierre Nicole, preface to Arnauld 1781, p. 10; see the discussion in Gardies 1984, p. 99.

107. In addition to Mancosu 1996, see Gardies 1984, pp. 99–106; Gardies 1982, 1995.

108. J309. Arnauld's dislike of *reductio* proofs was long-standing: see "Conclusiones philosophicae," [25.7.1641], ed. and trans. in Arnauld 2001, p. 13.

109. In so insisting, Pascal staked out positions close to those that Jesuit mathematicians maintained against Jesuit philosophers in the defense of their discipline. See Gardies 1984, pp. 98–99.

110. See Gardies 1984, pp. 122–23, on the contrast between the views of Pascal and Arnauld on infinite numbers.

111. See Descotes 1993 for a detailed analysis of the give-and-take in *Pensées.*

CHAPTER FOUR

1. Leibniz to Thomas Burnett of Kemney, 1/11.2.1697, A1,13:556–57. See Carraud 1986 for a good comparison of the apologetic strategies of Pascal and Leibniz.

2. "Récit," M I:651. This anonymous account confirms Leibniz's story: an "extraordinarily violent" toothache led Pascal to seek distraction in struggling with some unsolved and perhaps insoluble problems concerning the *roulette,* also known as the cycloid, the curve generated by a point on a rotating circle. In distracting himself from his pain, he discovered a number of results about this curve, including some important general methods. Having long since abandoned the quest for glory of the mathematician, he did not write up his results, which he dismissed as vain curiosities deflecting him from his real work. A friend "commented that God had perhaps permitted this encounter to allow him to procure a means of establishing and giving more force to the work he was pondering against Atheists and libertines." See also Mesnard 1965, pp. 646–47.

3. Pascal's *Pensées* is notoriously difficult to interpret due to its fragmentary quality and Pascal's avowed sense of the difficulties involved in interpreting and reading any text. While not denying the aporias of the text, in this chapter I aim to illuminate a central argumentative strategy that Pascal explicitly employed: the use of natural philosophy

and mathematics to lead the reader to seek the sorts of answers that Pascal maintained Christianity alone could provide.

4. Pascal's version of Christianity is distinguished theologically by a fairly negative view of human epistemic and moral capacity after the Fall of Adam and Eve. Very roughly, whereas the Jesuits stressed in general that human beings had been granted the capacity to know, choose, and perform good acts, Pascal and like-minded Jansenists stressed that humans had not in general been granted the capacity to know, choose, and perform good acts via their own power.

5. Le Moyne 1645, fol. ĩ3v. On Le Moyne, see Gouhier 1987, pp. 105–9, and the very critical but insightful Levi 1964, pp. 170–76.

6. Lemaistre de Sacy 1654, pp. 32–36.

7. Nicole in M I:1002; Mesnard identifies the work in question as "Ecrits sur la grâce" (c. 1655–56?), M III:594; the point holds for *Provincial Letters* and *Pensées*.

8. Numerous scholars have noted the deep problems with the notion of "popularization," especially the acceptance of hierarchy and the assumption of audience passivity. I use the term, nevertheless, because Le Moyne defined himself very much as a conduit for scholastic ideas in accordance with the Jesuits' disseminative vision. See now the very thorough study Secord 2000; as well as Stewart 1992; Terrall 2002; and the key attack on the notion of popularization in Cooter and Pumfrey 1994.

9. See Fumaroli 1980b, p. 45.

10. Le Moyne 1645, fol. ĩ4r.

11. Le Moyne 1645, fol. ĩ.

12. See Parish 1989, ch. 4, for a judicious account. Pascal himself noted, "I believe that it was necessary to write in a manner appropriate for making women and men of the world read my letters, so that they would know the danger" (1963, frag. 1002).

13. My analysis draws heavily on Viala 1985 and Parish 1989.

14. The sociologist Ludwik Fleck argued some years ago that "certainty, simplicity and vividness originate in popular knowledge" (1979, p. 115). I know of no general study of "commonsensical" or popular reason and philosophy, despite the constant interplay between academic philosophy and a more general popular or public philosophy throughout the early-modern period. For Kant and *"schönen Wissenschaften,"* see Zammito 2002, pp. 15–41, esp. pp. 36–39; and for Hume and his publics, Frasca-Spada 1999. For Descartes and a broader public, see Descartes to Girard Desargues, 19.6.1639, AT II:554–55; and the analysis in Cavaillé 1994.

15. For Pascal's deceptive use of the evidence, see Duchêne 1985.

16. See especially Viala 1985, pp. 174–75.

17. As we saw in the previous chapter, Pascal mocked the style of the Jesuit Noël as being regulated by reference to the rules of rhetoric, not the rules of nature or good reasoning; form, not meaning, generated his writing, just as, according to Pascal, form dominated the writings of the Jesuit theologians.

18. Miramion, *Discours de la conversation des femmes sçavoir si cest une chose utile à la jeunesse,* quoted in Maclean 1977, p. 146. Maclean provides a useful survey of attacks on and defenses of women in conversational culture (pp. 143–52).

19. *L'honneste femme, 1re partie,* 2nd ed. (1633), pp. 264–68, quoted in Zuber 1968, p. 222. The context concerns using proper French rather than Latin or Greek in serious conversation.

20. See Viala 1985, p. 173; Parish 1989, p. 51.

21. See Parish 1989, chs. 5–6, for a detailed study of Jesuit responses. For similar transformation in the hierarchy of disciplines in early-modern natural philosophy, see, e.g., Westman 1980; Biagioli 1993.

22. Pascal rejected a bifurcation of Christianity into formal authorities and simple believers. Following theologians like Cornelius Jansen, he condemned in particular speculative philosophy as insufficiently moored to the patristic tradition and revealed scripture in religion. See esp. M III:425–28; Ferreyrolles 1984, pp. 57–59; Descotes 1993, pp. 20–21. Pascal seems, at least rhetorically, to have rejected the need for an autonomous, formal class of intellectuals competent in scholastic logic, once held to be necessary for unraveling complex matters of faith and morality. Cf. Parish 1989, p. 16.

23. Suárez 1740, bk. 1, ch. 1, n. 1 (vol. 6, p. 205), n. 7 (vol. 6, p. 207). Suárez stressed that this knowledge is not trivially obtained at n. 8 (vol. 6, p. 207).

24. For a remarkable study of how Pascal's geometrical thought helped transform a somewhat banal comment from Montaigne on the "incapacity of man" into this fragment on disproportion, see Maeda 1964, esp. pp. 6–7.

25. The cases are slightly different. In mathematics, we have a certain knowledge of the possibility of always continuing to divide or augment and, therefore, according to Pascal, knowledge of the existence of infinitely small and infinitely large things; in natural philosophy, we have a certain knowledge that it is possible that natural mechanisms and structures could very well outstrip our ability to imagine or describe them; it is irrational to deny the existence of the geometric entities and irrational to deny the possibility of the natural ones.

26. Du Vair 1641, p. 3. See also Levi 1964, p. 84; and, for Du Vair's emphasis on the need for intense exercise to purge the soul and lead oneself to God, see Petey-Girard 1999. Carraud (1992, pp. 422–26) argues that Du Vair's *Saincte philosophie* is a key unacknowledged source for S230. For the importance of Du Vair in early-modern French elite culture, see Miller 2000, pp. 111–15.

27. See Levi 1964, ch. 3.

28. Fumaroli 1994a, pp. 370–91.

29. See Bellarmine's *De ascensione mentis in Deum per scalas rerum creatarum*, in Bellarmino 1870, vol. 8.

30. Binet 1624. Binet used, however, the rainbow as an example of the limitations of human knowledge.

31. In Joseph Filère's text, the readers literally achieve a proper perspective on all created things. See Filère 1636, p. 518; Fumaroli 1994a, p. 378. For mirrors and rational wonder as instruction, see chapter 5.

32. See chapter 1.

33. Vincent Carraud (1992, pp. 393–450) has emphasized how S230 shares some narrative movements with contemplative literature but fundamentally subverts the knowledge typically held to be gained through contemplative procedures. See also Michon 1996, pp. 97–119.

34. Hence the famous assertion: "It is an infinite sphere whose center is everywhere, the circumference nowhere" (S230). The sentence has a long history as a description of God; Pascal applied it to the universe.

35. See Ferreyrolles 1995.

36. See the complaint at M III:426.

37. Cf. Descotes 1993, pp. 20–21.

38. See S276, S694. For Pascal and Islam, see Wetsel 1994, pp. 182–211.

39. The attack on Muslim *soldiers* is particularly important for debates about salvation. For Pascal and his Jansenist anti-Pelagian allies, the Qur'an's guarantee of salvation to soldiers killed in defense of the faith offered an account of grace that was most antithetical to Christianity. Divine grace is completely scrutable in this case. The Qur'an offered the sort of salvation through works they accused the Jesuits of teaching. For this view and Pascal's mentor Lemaistre de Sacy, see Wetsel 1994, p. 209. For Pascal's considering philosophy as a game, see Michon 1996, p. 21.

40. Arnauld 1644, p. 189. The language of Arnauld's and Pascal's attacks on "philosophy" comes from the massive attack on "pure nature" in Jansen [1640] 1964, e.g., bk. 2, cols. 789–98.

41. Méré to Pascal, M III:353–54.

42. For the Cold War origins of rational-choice theory, see Amadae 2003.

43. For analyses, see Hacking 1975, 57–62; Thirouin 1991; Daston 1988, pp. 15–17.

44. For the letters, see M II:1132–58. For the background to the question, see Mesnard 1965, pp. 367–79.

45. Fermat and Pascal offered considerably different methods. Very briefly, Fermat counted combinations; Pascal considered equity and certain gain. See Daston 1988, pp. 16–17; Edwards 2002, 2003.

46. For a richly textured study of these business, personal, and religious connections between Pascal and the Roannez family, see Mesnard 1965. Mesnard argues that Pascal's friendships with Méré and Mitton were closer than with the more strictly business associates.

47. My discussion rests on Daston 1988, pp. 15–23. For the language of justice and the just, see, e.g., M II:1137, 1147.

48. Pascal elsewhere challenged the view that considerations of justice were integral to rationality itself and argued that communities stem, not from a sociable instinct, but from the workings of self-interest and imagination. Nevertheless, the former view of human rationality structured the answer to the gaming problem.

49. See Daston 1988, pp. 21–22.

50. The full flourishing of probability in Jesuit casuistry involved many economic questions. For a recent, well-documented account, see Stone and Van Houdt 1999.

51. Mitton, "Pensées sur l'*honnêteté*," edition in Grubbs 1932, p. 55.

52. The *honnête homme* has both, to draw on Pierre Bourdieu's account of habitus, "the capacity to produce classifiable practices and works and the capacity to differentiate and appreciate these practices and products (taste)" (1984, p. 170).

53. Mitton, "Pensées sur l'*honnêteté*," edition in Grubbs 1932, p. 55. See the discussion in Wetsel 1994, pp. 122–23. For *honnêteté* and community-making, see Mesnard 1976, pp. 106–10.

54. See the fine discussion in Wetsel 1994, pp. 119–29. Wetsel convincingly argues that the moderate, lukewarm belief exemplified by Mitton is a central target of Pascal's projected apology.

55. Mitton, "Description de la *honnête homme*," edition in Grubbs 1932, p. 56.

56. I take "metaphysical nonchalance" from Geertz 1973, p. 176.

57. Jones 2003.

58. I am consciously not dedicating much space to the famous wager, both because it has been extensively studied from many points of view and because its centrality to the "philosophical" parts of *Pensées* is easily overstated. Its relationship to Pascal's other arguments in *Pensées* is often disregarded. By using mathematics in the wager, he formalized the conundrum of being unable to eliminate metaphysical and physical possibility so central in the "disproportion of man."

59. F. Liceti, *De monstrorum caussis, natura, et differentiis libri duo* (Padua: G. Crivellari, 1616); French trans. (Leiden: Veuve B. Schouten, 1708), pp. 1–2, cited in Céard 1996, p. 443. A key source was Horace's account in his canon of artistic regulation; the poet's license does not go "so far that savage should mate with tame, or serpents couple with birds, lambs with tigers" (*De arte poetica*, ll.12–13; in Horace 1970, p. 451). For this account of monstrosity, see Westman 1990, pp. 178–84. For the interest in monsters more generally, see Daston and Park 1998, ch. 5.

60. For difficulties in working out whether monsters had supernatural or merely unusual and complex natural causes, see Daston and Park 1998, pp. 191–93, 201–10. For a more general survey of means for discerning natural, demonic, and truly supernatural causes, see Clark 1997, chs. 14–17, esp. pp. 268–77.

61. This is no small point, especially as Pascal seems here to open himself up to charges of invalid hypostatization, but I pass over it in the interest of sketching the major structure of Pascal's argument. Minimally, Pascal's argument holds that no secular philosophy, be it Stoicism, Epicureanism, Skepticism, or Cartesianism, can offer an explanation of these aspects of human nature through natural causes alone. Even if they could, they quite simply do not describe human beings as monstrous.

62. What seems to be connected here is "common sense" in our usage (commonly accepted general views on nature, humanity, and God) and "common sense" in Aristotelian usage (roughly, the faculty of synthesizing manifold sense impressions and deriving fundamental and apparently universal principles from their general, not accidental, characteristics). Pascal worked to show that common sense in our usage ought never to produce the standard results attributed to Aristotelian common sense. For analyses of Aristotelian common sense and its experimental uses, see Dear 1995a, ch. 2; Pérez-Ramos 1988.

63. Suárez 1740, bk. 1, ch. 1, n. 15 (vol. 6, p. 208). For the doctrine of "pure nature," see Lubac 1969, ch. 6.

64. Jansen [1640] 1964, "De statu purae naturae," cols. 677–980.

65. See Jones 2003.

66. Here again we see Pascal's insistence on the need to separate arguments about possibility from proofs of existence, discussed in chapter 3.

67. "What are our natural principles, if not the principles we are accustomed to?" (S158). The Ferreyrolles edition traces this to Montaigne 1992, I, 23, p. 115; noted in Pascal 2000, p. 111 n. 3.

68. "Ecrits sur la grâce" (c. 1655–56?), T1, §6, M III:767. Cf. S683.

69. Hadot 1995, p. 103 (italics in original).

70. Carraud makes a similar point about metaphysical proofs of God for Pascal: the pursuit of a metaphysical proof and concept of God precludes coming to more limited

knowledge of and love for God. "Dès lors, plus preuve métaphysique de l'éxistence de Dieu est performante, plus elle interdit l'accès à lui-même. Dieu connu, c'est-à-dire Dieu en tant que concept de Dieu, cache Dieu" (Carraud 1992, p. 382).

71. See Carraud 1992, ch. 2, for a thorough analysis of Pascal's classification of philosophies and philosophers.

72. Aristotelianism appears not to be something he felt necessary to refute in the *Pensées*, at least not for convincing the audience of *honnêtes gens* at whom he aimed his apology.

73. Le Moyne 1645, pp. 592–93. Such a polemical view offered a mistaken vision of the Stoics, particularly as they had been understood in the late Renaissance. See Tuck 1993, p. 54.

74. Pascal expressed such a point fairly clearly in S179: "Stoics. / They conclude that one can always do what one can do sometimes."

75. See "De l'art de persuader," M III:413.

76. For a thorough analysis of Pascal's various attacks on metaphysical proofs, see Carraud 1992, pp. 347–92, esp. pp. 356–57.

77. Descartes had similar concerns about the power of traditional logical demonstrations to provide psychologically satisfying and also truthful demonstrations. Both Pascal and Descartes illustrate the continuing power of the long humanist jeremiad against scholastic logic, and both illustrate the power of that set of critiques to provoke new forms of reasoning.

78. Pascal allowed that "la nature est une image de la grâce," but in S38 noted that natural things reveal God only "pour quelques âmes," and not for most people. See also S738; S306; Pascal to Mlle de Roannez, [29.10.1656], M III:1035–37.

79. Grotius 1632, I, 6, p. 23 (italics in original). For the importance of Grotius's book and his notion of *credo minimum*, see Miller 2000, pp. 103–6; Lagrée 1991, esp. pp. 184–86, 191–92.

80. In S644 Pascal moved immediately to a theological and metaphysical explanation, not just a psychological account, of why natural-theological proofs never could satisfy: God is a hidden god, a *deus absconditus*, and therefore not present in the world as the purveyors of such proofs hold.

81. For one thesis on the development of *sentiment* in Pascal's thought, see Jones 2001.

82. See Wetsel 1981, p. 126.

83. Force 1989, esp. pp. 50–51; see also the important remarks in Wetsel 1981, pp. 133–38, 166.

84. See Sellier 1999 for Pascal's account of the rhetoric of the Bible.

85. Pascal stressed both the epistemic and the rhetorical aspects of the necessity for Christ. See S224 and S225.

CHAPTER FIVE

1. Leibniz to Hermann Conring, 13.1.1678, A2,1:387.

2. Leibniz to Conring, 13.1.1678, A2,1:387.

3. Leibniz to Conring, 19.3.1678, A2,1:402; L191.

4. "La vraie méthode," [first half of 1677], A6,4:5.

5. Although this chapter ranges across several topics in his early work, it does not consider numerous key questions in his philosophy of mathematics in any depth, most

notably, his changing views concerning infinitesimals and the continuum, subjects other scholars have discussed with great care; see Richard T. W. Arthur's fine edition and translation LC, as well as Arthur, forthcoming a; Levey 1998; Bassler 1999; Knobloch 1994, 2002; Bertoloni Meli 1993.

6. This result was first published in 1682 (GM V:118–22), and Leibniz announced or proved the result in a number of places: Leibniz to Oldenburg, 15.8.1674, A3,1:120; to Mariotte, 10.1674, A3,1:139–41; to Huygens, 10.1674, A3,1:141–69; to Le Roque, late 1675, A3,1:336–55; to Gallois, late 1675, A3,1:355–63; to Arnauld, [after 1678], A3,2:779–82; and others. For overviews of the proof, see Hofmann 1974, pp. 54–62; Guicciardini 1999, pp. 139–42; Granger 1981; Mahnke 1926, pp. 11–13; Parmentier 2001; Knobloch 1989a and his introduction to K.

7. Leibniz's series was new, but both his contemporary David Gregory and Indian mathematicians of the fifteenth century (perhaps Mādhava, 1340–1425) had independently found formulations of the infinite series for the arctangent, of which Leibniz's result is a special case. See Roy 1990, reprinted in Berggren, Borwein, and Borwein 2004, pp. 92–107—a very useful collection of primary and secondary literature on π.

8. See Leibniz to Gallois, late 1675, A3,1:359. Cf. Leibniz to Le Roque, late 1675, A3,1:346; Granger 1981, pp. 23–24.

9. For the transmutation theorem, see the note at A3,1:115–16; Hofmann 1974, p. 55; Mahnke 1926, pp. 9–10, 34–35, depending on the still-unpublished documents Cc 545–46, among others. Although Leibniz certainly would not have missed the alchemical resonance of the term "transmutation," the term was in wide use among mathematicians. See H. Heurat's *Epistola de transmutatione curvarum linearum in rectas*, in Descartes 1659, pp. 517–20. See also Kircher and Schott 1669, bk. 8, ch. 3: "De transmutatione quadrangulorum in alias figuras planas," at p. 244, and pp. 247–50; Jungius n.d., p. 9.

10. See M IV:478, 563, fig. 26; Loeffel 1987, pp. 108–11; Descotes 2003, p. 50.

11. As *ds* becomes infinitely small, it becomes a straight segment rather than a curve.

12. "Fines geometriae," [summer 1673], Cc 552, A7,4 Vorausedition (hereafter VE), [24.3.2003], no. 25, p. 82. See Hofmann 1974, pp. 48–49; Leibniz to Tschirnhaus, 5.1679, A3,2:932–33.

13. The new curve, in notation Leibniz would soon invent, is equivalent to

$$z = y - x\frac{dy}{dx}$$

14. "Fines geometriae," [summer 1673], Cc 552, A7,4 VE, no. 25, p. 79.

15. "Fines geometriae," [summer 1673], Cc 552, A7,4 VE, no. 25, p. 80. See his more detailed discussion in "Schediasmatis de summis serierum et quadraturis figurarum pars X," 10.1674, A7,3:484–88. Compare similar remarks elsewhere, e.g., Leibniz to [?], [1679], A2,1:502–3. As we saw in chapter 1, Descartes maintained the traditional notion that "the proportion between straight lines and curves is not known" (AT VI:412). See Molland 1976, pp. 26, 36; Bos 1981, p. 314; Hofmann 1974, pp. 101–3. For the contrast with Descartes' limitations on allowable curves, see, e.g., Breger 1986.

16. For the quadratures of Wallis and Brouncker, see now Stedall 2002, pp. 156–65, 185–94. Wallis had found the product

$$\frac{\pi}{2} = \frac{2 \cdot 2 \cdot 4 \cdot 4 \cdot 6 \cdot 6 \cdot 8 \cdot 8 \cdots}{1 \cdot 3 \cdot 3 \cdot 5 \cdot 5 \cdot 7 \cdot 7 \cdot 9 \cdots}$$

and Brouncker the continuing fraction

$$\pi = \cfrac{4}{1 + \cfrac{1}{2 + \cfrac{9}{2 + \cfrac{25}{2 + \cdots}}}}$$

Newton drew heavily on the work of Wallis, but by 1674 Newton had surpassed him in rigor and generality. Leibniz maintained that Wallis's "expression" was good for approximations, "sed non pro exacta expressione per infinitam seriem considerata semel in universum" ("De serie Wallisiana," [1676], A7, 3:824).

17. "Fines geometriae," [summer 1673], Cc 552, A7,4 VE, no. 25, pp. 81–82. While trumpeting the extent of his personal contributions, Leibniz underlined the importance of the works of Descartes, Mercator, Pascal, Wallis, Wren, Heurat, Huygens, and Brouncker to his efforts.

18. Huygens to Leibniz, 6.11.1674, A3,1:170.

19. See Stedall 2002, pp. 165–72; Jesseph 1999.

20. For the varying accounts of mathematical knowledge during the period, see Mancosu 1996; see also the sources cited in chapter 1.

21. Leibniz noted that his series did not show anyone how to construct a square equal to a given circle using geometric techniques. Leibniz to Gallois, end of 1675, A3,1:356. See also GM V:97. For Leibniz's considerations about whether the ratio of the circumference of a circle to its diameter is transcendental, that is, not the root of any integer polynomial, see "Numeri infiniti," [4.1676], A6,3:497–98; trans. in LC 87, discussed by Arthur at p. lxxi, and especially Arthur 1999; for Leibniz's emphasis on and naming of transcendental curves, quantities, and numbers in mathematics, see Breger 1986.

22. In a draft preface to K, Leibniz offered a taxonomy of quadratures. See GM V:96. Compare with different categories of mathematical knowledge discussed in "Schediasmatis de summis serierum et quadraturis figurarum pars X," 10.1674, A7,3:486–88.

23. Leibniz to Oldenburg, 16.10.1674, A3,1:131 (my italics). The language of "exhibition" also figures in the summer 1673 "Fines geometriae," Cc 552, A7,4 VE, p. 81.

24. See also GM V:96. The vocabulary of "confessing" figures very prominently in "Schediasmatis de summis serierum et quadraturis figurarum pars X," 10.1674, A7,3:484–85.

25. Leibniz apparently rejected, as early as 1673, the existence of some single infinite number greater than all others (the categorematic infinite), as opposed to the assurance that, for any number, a still larger one exists (the syncategorematic infinite). While Leibniz countenanced the possible existence of actual infinitesimals—categorematic infinitesimals—in mathematics and natural philosophy through 1676, early in that year he began to consider them merely as fictions, as *"compendium loquendi,"* which Richard Arthur describes as "shorthand for the fact that finite variable quantities may be taken as small as desired, and so small that the resulting error falls within any present margin of error." See Arthur, forthcoming a, esp. "Phase 4"; his argument rests on "Numeri infiniti," [4.1676], A6,3:502–4; LC 97–101. Cf. Levey 1998; Bertoloni Meli 1993, pp. 61–65.

26. Earlier in the treatise he defended the beauty of his method of exhibition (K56).

27. As best as I can tell from the documents currently available, Leibniz began using the term "expression" in earnest as his general term for symbolic enunciations by fall 1674; see, e.g., "Expressio unius literae per multas," 4.9.1674, Leibniz 1976, p. 4; "De progressionibus et geometria arcana et methodo tangentium inversa," 12.1674, A7,3:557–58. The term figures ever more centrally as the general substantive for equations and formulas in the following years; see the numerous titles cited below; see, e.g., A7,3, nos. 48, 61, 73; as well as later examples, "Summa calculi anlytici fastigium," 12.1679, and "Mira Numerorum omnium expressio per 1 et 0," [17?.5.1696], in Zacher 1973, pp. 216–28. Compare the use of "expression" in one of his earliest papers explaining the quadrature of the circle: "De serie differentiae inter segmentum quadrantis et eius fulcrum," [fall 1673], A7,3:277; and see also Leibniz to Gallois, late 1672, A3,1:18. For the development and articulation of the metaphysical account of "expression," discussed below, see Mercer 2001, pp. 427–36. She argues that a passage in "De vera methodo philosophiae et theologiae ac de natura corporis," A6,3:157, is the first formal use of "expression" as a technical term "to describe the relation between an emanative source and its product" (2001, p. 405 n. 74); she dates the paper to 1672–73 (2001, p. 403).

28. Parmentier 2001, p. 285. Parmentier stresses that this claim of exactitude is one of Leibniz's key innovations here. See also Arthur, forthcoming a; Knobloch 2002, p. 63; Ferraro 2000, pp. 46–47, on Leibniz's new notion of equality; see also Breger 1986, p. 131.

29. In 1676, Leibniz explained, "Whenever it is said that a certain infinite series of numbers has a sum, I am of the opinion that all that is being said is that any finite series with the same rule [*regula*] has a sum, and that the error always diminishes as the series increases, so that it becomes as small as we would like" ("Numeri infiniti," [4.1676], A6,3:503; trans. in LC 99).

30. See Parmentier 2001, p. 286. For a fine discussion of Leibniz and infinite series throughout his mathematical work, see Ferraro 2000, esp. pp. 45–47, on these early works. For an account of the importance of his pre-Paris considerations of the continuum for his work with series and sequences, see Bassler 1999, pp. 173–80.

31. Cf. Beeley 1999, pp. 144–45.

32. See also GM V:120: "Et licet uno numero summa ejus seriei exprimi non possit, et series in infinitum producatur, quoniam tamen una lege progressionis constat, tota satis mente percipitur."

33. "Geometriae utilitas medicina mentis," [1676], A6,3:451.

34. Leibniz to Conring, 19.3.1678, A2,1:402; L191.

35. Leibniz to Gallois, late 1675, A3,1:358 (italics in original).

36. Laloubère 1651, pp. 13–14; trans. in Jesseph 1999, p. 313 n. 9; cited in Hobbes 1966, vol. 7, p. 320.

37. See, e.g., "Elementa juris naturalis," 1671, A6,1:484; "De conatu, et motu, sensu et cogitatione," [1671], A6,2:283. For the centrality and richness of harmony for Leibniz, see Antognazza 2003; Mercer 2001, passim, esp. pp. 208–20, 427–40.

38. "De constructione," [summer–fall 1674], A6,3:415. After using the cone to construct the various conic sections (circle, ellipse, hyperbola, parabola), Apollonius proved various theorems about each of them in the plane. He then noted analogies among the theorems that apply to the various sections. Taking for granted that the conic sections can be understood as projections of one another, Desargues demonstrated a number of

properties that did not change under projection. A proof about certain characteristics of one conic section thus applied to the others. Desargues thereby unified the various analogous theorems. See the fine exposition in Field 1997, pp. 183, 202, 205; for more detail, see Field and Gray 1987. See also Leibniz's remarks on how Pascal's method allows one to prove theorems about the circle and then generalize them at M II:1125; cf. Child 1920, p. 188.

39. "De la méthode de l'universalité," C98.

40. "De la méthode de l'universalité," C98.

41. For his sense of the importance of his discoveries as he was making them, see "Fines geometriae," [summer 1673], Cc 552, A7,4 VE, no. 25.

42. Pascal, Fragments on conics, as copied by Leibniz; M II:1127. See Costabel 1962, p. 259.

43. See "De la méthode de l'universalité," C98; "De constructione," [summer–fall 1674], A6,3:415.

44. See Leibniz to Gallois, late 1675, A3,1:359. Similar sentiments are expressed in K36; "De constructione," [summer–fall 1674], A6,3:415. See the starting point of Desargues 1639, fol. Ar; see Field and Gray 1987, p. 70.

45. Leibniz's notes are all that remain of Pascal's work, see M II:1120–31. René Taton, Jean Mesnard, and Javier Echevarría have detailed what Leibniz knew and when. See Taton 1978; Mesnard 1978; Echevarría 1983, 1994; see also Costabel 1962.

46. Leibniz to Bernard Le Bovier de Fontenelle, 12.7.1702, in Fontenelle 1989–97, vol. 3, p. 394.

47. Bosse 1973, *Traité*, p. 39. This text is suggested in Taton 1978, p. 120 n. 89; see also p. 124 n. 102. See Leibniz's annotations before the title pages of works of Bosse, quoted in Echevarría 1983, p. 195. Echevarría (1994) has illustrated Leibniz's knowledge of perspective and discussed the debates within and outside the French Academy of Painting concerning perspective. For these debates, see n. 203 below.

48. See Bredekamp 2004, p. 79.

49. "Drôle de pensée," 9.1675, A4,1:562–68, with annotations at A4,1:694–96 (the annotations are in A4,2 in some editions). See now the important study by Bredekamp (2004, esp. pp. 45–48, 79–80, etc., with an improved edition of the text at pp. 200–206 and a German translation with detailed annotations at pp. 237–46). See also Belaval 1958; Wiedeburg 1962, pt. 2, vol. 1, pp. 610–39; English trans. in Leibniz 1951, pp. 585–94.

50. "Drôle de pensée," 9.1675, A4,1:567.

51. "Drôle de pensée," 9.1675, A4,1:563–65.

52. "Drôle de pensée," 9.1675, A4,1:565.

53. "Drôle de pensée," 9.1675, in Bredekamp 2004, p. 202. The edition in A4,1:564 gives a blank for the second half of the line.

54. Though painted long before Leibniz's day, the greatest example of anamorphosis is Hans Holbein's *Ambassadors* in the National Gallery, London. A silver-white slash cuts diagonally across the lower third of the painting. If seen from an appropriately oblique point of view, this slash becomes a skull floating amid the two ambassadors, their sumptuously displayed wealth, and their instruments of knowledge.

55. See Baltrušaitis 1984; Kemp 1990, pp. 208–13; Bessot 1999. See also Pérez-Gómez and Pelletier 1995, covering both primary and secondary literature.

56. Quoted in Baltrušaitis 1984, p. 82.

57. For books of artificial and natural magic, see Eamon 1994. For governing arts in seventeenth-century Europe, see Donaldson 1988, ch. 4; Apostolidès 1981; Maravall 1986. For a sophisticated analysis of deception and governing in the seventeenth century, see Pascal's remarks at S59, 74, 116, 454. As explained in chapter 2, rhetoric was understood as a rational, passion-directing art on all sides of the confessional divide. For "scientific" spectacles, see Gorman 1999, 2001, 2002; Findlen 2003. For the use of demonstrations of natural phenomena to teach natural and political truths more broadly, see especially Schaffer 1983, as well as Schaffer 1994, 1995.

58. Brusati 1995, p. 172.

59. Niceron 1652, pp. 21–24.

60. See the fine discussion in Malcolm 2002, pp. 206–10.

61. See the editor's introduction to the facsimile reprint, Harsdörffer and Schwenter 1991, vol. 1, pp. xxi–xxiii.

62. Among many examples, see Harsdörffer and Schwenter 1991, vol. 1, pp. 302–6, vol. 2, pp. 246–47, vol. 3, pp. 239–41, 244–45. Many of the illustrations were simply copied from the works of Kircher, Niceron, and other authors.

63. For Leibniz's knowledge of Harsdörffer and Schwenter, see the editorial annotations to "Dissertatio de arte combinatoria," 1666, A6,2:548–50; introduction to Harsdörffer and Schwenter 1991, vol. 1, pp. xxxv–xxxvii. See also Westerhoff 1999. Harsdörffer is mentioned in the "Drôle de pensée" at A4,1:564; it is likely a reference to Harsdörffer 1968.

64. Leibniz was reading, for example, Honoré Fabri's advanced *Synopsis optica* (1667), which, among other things, focused on the power of perspective for understanding the varied appearances of celestial bodies, notably Saturn and its rings. See A7,1:106, for the use of Fabri, though not on this point. Echevarría has detailed considerations of the perspective treatises Leibniz probably or certainly read. See Echevarría 1983; Echevarría 1994, pp. 283–88; these important articles do not discuss Leibniz's reading in the optical literature or in sources such as Schwenter and Harsdörffer.

65. "Drôle de pensée," 9.1675, A4,1:565.

66. For an extraordinary list mixing works of mathematics and mathematical magic, see Sketch of a Bibliotheca Universalis Selecta, [1689], A1,6:442–43.

67. "Confessio philosophi," [fall 1672–winter 1672/73], A6,3:135 (italics in original); quoting Matthew 13:13, where Jesus quotes Isaiah 6:9. See Belaval's commentary at Leibniz 1970, p. 131 n. 119; and Mercer 2001, pp. 395–96.

68. "Drôle de pensée," 9.1675, A4,1:570.

69. For the sort of transformation Leibniz deemed necessary, see the remarkable text written for Duke Johann Friedrich: "But these same demonstrations cannot be exposed in a disordered speech; nor if they could, should they be. For they merit not a curious reading but the patience of the attention; neither are they set forth such that, flowing into the ears following a speech, they move one to a certain temporary rapture; rather, they should descend into the mind and indelibly implant themselves and be perpetually considered in acting: in the same way that the demonstrations of Euclid's geometry should not be run through but closely examined and resolved into the most basic Elements, until they attain clarity and can be neglected by none" ("De usu et necessitate demonstrationum immortalitatis animae," [1671], A2,1:114). See Mercer 2001, p. 384.

70. "Marii Nizolii...libri iv," [1670], A6,2:409. For some commentary, see Fenves 2003, p. 72.

71. "De vi persuadendi; de somnio et vigilia," [1669–summer 1670], A6,2:276. For Leibniz's views on the persuasive powers of language, see Gensini 1996, pp. 82–83, which examines "Modus examinandi consequentias per numeros," 4.1679, A6,4:230; Heinekamp 1972, pp. 482–83.

72. Cf. Piro 1999.

73. For the importance of the Jesuits as a model, see, e.g., "Societas philadelphica," [1669], A4,1:554.

74. "Nova methodus discendae docendaeque jurisprudentiae," 1667, A6,1:274–76.

75. "Drôle de pensée," 9.1675, A4,1:567.

76. Leibniz to Johann Friedrich, 9.1678, A1,2:76–77.

77. See the annotations at A4,1:696. Now see Bredekamp 2004, ch. 4.

78. Descartes, *Passions de l'âme*, §§70–77, AT XI:380–86. For Leibniz's reworking of *Passions*, see chapter 6.

79. Brusati 1995, pp. 90–91, 193–94, 291 n. 89. Published in 1678, Van Hoogstraten's *Inleyding tot de hooge schoole der schilderkonst* cannot be Leibniz's direct source.

80. "Origo regularum artis perspectivae quales sine libro ac magistro inveni"(LH XXXV, XI, 17, fols. 19–20); "Fundamentum perspectivae meo marte investigatum" (LH XXXV, XI, 17, fol. 21). Using Leibniz's choice of equal sign, Echevarría (1983, p. 198) dates them to c. 1678–85. See also Bredekamp 2004, pp. 79–80.

81. Niceron 1652, p. 133.

82. "Scientia perspectiva," [1695–], LH XXXV, XI, 1, fols. 1–2, partial edition in Echevarría 1983, pp. 198–99; cf. Niceron 1652, p. 128. For ease of reading, I have omitted a series of ellipses in this quotation; every ellipsis is an important set of generalizations and clarifications.

83. Echevarría 1983, p. 200.

84. "Recommandation pour instituer la science général," [1686], A6,4:708–9. See also the remarkable analogy of the characteristic and scenographical representation in "Characteristica geometrica," 10.8.1679, GM V:141.

85. See Mercer, 2001, pp. 303–19.

86. Leibniz to Johann Friedrich, [10].1671, A2,1:161. Belaval (1969, p. 60) remarks on this letter as potentially important in understanding Leibniz's increasing interest in the metaphor of point of view.

87. Leibniz 1671; edition in Leibniz 1768, vol. 3, pp. 14–15.

88. See Leibniz to Spinoza, 5.10.1671, A2,1:155; Spinoza to Leibniz, 9.11.1671, A2,1:184–85; Carcavy to Leibniz, 10.7.1671, A2,1:139; Leibniz to Arnauld, 11.1671, A2,1:180; Johannes Ott to Leibniz, 6 or 16.7.1671, A2,1:140–41. The pamphlet belongs within the seventeenth-century genre of announcements of devices that provided just enough detail for the author to claim priority if the invention should come to fruition but not enough to allow theft or replication of the central idea.

89. A few of these studies are printed in Ger. As with most of Leibniz's technological and experimental papers, they have been little studied. Just as I am completing this work, the firstfruits of the online edition of series 8 of the Deutsche Akademie der Wissenschaften edition of Leibniz's works have begun to appear, so more scholarly editions will soon be available. Although Gerland did not give dates for many of these

documents, there can be little doubt about the date of their composition: Leibniz's enthusiasm for the Jesuit Lana Terzi's work, discussed below, is steady but short-lived, from 1671 to 1674 or so. Other documents on optics and dioptrics are either dated later and/or involve notation and calculational techniques that Leibniz devised only later. Some of Leibniz's more theoretical optical concerns in the period are included among his "philosophical" papers; see, e.g., "Leges reflexionis et refractionis demonstratae," [second half of 1671], A6,2:309–23. For his theoretical optical work, especially his use of the principle of least action, see Hecht 1996.

90. Leibniz's references to Ott are obscure (e.g., Ger 94); he is almost certainly referring to Ott 1671. Before and after he went to Paris, Leibniz was reading many of the best books on optical instrumentation of his day: e.g., Schyrleus de Rheita 1645, vol. 1, pp. 336–59, on binoculars; see Thewes 1983. In the *Notitia*, Leibniz likewise refers to the important discussion of optical instruments in Hevelius 1647, pp. 1–31. Slightly later, Leibniz calls for a "Tabula Catoptrica. Tabula dioptrica. Nucleus Cherubini"—based on the best technical optics manual of the day: Cherubin d'Orléans 1671. See A6,4:88, line 13.

91. Lana Terzi 1670. For the cultural sphere in which Kircher, Lana Terzi, and Schyrleus de Rheita produced their works, see Evans 1979, esp. pp. 330–40. For Lana Terzi and the conditions for the production of his *Prodromo*, see Gorman 2000. For its importance among instrumental optical treatises in the seventeenth century, see Bedini 1994, p. 260; Bedini and Bennet 1999, p. 107.

92. Lana Terzi 1670, p. 10.

93. See Leibniz's reading notes on the optical sections of Lana Terzi's *Prodromo*: LH XXXVII, II, fol. 2r–v. High-resolution versions of these manuscripts are now available at http://echo.mpiwg-berlin.mpg.de/content/scientific_revolution/leibniz. Preliminary editions are available through http://www3.bbaw.de/forschung/leibniz/web/Leibniz_Reihe_8/index.php. Leibniz praises Lana Terzi's famous plans for flying ships in his "Hypothesis physica nova," [winter 1670–71], A6,2:233. For Leibniz's skepticism about some of Lana Terzi's claims, see A2,1:142n, noted in Gorman 2000, p. 421.

94. Lana Terzi 1670, p. 219.

95. One finds calls for improving optical distinctness in his other sources as well. Nevertheless, his key, much-repeated reference in these texts seems to have been Lana Terzi and, to a lesser extent, the even less well known Ott. For other uses, see, e.g., Hevelius 1647, p. 5; Schyrleus de Rheita 1645, vol. 1, p. 338.

96. Compare Leibniz's remarks in the *Notitia*. For the importance of concave mirrors as magnifying devices, especially in conjunction with lenses, see Gorman 2003. For the impossibility of using anything but flat mirrors for accurate observational purposes in the seventeenth century, see R. Wilson 2004, pp. 10–11.

97. See also Leibniz's considerations on distinctness and clarity with respect to telescopes in LH XXXVII, II, fol. 12r; and his notes and reflections on Lana Terzi in LH XXXVII, II, fol. 2r–v. "Confused" and "distinct" were, of course, good scholastic terms that Lana Terzi appears to have applied to questions in optics, as did other Jesuits, such as Honoré Fabri, not to mention Descartes. See, e.g., the discussion of divine and human knowledge of infinity in Coimbra 1984, vol. 1, p. 393.

98. Due to chromatic aberration, different colors focus at different places along the axis. Only with Newton's theory of light did the importance of chromatic aberration

slowly become understood. Leibniz was writing just before and as Newton's theories began to be debated. For this moment in relation to telescope design and optical improvements, see Dijksterhuis 2004, pp. 83–92.

99. See Wilson 1995, pp. 81–84; Hacking 1983, pp. 192–94.

100. See the fine discussion in Wilson 1995, ch. 7.

101. [10?.1675], Cc 1077, quoted in Hofmann 1970, p. 104. The quotation comes from a discussion of a mechanical, not an optical, instrument.

102. For a survey of then-current forms of tachygraphy, in a source we know Leibniz read, see Schott 1664, bk. 8, with discussions of proposals new and old.

103. "Totos libros explicare una figura," [1670–71], A6,2:477–78.

104. "De differentiis progressionis harmonicae," [1672–73], A7,3:126.

105. "Totos libros explicare una figura," [1670–71], A6,2:478. Cf. "De arte inveniendi in genere," [summer–fall 1678], A6,4:81.

106. "Demonstratio propositionum primarum," [fall 1671–beginning of 1672], A6,2:481. Leibniz was responding to Hobbes's skepticism about whether algebraic signs can truly produce new knowledge in the chapter of *De corpore* containing an attempted quadrature of the circle. See Hobbes 1999, ch. 20, pp. 220–21.

107. "Demonstratio propositionum primarum," [fall 1671–beginning of 1672], A6,2:481.

108. "Theoremata sunt cogitandi compendia," [summer 1674–fall 1676], A6,3:426–27.

109. Leibniz to Johann Friedrich, [10].1671, A2,1:160–61. For example, in the same 1671 letter to Johann Friedrich that mentioned his three optical discoveries, Leibniz promised an instrument called a "living Geometry" "through which the entirety of geometry, insofar as useful to life, will be perfected at once." The machine could mechanically perform addition, multiplication, and calculations involving proportions, "without any head breaking." Leibniz's machine offered him hope of bringing "the quadrature of the circle" into practice.

110. For some a and b, x is a mean proportional if $a/x = x/b$. Finding a mean proportional with straightedge and compass is one of the unsolved problems of antiquity; Descartes' compass produces mean proportionals.

111. Lana Terzi 1670, p. 252.

112. For another example, see the proportional circle as described and pictured in Harsdörffer and Schwenter 1991, vol. 3, pp. 129–30. For a survey of instruments in mathematics and elsewhere, see Bennett 1987; I. Schneider 1970.

113. Ger 107. In a roughly contemporary mathematical study, Leibniz referred to an example of using optics to construct ellipses taken from Honoré Fabri's important technical treatise *Synopsis optica* (1667). See "De geometria seu potius algebra mechanica," [spring–summer 1673], A7,1:106, citing Fabri 1667, p. 73. The editors of A7,1 note that Leibniz underlined the passage in his copy of Fabri.

114. There can be little doubt about the dating of this fragment. His enthusiasm for Lana Terzi begins in 1671 and wanes by 1674–75; Buot's problem, mentioned in the fragment, was a major focus of inquiry only after Leibniz arrived in Paris. See his proposed mechanical solution of the problem in A7,1:105. Compare the mention of optics in A7,3:560.

115. For his later considerations of these sorts of curves, see, e.g., A7,3:492; he drew on the review of Gregory, in *Philosophical Transactions* 3 (1668): 685–88.

116. For a survey of such machines, see Kemp 1990, pp. 167–88, esp. pp. 180–81.

117. See the detailed description in Kircher and Schott 1669; also discussed in Kircher 1641, bk. 2, pt. 2, ch. 3, which Leibniz had read, as shown by annotation to a work of Fabri; A6,2:216 and the reference in "Hypothesis physica nova," 1671, A6,2:239.

118. See Gorman 2002, pp. 5–9.

119. See, e.g., the instrument for solving algebraic problems discussed in Hofmann 1970, pp. 101–4; see also Leibniz to Huygens, [9.1675], A3,1:280; Huygens to Leibniz, 30.9.1675, A3,1:283 (draft), 284; see also "Calculus per instrumenta," [12.1675], A7,1:904–5; for his arithmetic machines, see Leibniz to Oldenburg, 8.3.1673, A3,1:40.

120. See also "De geometria seu potius algebra mechanica," [spring–summer 1673], A7,1:105.

121. "De geometria seu potius algebra mechanica," [spring–summer 1673], A7, 1:108.

122. "De arithmetica infinitorum perficienda," [spring–summer 1673], A6,3:408, 409. In 1673 Leibniz was carefully reading John Wallis's *Arithmetica infinitorum* (1656) and clearly sought to make its interpolative techniques more rigorous; see Hofmann 1974, pp. 52–54. He set out many times what the perfection of the art of progressions required; see, e.g., "De methodi quadraturarum usu in seriebus," [8–9.1673], A7,3:252–53, on the need to discern the rule underlying "a given series" that is "confused."

123. In spring 1673 Leibniz learned a crucial technique of dividing equations with binomials from Nicolaus Mercator's *Logarithmo-technia* (1668); this technique allowed him to convert the equation produced using his transmutation theorem into an infinite series.

124. Leibniz to La Roque, [late 1675], A3,1:346. He claimed that his combinatorial art allowed him quickly to run through many metamorphoses.

125. See A7,3 and Arthur, forthcoming b.

126. "Demonstratio propositionum primarum," [fall 1671–beginning of 1672], A6,2:482 (italics in original).

127. For lucid introductions to Leibniz's calculus, see Bos 1993; Bos 1980, pp. 60–70; Guicciardini 1999, pp. 136–45; Bertoloni Meli 1993, pp. 56–68; all drawing on the fundamental study Bos 1974, esp. pp. 13–35; see also Hofmann 1974, pp. 187–201.

128. "Extrapolation" is Bos's apt metaphor (1974, p. 13). For integration, Leibniz used the term "summation"; "integration" is Bernoulli's term, which Leibniz disliked (see Bos 1974, p. 21).

129. "Analyseos tetragonistae pars 2da," 29.10.1675, in Leibniz 1899, p. 155, modernized slightly.

130. Current mathematical usage reserves "series" for infinite sums, using "sequence" where Leibniz used "series" or *"progressio."* To bring out some of the connections in his treatment of mathematical objects and other objects of knowledge I sometimes use the term "series" in his sense. Bos (1974, p. 16) notes that Leibniz used "series" and *"progressio"* to mean "an ordered sequence of values."

131. See the mature treatments at A7,3:361, 368, etc. See also Hofmann 1974, pp. 17–18; Edwards 2002, pp. 104–7.

132. For the different uses of the triangles, see "Triangulum harmonicum et triangulum Pascalii," 2.1676, A7,3:708–11 and K84–88. For a facsimile of Leibniz's annotations on Pascal's printed triangle, see Loeffel 1987, p. 61.

133. He later promoted this as a good general practice; see "De combinatoria et usu serierum," [1680–84], A6,4:415–16.

134. See, e.g. "De methodi quadraturarum usu in seriebus," [8–9.1673], A7,3:252, as well as later papers.

135. Leibniz often referred to these small differences as infinitesimals, but after April 1676 he did not hold them to be actual infinitesimals; see above, p. 301 n. 25. Bertoloni Meli (1993, p. 63) helpfully calls them "incomparably smaller" quantities.

136. From April 1676 at the latest, Leibniz held that "infinitely sided" is shorthand (a compendious expression) for a polygon with a number of sides greater than any number we might choose, not for a polygon with an actually infinite number of sides. See "Numeri infiniti," [4.1676], A6,3:498; LC 89; see also Arthur, forthcoming a.

137. Initially these uses involved holding one of ds, dx, or dy constant. In the early works exploring his new procedures, he expressed the idea of holding dx or dy constant by saying that x or y increases as an "arithmetical progression"—betraying again the roots of his procedure in considerations about sequences of numbers. See, e.g., "Methodi tangentium inversae exempla," 11.11.1675, in Leibniz 1899, pp. 162–64. For the power of dividing the curve in various ways, see Bos 1974, pp. 25–26; Bos 1993, pp. 84–90; Bertoloni Meli 1993, pp. 66–68.

138. His late historical account underlines this continuity: "Historio et origo calculi differentialis," GM V:392–410.

139. "Elementa calculi novi pro differentiis et summis...," [before 9.1684], in Hess 1986, pp. 100–101; cf. Bos 1974, p. 20; Child 1920, p. 142.

140. See Bos 1974, pp. 16–17, 34–35.

141. Leibniz to Tschirnhaus, 5.12.1679, A3,2:925; trans. in Guicciardini 1999, p. 145.

142. See Hess 1986.

143. Leibniz to Tschirnhaus, 5.12.1679, A3,2:925; trans. in Guicciardini 1999, p. 145.

144. "Expressio quantitatis per seriem," [late 4.1676], A7,3:734. He claimed, by his theorem, that "any figure whatsoever is transformed in another rationally equivalent figure.... So it will come about that any figure can be reduced to infinite series" (Leibniz to Oldenburg, 27.8.1676, A3,1:569; trans. in Newton 1959–77, vol. 2, p. 65); on this, see Hofmann 1974, 234–35.

145. "De figurarum areis per infinitas series exprimendis regula generalis," 28.6.1676, A7,3:768–69 (italics in original). Compare his draft for Oldenburg two months later, A3,1:561–62.

146. See one version at GM V:408.

147. See the edition in A6,3; see the essential translation in Pk, as well as the translations and helpful annotations in Leibniz 1998. For starting points on the literature, see Parkinson 1986; Mercer and Sleigh 1995; S. Brown 1999; Mercer 2001; Dascal 1987a; Dascal 1978, ch. 7, esp. pp. 180–206.

148. "De mente, de universo, de deo," 12.1675, A6,3:463; trans. modified from Pk 5.

149. AT X:407–8.

150. "Demonstratio propositionum primarum," [fall 1671–beginning of 1672], A6,2:479. See Dascal 1987b, p. 70; Dascal 1978, esp. chs. 6–7.

151. Preparatory work on the Characteristic [second half of 1671–spring 1672], A6,2:493. Or again, in his introduction to Nizolius, he wrote, "a definition is nothing other than a signification expressed in words; that is, more briefly: a signified signification"

("Marii Nizolii . . . libri iv," [1670], A6,2:411). Some early texts include sharp criticisms of clarity and distinctness; see "Novus methodus discendi . . . ," 1667, §25, A6,1:280. Nevertheless, in such texts signs are supposed to lead one to cognitions of ideas themselves.

152. Leibniz to Gallois, late 1672, A3,1:17. One of his central concerns in this letter was to counter Hobbes's claims about the arbitrary nature of definitions; see above, pp. 218–19.

153. Leibniz to Gallois, late 1672, A3,1:18.

154. See M II:1203, discussed in chapter 3.

155. For example, Leibniz invoked Pascal's use of the sums of powers to find quadratures of parabolas in "De artibus resolvendi progressionem irreductam," [7–12.1672], A7,3:30–31. This interest in extending Pascal's discoveries was a central part of his mathematical practice; see, e.g., "Triangulum harmonicum et triangulum Pascalii," 2.1676, A7,3:708–11, among many others.

156. Leibniz to Gallois, late 1672, A3,1:17. The phrase "blind thinking" appears already in his "Dissertatio de arte combinatoria" of 1666; see Maat 2004, pp. 306–7. For signs as aids to "mnemonics" in Leibniz's early thought, see "Novus methodus discendi . . . ," 1667, §23, A6,1:277–78; L88; Dascal 1978, ch. 6.

157. See the discussion of the "Fines geometriae" of summer 1673 above. See also his doubts in 1673 about knowing anything clearly and distinctly in physics: "De imperfectione analyseos," [spring–summer 1673], A6,3:404.

158. "De mente, de universo, de deo," 12.1676, A6,3:463; trans. modified from Pk 5, 7. For this passage, see Dascal 1987c, p. 51. The semantic ground captured by the term "idea" was soon broken into a number of other terms, for example, "concept," "notion." In "Quid sit idea" [fall 1677], he articulated a new account of ideas: an idea, he argued, "consists, not in some act, but in the faculty of thinking" (A6,4:1370; L207).

159. It is perhaps useful to distinguish several claims here, in order to understand the movement of Leibniz's thought:

1. Symbolic reasoning is helpful in solving certain geometrical and arithmetical problems. It helps us obtain clear and distinct answers to them. We can nevertheless proceed without the use of symbols.

2. Symbolic reasoning is helpful in solving certain geometrical and arithmetical problems. It helps us obtain clear and distinct answers to them. We can nevertheless proceed without the use of symbols, although our limited life span will prevent us pragmatically from ever reaching answers in many cases. (This is the view Leibniz appears to have held around 1670.)

3. Symbolic reasoning is necessary in solving certain geometrical and arithmetical problems. It helps us obtain clear and distinct answers to them. For a wide variety, even the majority, of problems, we cannot proceed without the use of symbols, due both to our limited life span and to our cognitive makeup. (This is the view Leibniz appears to have held in late 1672.)

4. Symbolic reasoning is helpful in solving certain geometrical and arithmetical problems, but we do not thereby reach clear and distinct understanding about any (or most) of them. For a wide variety, even the majority, of problems, we cannot proceed without the use of symbols, due both to our limited life span and to our cognitive makeup. (This is the view Leibniz appears to have held by 1676.)

For a judicious recent survey of the disputes on these matters, see Maat 2004, pp. 346–56. On the literature, see especially Dascal 1987c; Dascal 1978, chs. 6–7.

160. "Numeri infiniti," [4.1676], A6,3:498–99; trans. modified from LC 89, 91. This fiction can mislead us into believing that a polygon with infinitely many sides exists: "if certain polygons are able to increase according to some law, and something is true of them the more they increase, our mind imagines some ultimate polygon." See also K69, discussed at LC 393 n. 5.

161. "Expressio seriei per numerum primum et ultimum," [10–12.1676], A7,3:834–35.

162. Leibniz to Tschirnhaus, 12.1679, A3,2:940.

163. "Quadratura arithmetica circuli et hyperbolae," 3.5.1676, A7,3:751–53.

164. "De l'usage de la méditation," [1676], A6,3:663, note to lines 10–11, variant (bb). Although focused primarily on Leibniz's theological work, see Antognazza 1994 (esp. pp. 64–66) on the Incarnation, human embodiment, and human ability to have knowledge other than clear and distinct knowledge.

165. "Quid sit idea?" [1677], A6,4:1370.

166. "Quid sit idea?" [1677], A6,4:1370, lines 17–18, variant (2).

167. "Quid sit idea?" [1677], A6,4:1370.

168. "Specimen inventorum de admirandis naturae generalis arcanis," [1688], A6,4:1618 n. 2; see LC 309.

169. "Monodologie," §57, GP VI:616. See the remarks in Elkins 1994, p. 20. For a zeitgeist-style argument relating Leibniz's cityscapes to Baroque aesthetics and the state, see Assunto 1979, pp. 61–69.

170. Exactly how close Leibniz was to his mature philosophical views before, during, and after the Paris period is a question of contention among historians of Leibniz's philosophy, but need not detain us here. For some current views, see S. Brown 1999. For a powerful argument that Leibniz held nearly all of the core doctrines of his mature metaphysics by the Paris period, except for the containment theory of truth, see Mercer 2001.

171. "Specimen inventorum de admirandis naturae generalis arcanis," [1688], A6,4:1618 n. 2.

172. "Conspectus libelli elementorum physicae," [summer 1678–winter 1678/79], A6,4:1989.

173. Even in late examples, Leibniz worked with the concreteness of the metaphor; see Leibniz to de Volder, 20.6.1703, GP II:252.

174. Devices such as these have received considerable attention of late thanks to the controversy over David Hockney's thesis concerning optical devices in Renaissance and early-modern art. See Hockney 2001. For pointers to the literature and an important corrective contribution, see Gorman 2003. For the history of these projections in technical drawing, see Camerota 2004; Lefèvre 2004.

175. Harsdörffer and Schwenter 1991, vol. 2, pp. 198–99.

176. Alsted 1649, vol. 3, p. 682.

177. Kircher and Schott 1669, pp. 119–20, 220–21; Kircher 1654, pp. 174–89.

178. See *Notitia*, 1671, in Leibniz 1768, vol. 3, pp. 14–15; Leibniz to Johann Friedrich, [10].1671, A2,1:161.

179. For another early example, but involving mirrors *and* lenses, see "Elementa juris naturalis," 1671, A6,1:438. For the importance of images and metaphors of mirrors in

Leibniz's thought from 1671 onward, see Mercer 2001, pp. 217–20, 248–50; she follows L138 n. 9 in underscoring the deficiencies of seventeenth-century mirrors. For a more technical discussion of problems with lenses and mirrors in accurate astronomical work, see R. Wilson 2004, pp. 1–11. Compare Newton's contemporaneous optimism about overcoming the technical difficulties in polishing and metallurgy necessary to make mirrors into useful instruments. See, e.g., Newton to Oldenburg, 20.3.1672, in Newton 1959–77, vol. 1, p. 127; and Newton's manuscript on mirrors, in Newton 1959–77, vol. 1, pp. 85–88.

180. "De modo perveniendi ad veram corporum analysin et rerum naturalium causas," 5.1677, A6,4:1974.

181. Harsdörffer and Schwenter 1991, vol. 3, p. 259.

182. Harsdörffer and Schwenter 1991, vol. 3, pp. 229–30; vol. 1, p. 270 ("Durch Perspectivische Instrument vergleichen in Werck zu richten"); among many other discussions.

183. Harsdörffer and Schwenter 1991, vol. 3, p. 260.

184. I do not want to claim that these scientific and mathematical practices by themselves caused the development of his doctrines, so much as that they offered languages and heuristics that played a central and essential constitutive role in their development. Failing to recognize the practical domain whence stemmed these metaphors threatens to sever Leibniz's philosophical development from the many sources he drew upon for this articulation and development, many of which seem to have little serious "philosophical" significance today.

185. Mercer (2001, esp. pp. 303–5, 427–36) studies Leibniz's changing use of the town analogy in some detail; she argues that as Leibniz became more committed to preestablished harmony among minds "he became more attached to the town image and became more inclined to think of God both as the ultimate source of these projections and as something to be accessed by these means" (p. 429).

186. Leibniz to Jacob Thomasius, 4.1669, A6,2:437; L97.

187. "Specimen demonstrationum de natura rerum ... ," [late 1671], A6,2:304; L142.

188. An important moment of transition can be see in Leibniz to Gallois, late 1672, A3,1:17–8.

189. "De plenitudine mundi," [early] 1676, A6,3:524; Pk 85. Mercer (2001) documents well the Platonist background for Leibniz's account of creation as the emanation of limited expressions of God's essence. I elide the Platonist background to focus on the sharpness Leibniz gave his metaphysics *following* his expanded knowledge of the mathematics of series and of quadratures and his efforts in optics and catoptrics.

190. "De formis simplicibus," 4.1676, A6,3:523; Pk 83; on this, see Kulstad 1999, pp. 78–80.

191. "De origine rerum ex formis," [4.1676], A6,3:518; Pk 77. While he continued to use the town analogy throughout his life, Leibniz soon stopped using the arithmetical analogy; this may have been because he was worried that this articulation of the relationship between God and created things appeared (or may actually have been) dangerously close to Spinoza's views. See Malcolm 2003, pp. 239–40.

192. See the examples above and many others, e.g., "Expressio unius literae per multas," 4.9.1674, Leibniz 1976, p. 4.

193. "De inventione theorematum elegantium," 11.1675–2.1676, A7,3:697.

194. "De formis simplicibus," 4.1676, A6,3:523; trans. modified from Pk 85.

195. The analogical connections between various aspects of Leibniz's work, upon which Michel Serres insisted, seem instantiated within Leibniz's philosophy itself. See Serres 1982.

196. See Rauzy 2001, pp. 63–69; Krämer 1991, pp. 298–305; Swoyer 1995; Lamarra 1982, 1991; Kulstad 1977. As Swoyer and Rauzy argue, expression does not always entail a strict bijection between a thing and its expression; that is, not every relation or quality of the thing needs to be expressed, and not every attribute of the expression needs to correspond to some set of relations or properties in the thing expressed. Some mathematical examples of expression evidently involve bijection, but many of Leibniz's examples quite clearly do not, for example, a *scenographia* of a building.

197. See, e.g., Leibniz to de Volder, 20.6.1703, GP II:253; L531.

198. Mercer has illuminated the genealogy of this account of the creation of beings, from its origins in Platonist doctrines of emanation of all beings from God, up through its full articulation as the doctrine of expression. See Mercer 2001, esp. pp. 427–36 for his 1676 work; see also Parkinson's introduction to Pk.

199. "Dialogue entre Theophile et Polidore," [1679], A6,4:2237. There are numerous other examples, e.g., "Discours de métaphysique," 1686, §9, A6,4:1542.

200. While excellent on projective geometry as a key model for expression, Swoyer 1995 offers a less convincing account of equations and other written expressions.

201. Elkins 1994, pp. 1–44 (on Leibniz, pp. 19–20). Elkins sees approaches like those of Panofsky 1991 that unearth a totalizing worldview of the "Renaissance" as the projection back of this later *philosophical* understanding of perspective. For Leibniz's key role in moving the terms "viewpoint" and "perspective" into philosophical and quotidian language, see Guillén 1971, pp. 318–25. For a brilliant reading placing point of view in a history of rhetorical accommodation and its hermeneutic implications, see Ginzburg 2001, pp. 152–56.

202. Alpers 1983, pp. 53–70.

203. See Huret 1670, p. 86, among many examples; for consideration of the Huret-Bosse debate, see Heinich 1993; Kemp 1990, pp. 119–31. Leibniz probably read Huret; see Echevarría 1994, pp. 285–86; A3,1:55, 61, 71.

204. Cf. Mercer 2001, p. 434.

205. "De characteribus et compendiis," [26?.3.1676], A6,3:433–34.

206. For confusion and multiple acts of the mind, see, e.g., "De plenitudine mundi," [3].1676, A6,3:524; Pk 85; LC 59. For his understanding in 1676 of the process of abstraction involved when cognizing characters or images, see "Numeri infiniti," [4.1676], A6,3:498–99; LC 89, 91.

207. "De veritatibus, de mente, de deo, de universo," 15.4.1676, A6,3:512; trans. modified from Pk 65.

208. Notes on Metaphysics, 12.1676, A6,3:400; Pk 115.

209. "Dialogue entre Theophile et Polidore," [1679], A6,4:2234. For discussions of the proportionality of intellects, see Riley 1996, p. 51; Hunter 2001, pp. 105–7.

210. "Aphorismi de felicitate, sapientia, caritate, justitia," [1678/79], A6,4:2801.

211. "Elementa verae pietatis, sive de amore dei super omnia," [early 1677–early 1678], A6,4:1359–60.

212. Leibniz to Mariotte, [first half of] 1676, A2,1:270 (my italics).

213. Leibniz to Mariotte, [first half of] 1676, A2,1:271 (my italics).

214. Leibniz to Mariotte, [first half of] 1676, A2,1:271.

215. For the purposes of this chapter, I skip over Leibniz's account of "real definition," a definition that includes a proof of the possibility of the thing defined.

216. The validity of Leibniz's reading of Hobbes need not concern us here. For Hobbes's account, see, e.g., *Leviathan*, 1.4; Couturat 1901, pp. 457–72; Dascal 1987b; Maat 2004, pp. 332–33.

217. Leibniz had been concerned with Hobbes's claims for some time. E.g., see Leibniz to Gallois, 1672, A3,1:13–14.

218. "Dialogus," 8.1677, A6,4:24; trans. modified from L184. For an important translation of and commentary on this text, see Gaudin 1993, esp. pp. 88 and 98; Dascal 1987b; clear and very useful is Maat 2004, pp. 335–40; Gensini 1996, pp. 87–91.

219. Mariotte 1678, fols. à2v; see pp. 57, 67–70.

220. Mariotte 1678, p. 68.

221. Leibniz knew Pascal's *L'esprit de géométrie* by the 1680s at the latest, and its doctrines of nominal definitions much earlier, through the *Port-Royal Logic*.

222. "Meditationes de cognitione, veritate et ideis," 1684, A6,4:585–92. On the authoritative status of this publication, see, e.g., Leibniz to Burnett, 20/30.1.1699, A1,16:510–11; NE II.29.2, A6,6:254–55; Rutherford 1995, p. 74; Maat 2004, p. 351.

223. "Meditationes de cognitione, veritate et ideis," 1684, A6,4:590–91. See also "De analysi veritatis et judicorum humanorum," [late 1684–early 1685], A6,4:593. Leibniz likewise condemned Pascal for his dependence on a subjective "natural light," which was no better than Descartes' clarity and distinctness ("Projet et essais pour avancer l'art d'inventer," [8.1688–10.1690], A6,4:970).

224. "Meditationes de cognitione, veritate et ideis," 1684, A6,4:586; cf. Leibniz to Sophie, [1690], A1,6:76.

225. "Meditationes de cognitione, veritate et ideis," 1684, A6,4:587, 587–88 (italics in original).

226. "Meditationes de cognitione, veritate et ideis," 1684, A6,4:587. The vocabulary here follows Descartes but is an older scholastic vocabulary. For Descartes on "adequate" knowledge, see AT VII:365; cf. the discussion of "entire and perfect knowledge" at AT IX/1:171.

227. In more distinct definitions, we have analyzed more of the defining marks: "Distinct knowledge has degrees, for usually the notions in use in a definition will have themselves need of definition" ("Discours de métaphysique," 1686, §24, A6,4:1568). Experimental discovery makes knowledge more distinct: with Newton's book on colors, "we will understand them more distinctly" (Leibniz to Thomas Burnett, 20/30.1.1699, A1,16:510). See also the numerous examples in "Praefatio ad libellum elementorum physicae," [summer 1678–winter 1679], A6,4:1992–2009; "Revocatio qualitatum confusarum ad distinctas," [1677?], A6,4:1961–62.

228. Fichant 1998b, pp. 274–75, discussing "De affectibus," 20–22.4.1679, A6,4:1435.

229. Compare Mercer on "partial cognition" in the "expression relation" (2001, pp. 432–36).

230. In investigating the world, different researchers will often produce different nominal definitions of things. Nominal definitions suffice to distinguish the thing defined but need not be the same. They must, however, not contradict one another.

231. "Paraenesis de scientia generali," [8–10.1688], A6,4:974; cf. A6,4:970; Couturat 1901, p. 183. Leibniz worked on a philosophical language that began with the elements of human thinking, and not the elements of reality, whose constitution was the never-attainable endpoint of the project, not its beginning. See Maat 2004, pp. 324–26, 364.

232. "De mente, de universo, de deo," 12.1675, A6,3:462 (my italics); cf. Pk 3.

233. Leibniz to [Elisabeth?], 1678, A2,1:437 (italics in original).

234. Leibniz to [Elisabeth?], 1678, A2,1:437.

235. See Locke 1975, 4.2.6, p. 533. See Schuurman 2001 for Locke's indebtedness to the Cartesian logical tradition.

236. Leibniz to [Jean Berthet?], 1677, A2,1:384.

237. See the quick summary at "De vita beata," [early 1676], A6,3:636, 668–69.

238. A6,3:636 n. 1. Leibniz repeated versions of this claim from time to time. See "Le cartesianisme l'antichambre de la véritable philosophy," [1677], A6,4:1356; Leibniz to H. Fabri, 17.5.1677, A3,2:144.

239. Leibniz quotes the Latin version of the *Passions*; "De vita beata," [early 1676], A6,3:643; see Descartes, *Passions*, §153, AT XI:445–46.

240. "De la sagesse," [1676], A6,3:673 (italics in original).

241. Descartes was far from Leibniz's only source; besides his wide philosophical background, Leibniz also was reading Epictetus, Robert Boyle, and others around the same time. Many of his reading notes and annotations from this period appear in A6,3. For his ethical thinking and reading up to the Paris period, see Piro 1999; Mercer 2001, pp. 215–20, 310, etc.; Mercer 2004.

242. "De vita beata," [1676], A6,3:637.

243. "De la sagesse," [1676], A6,3:671–73 (quotation on p. 671), taken from his accounts of what is necessary to discover and remember.

244. "De vita beata," [1676], A6,3:637 n. 3. See the similar complaint in "De veritatibus, de mente, de deo, de universo," [15.4.1676], A6,3:509; Pk 59.

245. "Geometriae utilitas medicina mentis," [1676], A6,3:453. For Leibniz's views on the purpose of mathematics, see Beeley 1999, pp. 141, 144–45.

246. Leibniz to Conring, 13.1.1678, A2,1:387.

247. "De l'usage de la méditation," [1676], A6,3:665 n. 1.

248. Leibniz to Ernst August, [1685–87], A1,4:314; see Belaval 1960, p. 49.

249. "De l'usage de la méditation," [1676], A6,3:665.

250. "Optima ordinatio agendorum secundum tempora," [spring 1680], A4,3:907. Such was Leibniz's own practice. "Sometimes so many thoughts come to me in the morning, in the hour while I am still in bed, that I need to use the entire morning and sometimes the entire day and beyond to put them distinctly into writing" (Bodemann 1966, p. 338). See the more detailed advice in the contemporaneous "Conversation du Marquis de Pianese et de Père Emery Eremite," [1679–81], A6,4:2277–79.

251. "Optima ordinatio agendorum secundum tempora," [spring 1680], A4,3:907. In an earlier text Leibniz notes, "Hence the Pythagoreans held that no one should fall asleep before 'he had reviewed all his deeds for the whole day'" ("Novus methodus discendae...," 1667, §12, A6,1:272; L86). For these Pythagorean procedures, see Thom 1995, pp. 163–67 and trans. at p. 97.

252. "De l'usage de la méditation," [1676], A6,3:662–63 (italics in original).

253. "De l'usage de la méditation," [1676], A6,3:666–67.

254. "Confessio philosophi," [1672–73], A6,3:135 (italics in original). See the discussion in Mercer 2001, pp. 395–96.

255. "De l'usage de la méditation," [1676], A6,3:665.

256. "Geometriae utilitas medicina mentis," [1676], A6,3:450.

257. Memo for Johann Friedrich, "De usu et necessitate demonstrationum immortalitatis animae," [1671], A2,1:114.

258. "De usu geometriae," [1676], A6,3:448.

259. Cf. Rutherford 2003, p. 79.

260. "De usu geometriae," [1676], A6,3:448–49.

261. In this period, Leibniz responded positively to Robert Boyle's *Excellency of Theology Compar'd with Natural Philosophy* (London, 1674); in Hunter and Davis 1999–2000, vol. 8. Boyle attacked the sufficiency of a Cartesian natural-philosophical way of life, stressed the dangers of natural philosophy, and insisted on the need for revealed theology. See Leibniz's reading notes on Boyle, [12.1675–]2.1676, A6,3:218–31, esp. 221–22.

262. Leibniz to [?], [1679], A2,1:500, 501. For a more nuanced account of Leibniz's views on Stoicism, particularly good on his account of hope, see Rutherford 2003, esp. pp. 74–83.

263. Leibniz to [?], [1679], A2,1:502. Cf. GP VI:606; and see the important discussion in Rutherford 1995, p. 54.

264. See, in his later work, NE IV.10.11, A6,6:440–41, on the need for proper metaphysics to provide more than patience.

265. "De vita beata," [1676], A6,3:639 n. 10: "non φιλορωμαῖος sed φιλάνθρωπος" (Leibniz's comments on one of Descartes' letters to Elisabeth, 15.9.1645, AT IV:293).

266. "De vita beata," [1676], A6,3:639 n. 9.

267. "De vita beata," [1676], A6,3:639 n. 11.

268. "Aphorismi de felicitate, sapientia, caritate, justitia," [1678/79], A6,4:2801.

CHAPTER SIX

1. "Double infinité chez Pascal et monade," [after 1695], in Leibniz 1948, vol. 2, pp. 553–54. For this fragment, see Naërt 1985; Riley 1996, pp. 61–64; and especially Carraud 1986. See also Leibniz to Burnett, 1/11.2.1697, A1,13:557.

2. For Leibniz's account of the power of metaphysics to motivate the belief that this is the world of greatest perfection, see, above all, Rutherford 1995, e.g., at p. 9; and Mercer 2001.

3. "Initia et specimina scientiae generalis de nova ratione instaurationis et augmento scientarum," [1679], A6,4:353.

4. "Common Conception of Justice," [1702], in Leibniz 1988, p. 52. The comment on the novel is probably taken from Robert Boyle; see Leibniz's reading notes at A6,3:225; Hunter and Davis 1999–2000, vol. 8, p. 55.

5. For a balanced view of Leibniz's dissatisfaction and career in Hanover, see Rescher 1992.

6. For scholarly work, see, e.g., Jardine and Grafton 1990; Zedelmaier 2001. For writing and inscription in the history of science, see Lenoir 1998; Siegert 2003; Kaiser 2005. For connecting the two, see the work of Ann Blair, e.g., Blair 1997, 2003, 2004.

7. Although many of the considerations discussed in this chapter appear prominently in his later works, I have focused on a more temporally narrow band in order to make more apparent the fecundity of his efforts when working in many disciplinary domains more or less simultaneously.

8. "La vraie méthode," [early 1677], A6,4:7.

9. "La vraie méthode," [early 1677], A6,4:1.

10. "Introductio ad scientiam generalem modum inveniendi demonstrandique docentem," [1679], A6,4:370 (my italics).

11. "Initia et specimina scientia generalis ad instauratione et augmentis scientiarum," [summer–fall 1679], A6,4:359.

12. "Initia scientiae generalis. Conspectus speciminum," [summer–fall 1679], A6,4:362.

13. "De numeris characteristicis ad linguam universalem constituendam," [spring–summer 1679], A6,4:264; cf. L222. For a major revisionist account of Dalgarno, Wilkins, and Leibniz, one extremely critical of the well-known accounts of Umberto Eco and Paolo Rossi, among others, see Maat 2004. For a recent survey of scholarly opinions on the relationships between the characteristic and universal languages, see Nef 2000, pp. 86–88. For the universal characteristic in relation to early-modern drives for *pansophia*, see Rossi 2000, pp. 180, 317–18 n. 55; and the much more up-to-date Moll 2002. For a good appreciation of the connection between natural languages and the characteristic, see Heinekamp 1972 (but see Maat 2004, p. 328) and especially Gensini 1996, now in Gensini 2000, ch. 1.

14. Dascal underscores that these tools do not simply augment existing human epistemic powers but rather "redefine the natural limits of human reason" (1978, p. 210).

15. The characteristic provides means for expressing logical truths, but those truths might be knowable through other means. See Kauppi [1960] 1985, p. 64; Peckhaus 1997, pp. 33–34.

16. 1679? GP VII:25; trans. modified from Pasini 1997, p. 40; nearly verbatim in Leibniz to Ernst August, [1685–87], A1,4:314.

17. "Sufficient givens" are discussed in many places; e.g., "Introductio ad scientiam generalem modum inveniendi demonstrandique docentum," [1679], A6,4:370–71.

18. Leibniz to [Jean Berthet?], 1677, A2,1:384. For the efforts to create a characteristic for the estimation of probabilities, see the introduction to Leibniz 1995, pp. 39–42.

19. Far more important here than documenting the variations and fine structure of Leibniz's accounts of the general science over time is to establish his insistence on its palpability.

20. "Initia et specimina scientia generalis ad instauratione et augmentis scientiarum," [summer–fall 1679], A6,4:359. For a detailed and lucid account of the shifting goals and status of Leibniz's logical efforts, see now Rauzy 2001 and his extremely helpful edition of translations in Leibniz 1998. Rauzy (2001) shows well the increasing autonomy Leibniz afforded the study of symbolic logic, but also the limits of that autonomy. In English, see Leibniz 1966.

21. On sensibility, see especially Dascal 1978, pp. 96–98, 210–12. Many accounts of the general science and universal characteristic pass too quickly over Leibniz's repeated stress on the concreteness of writing: the power of good notation to make demonstrations

and inventories appear distinctly, to make them visible all at once, and to help human beings grasp order and interconnection among them. Scholars have at times neglected the considerable role those logics and notations were to have as a concrete means of discovery, a method of invention as well as of judgment. As Christia Mercer (2001) has now thoroughly demonstrated, the formerly dominant logical emphasis in Leibniz studies, stemming from the works of Bertrand Russell and Louis Couturat, has systematically obscured much in Leibniz that is foreign to such logical drives, has validated neglect of much of his philosophical motivation as epiphenomena, and has falsely projected his logical innovations of the 1670s and 1680s into his earlier work.

22. Leibniz to [Jean Berthet?], 1677, A2,1:384. Couturat (1901, pp. 89–95) stressed the materiality intended in Leibniz's universal characteristic; see especially Elster 1975, p. 108, on NE IV.1.9, A6,6:360–61.

23. "Examen religionis christianae," [4–10.1686], A6,4:2387. For a major Jesuit account of the corporeal impediments to knowledge, see Suárez 1740, bk. 1, ch. 1, nn. 9–11 (vol. 6, pp. 206–7).

24. Hotson 2000, pp. 68–70. For Leibniz's comments on Alsted and perfecting the encyclopedia, see A6,2:394–97, though not on these points. For Leibniz and Alsted more generally, see Antognazza and Hotson 1999.

25. Comments to a copy of Descartes' *Regulae ad directionem ingenii*, [1678–83], A6,4:1033, lines 5–11, n. 14; comments on the passages in AT X:383, line 23, to 384, line 8.

26. "Geometriae utilitas medicina mentis," [1676], A6,3:452.

27. "De mente, de universo, de deo," 12.1676, A6,3:463.

28. See "Analysis linguarum," 11/21.9.1678, A6,4:102.

29. "Elementa rationis," [4–10.1686], A6,4:715. See comments in Leibniz 1998, p. 177 n. 66.

30. "Elementa rationis," [4–10.1686], A6,4:725.

31. Lamarra (1978, pp. 60–62) sees the change as occurring in 1673. See Descartes to Mersenne, 20.11.1629, AT I:80–82; and Leibniz's excerpts from and comments on the letter (1678–80) in A6,4:1028–30.

32. As discussed in the previous chapter, Dascal (1978, 1987c) has stressed that Leibniz moved from a view of language as a mnemotechnic art useful for remembering toward seeing it as constitutive of thought itself. See also Lamarra 1978; Gensini 1996.

33. "Fundamenta calculi rationatoris," [1688], A6,4:920; trans. in Rutherford 1995, p. 103.

34. As noted in the previous chapter, with such forms of writing, human beings could narrow the gap between human and divine knowledge. Pascal maintained that an infinite disproportion separates the human intellect from God's. Leibniz denied such a disproportion. Although human beings share the same truths as God, they do not know them in the same way. Just as many human beings know simple arithmetical relations such as $12 \times 12 = 144$ without resorting to calculation, God has immediate knowledge of all truths without requiring artificial modes of reasoning such as syllogistic logic or infinite series, which are needed by human beings and, he added, probably angels. See "Dialogus inter theologum et misosophum," [1678–79], A6,4:2215. For scholastic concerns about God's knowledge of infinity, see the discussion of the Coimbra commentators above, in chapter 3.

35. His work on combinations involved both theorems about combinations taken abstractly and cognitive concerns about the display and order of thinking through things; e.g., "*Dispositions* are the varieties of ways of thinking about certain things, that is, of placing them in the mind" ("Problème général," [before 8.1673], in Leibniz 1976, p. 8; italics in original).

36. "De arte characteristica inventoriaque analytica combinatoriave in mathesi universali," [1679–80], A6,4:324 (italics in original).

37. "De arte inveniendi in genere," [summer–fall 1678], A6,4:81; and see the contemporaneous list of many useful tables: "Atlas universalis," [summer–fall 1678], A6,4:86–90. The sources of Leibniz's account of tables need serious study. See Bredekamp 2004, pp. 91–92; Rutherford 1995, ch. 5, and esp. p. 127 n. 17, where he rightly notes that one finds more tables than deductions in Leibniz's papers. Serres (1982) gives Leibniz's tables a prominent place. For a hint about Ramist and rhetorical sources, see Varani 1995, p. 154 n. 117. See Dascal 2002 for important considerations on Leibniz's various "cognitive technologies," including tables. For an important study of tables and scientific discovery, see Gordin 2004, ch. 2.

38. "De arte inveniendi in genere," [summer–fall 1678], A6,4:82.

39. As Eberhard Knobloch notes, "Leibniz is a master of inductive mathematical research, meaning that he made important discoveries in observing, in divining in an audacious manner, [and] in verifying in a sagacious manner" (1989b, p. 165). We owe our knowledge of these efforts in great part due to Knobloch's editorial work: notably Leibniz 1976 and 1980. For the work on series, especially on the quadrature of the circle and the problems associated with it, see A7,3. For an older stress on induction, see Couturat 1901, p. 261.

40. See Knobloch 1973, p. 241. Knobloch stresses that tables were both for inductive reasoning and to help reduce the dependence on long calculations. For the most thorough recent study of Leibniz's method of discovery in physics, see Duchesneau 1993, pt. 1; on the limitations of inductive practice see pp. 51ff.

41. For a remarkable example, see "Numeri progressionis harmonicae et proprietates quaedam in earum progressione observatae et summa circiter vera," 8.2.1676, A7,3:715–30, esp. 717–20.

42. See the nearly verbatim discussions in "Consilium de encyclopaedia nova conscribenda methodo inventoria," 15/25.6.1679, A6,4:340; "Concilium de scribenda historia naturali," [1679], A4,3:856–57. This example figures in Lana Terzi 1670, p. 131; compare the slightly more sophisticated example in "De arte characteristica inventoriaque analytica combinatoriave in mathesi universali" [1679–80], A6,4:325–26, inter alia.

43. "Specimen analyseos novae quae errores vitantur, animus quasi manu ducitur, et facile progressiones inveniuntur," 6.1678, in Leibniz 1980, pp. 5–12.

44. Cramer's rule holds that, given a set of linear equations $\mathbf{Dx} = \mathbf{A}$, where $\mathbf{D}$ is an $n \times n$ matrix of coefficients d_{ij}, $\mathbf{x}$ is a column of n variables $x_1, \ldots, x_n$, and $\mathbf{A}$ is a column of n constants $a_1, \ldots, a_n$, then $x_i = |\mathbf{D}_i| / |\mathbf{D}|$, where $\mathbf{D}_i$ is $\mathbf{D}$ with the ith column replaced by $\mathbf{A}$, and $|\mathbf{D}|$ is the determinant of $\mathbf{D}$.

45. "Specimen," 6.1678, in Leibniz 1980, p. 5.

46. "Specimen," 6.1678, in Leibniz 1980, p. 5.

47. "Specimen," 6.1678, in Leibniz 1980, pp. 6–7.

48. In modern notation, this is the determinant of the matrix of coefficients

$$\begin{vmatrix} 12 & 13 & 14 & 15 \\ 22 & 23 & 24 & 25 \\ 32 & 33 & 34 & 35 \\ 42 & 43 & 44 & 45 \end{vmatrix}$$

The top line of Leibniz's expression is the working out of the determinant for the 12 coefficient, that is,

$$12, \begin{vmatrix} 23 & 24 & 25 \\ 33 & 34 & 35 \\ 43 & 44 & 45 \end{vmatrix} = 12, [23, 34, 45 + 24, 35, 43 + 25, 33, 44 - 23, 35, 44 - 24, 33, 45 - 25, 34, 43].$$

49. "Specimen," 6.1678, in Leibniz 1980, p. 5. See Knobloch 2001, pp. 154–55.

50. "De combinatoria et usu serierum," [1680–84], A6,4:415, 416.

51. "De arte characteristica inventoriaque analytica combinatoriave in mathesi universali," [1679–80], A6,4:326 (italics in original).

52. He distinguished disciplines subject to deductive proof from those allowing only probable reasoning. See, e.g., "Concilium de scribenda historia naturali," [1679], A4,3:863.

53. Leibniz to [Jean Berthet?], 1677, A2,1:384.

54. Recent scholarship on Leibniz has rightly underlined his pursuit of "softer reason" as well as deductive reasoning. See, for a general overview, Dascal 2001 and, for a fine example, Antognazza 2001. See the heated discussion Dascal 2003; Schepers 2004; Dascal 2004. These treatments offer, in my opinion, too narrow an account of how "hard" reasoning based on mathematics had to be for Leibniz. Leibniz's "softer" forms of inductive reasoning were often as grounded in his mathematical practice as were his formal logical deductions. For law, see "De legum interpretatione, rationibus, applicatione, systemate," [1678–79], A6,4:2785. For discerning the *"rationes"* behind laws and the intentions of lawmakers, see Varani 2003, esp. pp. 643–46. Medicine is like law in its dependence on empirical data. See "Methodus jurisconsultorum exemplar methodi medicinae," [1677–78], A6,4:2755–56. For judicious remarks on Leibniz's empirical work in linguistics, see Gensini 2000, pp. 11–12. For his biblical interpretation in this period, see Duchesneau 2001, pp. 276–79; for the context and importance of apocalyptic interpretation, see Antognazza and Hotson 1999, pt. 2.A, introduction. For the need for a logic of probability using inductive means in history, see Leibniz to Eisenhardt, [2].1679, A1,2:426–28; discussed in Eckert 1971, pp. 2–5. For examples of discerning the "reasons" behind series of historical facts and actions, see Davillé [1909] 1986, pp. 426, 434, etc.; Belaval 1960, pp. 116–17, 125.

55. Leibniz to [François de la Chaise], [4–5].1680, A3,3:192. For indications on finding physical principles, see, e.g. "Praefatio ad libellum elementorum physicae," [1678–79], A6,4:2000; see generally Duchesneau 1993.

56. "Recommandation pour instituer la science générale," [1686], A6,4:691.

57. Leibniz to [Jean Berthet?], 1677, A2,1:383.

58. "Recommandation pour instituer la science générale," [1686], A6,4:691, 696.

59. See Fichant's fine edition with French translation and commentary: Leibniz 1994b.

60. See, e.g., "Pacidius Philalethi," [29.10–10.11.1676], A6,3:531–32, 570; LC 135, 137, 219.

61. A more thorough study of these documents in relation to tabularity and expression would prove quite interesting. In the course of the investigation, he advocated enumeration of all possible mathematical "expressions" of a measured quantity, to direct further experiments. See "De corporum corsu," 1–2.1678, in Leibniz 1994b, pp. 104–5, 226–28. For citations to the older literature on *vis viva*, see nn. 151 and 153, below.

62. "De corporum corsu," 1–2.1678, in Leibniz 1994b, p. 134; see also table 2, pp. 132–33.

63. "De corporum corsu," 1–2.1678, in Leibniz 1994b, pp. 138–58. He drafted, in the months afterward, a number of statements of the value of conjecture in the process of discerning causes. See "Veritates physicae," [1678–80/81], A6,4:1985; "Praefatio ad libellum elementorum physicae," [1678/79], A6,4:1998–2001. For the role of such conjectures, see especially Duchesneau 1993.

64. See Descartes, *Principia philosophiae*, II, 36, AT VIII/1:61–62.

65. For the striking and repeated use of *"reformatio,"* see Leibniz 1994b, p. 15, 15 n. 1.

66. See "De corporum corsu," 1–2.1678, in Leibniz 1994b, table 3 on p. 135; see Fichant's discussion on pp. 269–71.

67. See Fichant 1998a.

68. For the limits of his empiricism, see Duchesneau 1993, p. 51, etc.

69. In the same period that he was pursuing his experimental researches into dynamics, Leibniz worked out new philosophical justifications for accommodating empirical generalizations within a deductive epistemology. See Duchesneau 1993, pp. 171–92, 254–58.

70. "De arte inveniendi in genere," [summer–fall 1678], A6,4:82. See Couturat 1901, p. 261. See also, among many other examples, A6,4:416.

71. This chapter underscores the claims of Mercer and Garber (Mercer 2001, p. 57; Garber 1985, pp. 72–75) that the development of Leibniz's thinking can hardly be characterized as a deductive process.

72. See the judicious remarks of Beeley (1999, pp. 141–45). Leibniz underlined the need to define carefully the limits to the use of reason when studying the infinite. But since he had endured the pain of the necessary rigor, future mathematicians using his rules, definitions, and tools could be spared it and could study infinities and infinitesimals safely. See K33. For the importance of the *De quadratura arithmetica* treatise in clarifying accounts of the infinite, see, e.g., Knobloch 1994, 2002; Arthur, forthcoming a.

73. Leibniz to Tschirnhaus, [5–6].1679, A3,2:450; L193 (italics in original).

74. "Elementa Rationis," [4–10.1686], A6,4:716.

75. "Initia scientiae generalis. Praefatio," [summer–fall 1679], A6,4:367–68.

76. Scheel 1983, pp. 749–50.

77. Before going to Hanover, Leibniz believed that he had been offered an elevated post as a privy counselor (*geheimer Rat*), not simply as a court counselor (*Hofrat*). For a fine account of Leibniz's attempts to get a post he deemed appropriate, see Rescher 1992, esp. pp. 30–32 on his rank. For the failure of many of his plans of reorganization, see Reese 1967, pp. 1–25. For Leibniz's work as justifying consultation with a metaphysical sage for proper and legitimate governance, see the important study Hunter 2001. For his political thought and work, see Riley 1996; Robinet 1994.

78. Among political thinkers, Leibniz stressed a potent perfectionism, which itself would be important for later political thinkers, notably Christian Wolff. For Cameralism

as more than an economic doctrine, see Raeff 1983, pt. 2; Hull 1996, ch. 4; Stolleis 1988, esp. pp. 334–65; the still useful Small 1909; Wakefield 2005, esp. pp. 317–19.

79. On these efforts, see Knabe 1956; H.-P. Schneider 1999, pp. 27–28; Siegert 2003, pp. 166–71; Lodolini 1998, esp. pp. 247–50. On the organization of the archive for the state, see two memoranda for Duke Johann Friedrich, 1678, A1,2:75, 77–78, discussed in Eckert 1971, p. 6.

80. Mostly printed in A4,3; but to be consulted with memos in A1,2; A1,3; as well as more "philosophical" writings in A6,4.

81. Leibniz sought a new version of the *Corpus juris* that would allow readers to discern quickly what they needed in *"uno obtutu"*—one of his earliest calls for such a help; see A6,1:327; Varani 2003, p. 641 n. 11.

82. See "Bestellung eines Registratur-Amts," 1680, A4,3:376–81, quoted at 380, 376.

83. "Entwurff gewisser Staats-Tafeln," [spring 1680], A4,3:341, 348. Cf. Leibniz to Ernst August, 1680, A1,3:24. See also A4,3:376; A6,4:326; Aiton 1985, pp. 101–2.

84. In focusing on the reorganization of the bureaucracy and the tools for accounting needed for it, Leibniz was following the tradition of important German writers on bureaucracy, notably Veit von Seckendorff; on whom, see Johnson 1964, p. 383; Stolleis 1987. In particular, Seckendorff focused heavily on proper accounting procedures and the flow of information among different levels and kinds of bureaucrats; Leibniz included him on the list of sources of inventorying practices at A6,4:261. See Seckendorff 1660. For the long process of creating this sort of information in France, see Barret-Kriegel 1988, pp. 111–59. For Leibniz as projector, see Elster 1975; and for the culture of projection in early-modern German states, see Smith 1994; Schaffer 1995.

85. See, e.g., "Unvorgreiflicher Vorschlag, Einen continuirlichen Extractum Actorum betreffend," for Ludolf Hugo, 3.1678, A1,2:49; for the need for improved scripts, see p. 51.

86. Hacking 1990, p. 18. For extracting and making repertories, see "Bestellung eines Registratur-Amts," [1680], A4,3:377–78.

87. "Epargne d'un prince," [fall 1678], A4,3:329.

88. "Denkschift betr. die allgemeine Verbesserung des Bergbaues in Harz," 20–2.2.1682, A1,3:159; followed by the call for several sorts of tables (159–62). Cf. Hamm 1997. Compare the accounting system of his mines discussed in NE IV.1.9, A6,6:360–61. For the mining project, see Aiton 1985, pp. 87–90, 107–14; Elster 1975, p. 108.

89. "La place d'autruy," [1679], A4,3:903; cf. Leibniz 1988, pp. 81–82. For this text, see Gil 1984, pp. 162–63; Riley 1996, pp. 189–93, Dascal 1995; Salas Ortueta 1991. Leibniz used this image elsewhere; see, e.g., A6,4:2243 and "Meditation on the Common Concept of Justice," [1702–3], in Leibniz 1988, p. 56.

90. See also the use of "fictions" to aid the memory in physics, grammar, and so forth. Leibniz often discussed the power of fictions to transform someone's life using his favorite example of the fictions used to train assassins, ultimately stemming from Marco Polo's descriptions; see now Frémont 2003. For the story of the assassins, see the early indication in Leibniz to Simon Foucher, 1675, A2,1:248–49; and the important later uses: NE II.21.37–38, A6,6:190–91.

91. "Agenda," [1679], A4,3:899.

92. For all his ecumenicism, Leibniz had his share of prejudice; see, e.g., A4,1:408. See Dascal 1993, pp. 390–94.

93. Notes on Jean Domat, [end of 1695], in Leibniz 1948, vol. 2, p. 648.

94. I have largely elided the extent to which Leibniz's account of knowledge production involves integrating the viewpoints of others, both in institutions and otherwise. For Leibniz and epistemological pluralism (in various senses), see Dascal 2000. For his insistence on institutions, see, e.g., Ramati 1996; among numerous other studies. For China, see Perkins 2002. For his eclecticism as method and as rhetorical practice, see Mercer 2001, ch. 1, esp. pp. 49–59.

95. See the remarks of Maat 2004, p. 323.

96. "De controverses," [1680], A4,3:210, 212.

97. "Methodus disputandi usque ad exhaustionem materiae," [1683–85/86], A6,4:578, 577. For the legal background to these methods, see Olaso 1975, esp. pp. 212–14.

98. "De numeris characteristicis ad linguam universalem constituendam," [1679], A6,4:269. This account, like many others, stresses the use of probabilities in "weighing arguments." As Dascal often emphasizes, the metaphor of the balance figures prominently in Leibniz's considerations of resolving controversies. See Dascal 2003; Leibniz 1998, pp. 177–78 n. 69.

99. "The single means for fixing our reasoning is to make it as sensible as that of mathematicians, so that one can find one's errors in the blink of an eye, and when there are disputes among people, one can say simply: let's compute, without any ceremony, to see which of us is right" ("Projet et essais . . . pour avancer l'art d'inventer," [1688–90], A6,4:964). The certainty of mathematics needs to be imitated; that certainty comes in no small part from its ability to be examined or surveyed by the eye.

100. "Grundriss eines Bedenkens von Aufrichtung einer societät in Deutschland," [1671], A4,1:533; trans. and discussed in Riley 1996, p. 216.

101. "De summa juris regula," [1680], A6,4:2846–47. For Leibniz's vision of the obligations of the state, see H.-P. Schneider 1999, pp. 27–28.

102. "De summa juris regula," [1680], A6,4:2846–47. Such a set of strong positive goals for the state increasingly characterized much of German political thought in the late seventeenth and eighteenth centuries. See Raeff 1983, pt. 2. With Leibniz one sees an extremely clear relationship between two projects of Foucault's later works: philosophical exercise and the discipline necessary for improving production and permitting "liberal" citizenship.

103. The state's obligations are stressed in the remarkably contemporaneous "Kammergefälle" of 1680, A4,3:353–59. Cf. Robinet 1994, pp. 274–75.

104. "De summa juris regula," [1680], A6,4:2847.

105. "Erfordernisse einer guten Landesregierung," [1680], A4,3:369.

106. "De scientia juris tradenda," [1680], A6,4:2848; see also Leibniz 1994a, pp. 154–55. For a discussion of Leibniz's account of thinking before the Paris period, see Mercer 2001, pp. 319–29.

107. "De scientia juris tradenda," [1680], A6,4:2848. By 1671, Leibniz glossed thought as follows: "Thinking is nothing other than the perception of the relation, or more briefly, the perception of the many things at the same time, or the one in the many" ("De conatu, et motu, sensu et cogitatione," [1671], A6,2:282; trans. in Mercer 2001, p. 321). In 1671 Leibniz was pursuing the goal of producing techniques that would allow many things to be seen all at once, as shown in his optical projects of the period. He did not yet use

the language of representation and expression that he developed in part through his mathematical work discussed in chapter 5.

108. "De scientia juris tradenda," [1680], A6,4:2848–49. Here again we can see that Leibniz does not insist on bijection when discussing expression, as noted in chapter 5. See also "De affectibus," 20–22.4.1679, A6,4:1427.

109. "De scientia juris tradenda," [1680], A6,4:2849.

110. "De scientia juris tradenda," [1680], A6,4:2849.

111. "De scientia juris tradenda," [1680], A6,4:2849.

112. Hume 1998, §1.7, p. 75.

113. For his ethics before 1672, see Piro 1999; Mercer 2004.

114. "De affectibus," 20–22.4.1679, A6,4:1426; repeated several times, see pp. 1430, 1434, etc. Thinking itself "involves a certain order of thinking, since it is the thinking of a series of certain things" (1424). For this text, see Giolito 1996, with partial French translation of the earlier Grua edition (Leibniz 1948).

115. "De affectibus," 20–22.4.1679, A6,4:1433; see *Passions de l'âme*, §73, AT XI:383, and Leibniz's remarks at 1426.

116. "Memoire pour des Personnes éclairées et de bonne intention," [c. 1692], A4,4:615.

117. "De affectibus," 20–22.4.1679, A6,4:1433; similar remarks at 1427, 1434–35, and n. 12.

118. See his roughly contemporaneous denial of the liberty of indifference: "Du Franc Arbitre," [1678–81], A6,4:1409: "true liberty of the mind consists in recognizing and choosing the best." See also NE II.21.50, A6,6:199.

119. Compare his earlier account of the relation of the will and thinking in "Novus methodus discendae . . . ," 1667, §31 (and later revisions in note), A6,1:284; L88, 91 n. 11.

120. These texts underscore Riley's emphasis on the wisdom and knowledge integrated within Leibniz's account of love and charity. See Riley 1996, ch. 4, esp. pp. 142–43, 152–54, 156–59, 177–82. Cf. Hunter 2001, p. 110.

121. "Felicity," [c. 1694–95], in Riley 1988, p. 83; cf. early writings: A4,1:34–36; A6,1:465; for late 1670s: A6,4:2793, 2798, 2806. For Leibniz's earlier reflections on love, pleasure, and morality, see Piro 1999. For the relationship between happiness, the sense of perfection, and one's own perfection, see Rutherford 1995, p. 51; Rutherford 2003, esp. pp. 76–80.

122. Hadot 1995, pp. 179, 201–2.

123. For a fine analysis of Leibniz using Hadot, see Hunter 2001, ch. 2.

124. "Aphorismi de felicitate, sapientia, caritate, justitia," [1678/79], A6,4:2800, 2805. For the active, as well as highly intellectual, qualities of charity in Leibniz, see Riley 1996, pp. 17–21, 156–59, etc.

125. See the discussion and letters quoted in Antognazza and Hotson 1999, pp. 169 n. 149, 188 n. 215; Leibniz to Ernst von Hessen-Rheinfels, 15/25.3.1688, A1,5:66. For a balanced consideration of Leibniz's approaches to enthusiasm, see Cook 1998.

126. While Leibniz was remarkably reticent about publishing his works, over time he published or made available in manuscript numerous decorous accounts of his philosophy.

127. See Rutherford 1995, pp. 58, 61–62.

128. "Discours de métaphysique," [1686], §37, A6,4:1588.

129. Among other major sources, Leibniz knew and cited Stefano Guazzo's *Civile conversatione* and Harsdörffer's *Frauenzimmer Gesprächspiele*; see, e.g., "De Collegiis," [1665], A6,2:5. In addition to treating emblems and paradoxes, Harsdörffer's conversations include arithmetical questions. See Harsdörffer 1968, vol. 2, pp. 75–68. On Harsdörffer's importance for promulgating French notions of *honnêteté* in German-speaking areas, see Krebs 1991.

130. "Lebensregeln," [1679], A4,3:888. Leibniz mocked himself for his lack of social graces. See Mates 1986, p. 32, discussing Bodemann 1966, p. 339. In an autobiographical piece, he later noted: "*Conversationis* appetentia non multa; major meditationis et lectionis solitariae" (Guhrauer 1966, vol. 2, Beilage, p. 60). But we should probably read these statements as a bit of self-fashioning of himself as the solitary intellectual; cf. Shapin 1991. For his relationships with the educated women in his courtly circles, see now G. Brown 2004. Leibniz immodestly noted that rhetorical styles were so easy for him as a student that "my teachers feared that I was attaching myself to these delights" ("De numeris characteristicis ad linguam universalem constituendam," [1679], A6,4:265). Just as we no longer take Hobbes's criticisms of rhetoric as calling for a complete abandonment of figurative language, we need to come to terms with Leibniz's account of the proper uses of rhetoric. For Hobbes, see Skinner 1996. For revised accounts of Leibniz's relation to rhetoric, see Gensini 1996, pp. 82–83; Fenves 2003; Feldman 2004.

131. "Lebensregeln," [1679], A4,3:888. Later pages have hints for mathematicians (890) as well as: "Semper aliqua nova habere, moden, horologia, machinulas etc." (892). See also the skills included in the account of the complete formation of a gentleman in "Agenda," A4,3:899 (rhetorical skills of discovery and judgment), 901–2 (mathematics, curiosities), etc.

132. "Nova methodus discendae docendaeque jurisprudentiae," 1667, §17, A6,1:274; L87.

133. See his revisions for a new edition of the "Nova methodus discendae . . . ," [after 1676], A6,1:275–76, notes to lines 13, 16.

134. "Erfordernisse einer guten Landesregierung," [1680], A4,3:366. Another example: "Seria ex Ludo. Oder Concept einer Spiel-Cassa, darauß ein Banco zu machen und damit sowohl denen participanten als dem publico Nuz zu schaffen" ([1688–89], A4,4:102–11).

135. "Lettre sur l'éducation du Prince," 1685–86, A4,3:549, 553 (italics in original). See Riley 1996, pp. 225–26. For machines used to teach fundamental physical and political principles in Georgian Britain, see Schaffer 1994.

136. Like the Jesuits, Leibniz was aware of the need both for methods of teaching appropriate to polite conversation and for methods of teaching appropriate to coming to higher knowledge of things: "Methodus docendi una popularis altera scientifica perfectior," [1683–85/86], A6,4:579.

137. Leibniz for Johann Friedrich, 9.1678, A1,2:76–77. Leibniz drafted a set of rules for the society; see "Societas sive ordo caritatis" and "Societas theophilorum ad celebrandas laudes Dei," [fall 1678], A4,3:847–52.

138. Leibniz for Johann Friedrich, 9.1678, A1,2:77.

139. "Pacidius Philalethi," [29.10–10.11.1676], A6,3:529; LC 129.

140. See "Conversation du Marquis de Pianese et de Père Emery Eremite," [1679–81], A6,4:2240–83; henceforth cited within the text by page number. Dascal has done the most to underscore the importance of this text; see especially Dascal 2000, pp. 26–29.

141. After his "reformation" of the rules of impact in 1678, Leibniz carefully went through his notes, marking his previous errors and attempting to understand the genealogy of his own mistakes. See Leibniz 1994b, e.g., pp. 94, 114, 118–19, 134.

142. For this theme, see Dascal 2000.

143. "Remarques sur la doctrine cartesienne," [1689], A6,4:2045–46. Leibniz maintained such a view from early on. See Mercer 2001, p. 53.

144. For civil conversation and early-modern natural philosophy, see the citations in chapter 3; for Leibniz's knowledge of the conversational literature of his day, see above.

145. Montaigne 1992, III, 8, p. 926.

146. This sense of how proper knowledge of nature can both "inflame" and elevate one is stressed in another great dialogue focused on natural philosophy: "Pacidius Philalethi," [29.10–10.11.1676], A6,3:570; LC 219.

147. "Discours de métaphysique," [1686], A6,4:1529–88; henceforth cited within the text by section and page number.

148. For Leibniz's important relationship with Landgrave Ernst von Hessen-Rheinfels, see Sleigh 1990, ch. 2.

149. Leibniz to Ernst von Hessen-Rheinfels, 28.11/8.12.1686, A1,4:410–11.

150. Leibniz scholars do not agree about when Leibniz accepted substantial forms and when (or if) he did not. As it matters not to my argument here, I pass over the controversy.

151. For this argument, see Garber 1995, pp. 309–13, 340 n. 34, 348 n. 107 and the literature cited therein; see also the important remarks in Gabbey 1985, pp. 25–26. For an early form of this argument, see above and Leibniz 1994b, pp. 152–53.

152. See *Principia philosophiae*, II, 36, AT VIII/1:61–62. Conservation of momentum, in contrast, includes **v** as a vector quantity: $\sum m\mathbf{v}$ is conserved. Confusion about the sort of quantities involved–scalar or vector–played no small role in the subsequent acrimonious debate.

153. Leibniz here argued only that force and quantity of motion are distinct, and not that the quantity conserved is in fact mv^2 (noted in Garber 1995, p. 348 n. 107, drawing on G. Brown 1994, esp. pp. 183–88).

154. See Schaffer 1995.

155. More properly, this is the case in any universe in which Galileo's law of free fall holds. See Leibniz 1994b, pp. 134, 270. Leibniz worked to show that his conservation law was not contingent on this aspect of empirical reality. See Garber 1995, pp. 313–14.

156. See Hutchison 1991; Des Chene 1996, pp. 177–86. See also Jones 2003.

157. Mercer 2001 offers a rich account of Leibniz's understanding of his role as a philosophical mediator. See esp. pp. 49–59.

158. "Praefatio ad libellum elementorum physicae," [1678–79], A6,4:2009. See Fichant 1998a on the immediate political, religious, and metaphysical concerns connected to Leibniz's physical innovations of the late 1670s.

159. Schaffer 1987; cf. Henry 1986. Leibniz did not simply alter the principles of motion willy-nilly to suit his metaphysical, political, moral, and ecclesiastical agendas;

his metaphysical concerns ended up helping him interpret the misfit between his theory and practice. See Leibniz 1994b, pp. 57–58, 269–71, 409–16.

160. See, e.g., Ficino 2001, pp. 10–11.

161. "Praefatio ad libellus elementorum physicae," [1678–79], A6,4:1994.

162. "Elementa juris naturalis," 1671, A6,1:481; see Mercer 2001, p. 248.

EPILOGUE

1. Kant 2000, §83, Ak 5:433.

2. Kant 1996, Bxiii.

3. Kant 1996, A752/B780 (italics in original). The context here is the need for a critique of dogmatic reason.

Bibliography

Adam, Antoine, ed. 1986. *Les libertins au XVIIe siècle*. Paris: Editions Buchet/Chastel.

Aiton, E. J. 1985. *Leibniz: A Biography*. Bristol and Boston: A. Hilger.

Alanen, Lilli. 1999. "Intuition, Assent and Necessity: The Question of Descartes' Psychologism." In *Norms and Modes of Thinking in Descartes*, edited by Tuomo Aho and Mikko Yrjönsuuri, pp. 99–121. Acta philosophica Fennica, vol. 64. Helsinki: Societas philosophica.

Alanen, Lilli, and Mikko Yrjönsuuri. 1997. "Intuition, jugement et évidence chez Ockham et Descartes." In *Descartes et le moyen âge*, edited by Joël Biard and Rushdi Rashid, pp. 155–74. Paris: J. Vrin.

Alpers, Svetlana. 1983. *The Art of Describing: Dutch Art in the Seventeenth Century*. Chicago: University of Chicago Press.

Alsted, Johann Heinrich. 1649. *Joan. Henrici Alstedii Scientiarum omnium encyclopaediae*. 4 vols. Lyon.

Altman, Joel B. 1987. "'Preposterous Conclusions': Eros, Enargeia, and the Composition of Othello." *Representations*, pp. 129–57.

Amadae, S. M. 2003. *Rationalizing Capitalist Democracy: The Cold War Origins of Rational Choice Liberalism*. Chicago: University of Chicago Press.

Andersen, Kristi. 1986. "The Method of Indivisibles: Changing Understandings." In *300 Jahre "Nova Methodus" von G. W. Leibniz*, edited by Albert Heinekamp, pp. 14–25. Studia Leibnitiana, Sonderheft, vol. 14. Stuttgart: Franz Steiner Verlag.

———. 1994. "Precalculus, 1635–1665." In *Companion Encyclopedia of the History and Philosophy of the Mathematical Sciences*, edited by I. Grattan-Guinness, pp. 292–307. London: Routledge.

Antognazza, Maria Rosa. 1994. "Die Rolle der Trinitäts- und Menschwerdungdiskussion für die Entstehung von Leibniz' Denken." *Studia Leibnitiana* 24:56–75.

———. 2001. "The Defence of the Mysteries of the Trinity and the Incarnation: An Example of Leibniz's 'Other' Reason." *British Journal for the History of Philosophy* 9:283–309.

———. 2003. "Leibniz and the Post-Copernican Universe: Koyré Revisited." *Studies in History and Philosophy of Science* 34:309–27.

330 *Bibliography*

Antognazza, Maria Rosa, and Howard Hotson. 1999. *Alsted and Leibniz: On God, the Magistrate and the Millennium.* Wiesbaden: Harrassowitz.

Apostolidès, Jean-Marie. 1981. *Le roi-machine: Spectacle et politique au temps de Louis XIV.* Paris: Editions de Minuit.

Ariew, Roger. 1999. *Descartes and the Last Scholastics.* Ithaca: Cornell University Press.

Armogathe, Jean-Robert, Vincent Carraud, and Robert Fernstra. 1988. "La licence en droit de Descartes: Un placard inédit de 1616." *Nouvelles de la république des lettres,* 123–45.

Arnauld, Antoine. 1644. *Apologie de monsieur Jansenius evesque d'ipre.* N.p.

———. 1781. *Nouveaux éléments de géometrie, contenant des moyens de faire voir quelles lignes sont incommensurables.* . . . Vol. 42 of *Œuvres de messire Antoine Arnauld.* Paris and Lausanne.

———. 2001. *Textes philosophiques.* Edited by Denis Moreau. Paris: Presses universitaires de France.

Arnauld, Antoine, and Pierre Nicole. 1992. *La logique ou L'art de penser.* Edited by Charles Jourdain. Paris: Gallimard. (Abbreviated as J.)

Arthur, Richard T. W. 1999. "The Transcendentality of π and Leibniz's Philosophy of Mathematics." *Proceedings of the Canadian Society for History and Philosophy of Mathematics* 12:13–19.

———. 2003. "The Enigma of Leibniz's Atomism." In *Oxford Studies in Early Modern Philosophy,* vol. 1, edited by Daniel Garber and Steven Nadler, pp. 183–227. Oxford: Clarendon Press.

———. Forthcoming a. "From Actuals to Fictions: Four Phases in Leibniz's Early Thought on Infinitesimals." In a special volume of *Studia Leibnitiana* on the Philosophy of the young Leibniz, edited by Mark Kulstad and Mogens Laerke.

———. Forthcoming b. "The Remarkable Fecundity of Leibniz's Work on Infinite Series." *Annals of Science.* (Review of A7,3 and A3,5.)

Assunto, Rosario. 1979. "Un filosofo nelle capitali d'Europa: La filosofia di Leibniz tra Barocco e Rococò." In *Infinita contemplazione: Gusto e filosofia dell'Europa barocca,* by Rosario Assunto, pp. 9–88. Naples: Società editrice napoletana.

Baltrušaitis, Jurgis. 1984. *Anamorphoses, ou Thaumaturgus opticus.* 3rd ed. Paris: Flammarion.

Barret-Kriegel, Blandine. 1988. *La république incertaine.* Vol. 4 of *Les historiens et la monarchie.* Paris: Presses universitaires de France.

Bassler, O. Bradley. 1999. "Toward Paris: The Growth of Leibniz's Paris Mathematics out of the Pre-Paris Metaphysics." *Studia Leibnitiana* 31:160–80.

Batllori, Miguel. 1986. "Las obras de Luis Vives en los colegios jesuìticos del siglo XVI." In *Erasmus in Hispania; Vives in Belgio,* edited by J. IJsewijn and A. Losada, pp. 121–45. Louvain: Peeters.

Baxandall, Michael. 1971. *Giotto and the Orators: Humanist Observers of Painting in Italy and the Discovery of Pictorial Composition, 1350–1450.* Oxford: Clarendon Press.

Bayley, Peter. 1980. *French Pulpit Oratory, 1598–1650: A Study in Themes and Styles, with a Descriptive Catalogue of Printed Texts.* Cambridge: Cambridge University Press.

Bedini, Silvio A. 1994. "Citadels of Learning: The Museo Kicheriano and Other 17th-Century Italian Science Collections." In *Science and Instruments in Seventeenth-Century Italy,* by Silvio A. Bedini, pp. X:249–67. Aldershot, UK: Variorum.

Bedini, Silvio A., and Arthur G. Bennet. 1999. "'A Treatise on Optics' by Giovanni Christoforo Bolantio." In *Patrons, Artisans and Instruments of Science, 1600–1750,* by Silvio A. Bedini, pp. IX:103–26. Aldershot, UK: Variorum.

Beeley, Philip. 1999. "Mathematics and Nature in Leibniz's Early Philosophy." In *The Young Leibniz and His Philosophy (1646–76),* edited by Stuart Brown, pp. 123–45. Dordrecht: Kluwer Academic Press.

Belaval, Yvon. 1958. "Une 'Drôle de pensée' de Leibniz." *La nouvelle revue française,* no. 70:754–68.

———. 1960. *Leibniz: Critique de Descartes.* Paris: Minuit.

———. 1969. *Leibniz: Initiation à sa philosophie.* 3rd ed. Paris: J. Vrin.

Bellarmino, Roberto. 1870–74. *Opera omnia.* Edited by Justin Louis Pierre Fèvre. 12 vols. Paris: Ludovicum Vivès.

Bennett, J. A. 1986. "Mechanics' Philosophy and Mechanical Philosophy." *History of Science* 24:1–28.

———. 1987. *The Divided Circle: A History of Instruments for Astronomy, Navigation, and Surveying.* Oxford: Phaidon Christie's.

Bensoussan, David. 2000. "La moraliste et les philosophes (Descartes, Gassendi, Hobbes): Quelques aperçus épistémologiques." In *Entre Baroque et Lumières: Saint-Evremond (1614–1703),* edited by Suzanne Guellouz, pp. 173–92. Caen: Presses universitaires de Caen.

Berggren, Lennart, Jonathan M. Borwein, and Peter B. Borwein. 2004. *Pi, a Sourcebook.* 3rd ed. New York: Springer.

Bertoloni Meli, Domenico. 1993. *Equivalence and Priority: Newton versus Leibniz.* Oxford: Clarendon Press.

Bessot, D. 1999. "Drôles des visions, autour des anamorphoses." In *L'actualité de Leibniz: Les deux labyrinthes,* edited by Dominique Berlioz and Frédéric Nef, pp. 235–76. Studia Leibnitiana supplementa, vol. 34. Stuttgart: Franz Steiner Verlag.

Beugnot, Bernard. 1999. "La précellence du style moyen." In *Histoire de la rhétorique dans l'Europe moderne: 1450–1950,* edited by Marc Fumaroli, pp. 539–99. Paris: Presses universitaires de France.

Biagioli, Mario. 1989. "The Social Status of Italian Mathematicians, 1450–1600." *History of Science* 27:1–75.

———. 1993. *Galileo, Courtier: The Practice of Science in the Culture of Absolutism.* Chicago: University of Chicago Press.

———. 1996. "Etiquette, Interdependence, and Sociability in Seventeenth-Century Science." *Critical Inquiry* 22:193–238.

Biard, Joël, and Rushdi Rashid, eds. 1997. *Descartes et le moyen âge.* Paris: J. Vrin.

Binet, Etienne. 1624. *Essay des merveilles de nature, et des plus nobles artifices. Pièces très necessaire, à tous ceux qui font profession d'éloquence.* 4th ed. Rouen.

Blair, Ann. 1992. "Humanist Methods in Natural Philosophy: The Commonplace Book." *Journal of the History of Ideas* 53:541–51.

———. 1997. *The Theater of Nature: Jean Bodin and Renaissance Science.* Princeton: Princeton University Press.

———. 2003. "Reading Strategies for Coping with Information Overload ca. 1550–1700." *Journal of the History of Ideas* 64:11–28.

———. 2004. "[Scientific Readers:] An Early Modernist's Perspective." *Isis* 95:420–30.

Blay, Michel. 1992. *La naissance de la mécanique analytique: La science du mouvement au tournant des XVIIe et XVIIIe siècles*. Paris: Presses universitaires de France.

Blum, Paul Richard. 1985. "*Apostolato dei collegi:* On the Integration of Humanism in the Educational Programs of the Jesuits." *History of Universities* 5:101–15.

Bodemann, Eduard. 1966. *Die Leibniz-Handschriften der Königlichen öffentlichen Bibliothek zu Hannover*. Hildesheim: G. Olms.

Boileau, Nicolas. 1666. *Recueil contenant plusieurs discours libres et moraux en vers et un jugement en prose sur les sciences où un honneste homme peut s'occuper.* N.p.

Bold, Stephen C. 1996. *Pascal Geometer: Discovery and Invention in Seventeenth-Century France.* Geneva: Librarie Droz.

Bos, H. J. M. 1974. "Differentials, Higher-Order Differentials and the Derivative in the Leibnizian Calculus." *Archive for History of Exact Sciences* 14:1–90.

———. 1980. "Newton, Leibniz and the Leibnizian Tradition." In *From the Calculus to Set Theory, 1630–1910: An Introductory History,* edited by I. Grattan-Guinness, pp. 49–93. London: Duckworth.

———. 1981. "On the Representation of Curves in Descartes' *Géométrie.*" *Archive for History of Exact Sciences* 24:295–338.

———. 1984. "Arguments on Motivation in the Rise and Decline of a Mathematical Theory: The 'Construction of Equations,' 1637–ca. 1750." *Archive for History of Exact Sciences* 30:331–80.

———. 1990. "The Structure of Descartes' *Géométrie.*" In *Descartes: Il metodo e i saggi,* vol. 2, edited by Giulia Belgioso et al., pp. 349–69. Rome: Istituto della Enciclopedia italiana.

———. 1992. "Descartes, Pappus' Problem and the Cartesian Parabola: A Conjecture." In *The Investigation of Difficult Things: Essays on Newton and the History of the Exact Sciences in Honour of D. T. Whiteside,* edited by P. M. Harman and Alan E. Shapiro, pp. 71–96. Cambridge: Cambridge University Press.

———. 1993. "The Fundamental Concepts of the Leibnizian Calculus." In *Lectures in the History of Mathematics,* ed. H. J. M. Bos, pp. 83–99. Providence, RI: American Mathematical Society and London Mathematical Society.

———. 2001. *Redefining Geometrical Exactness: Descartes' Transformation of the Early Modern Concept of Construction.* New York: Springer.

———. 2003. "Descartes, Elizabeth and Apollonius' Problem." In *The Correspondence of René Descartes 1643,* edited by Theo Verbeek, Erik-Jan Bos, and Jeroen Van de Ven, pp. 202–11. Quaestiones infinitae, vol. 45. Utrecht: Department of Philosophy, Utrecht University.

Bosse, Abraham. 1973. *Traité des pratiques géométrales et perspectives; Discours sur la manière universelle de Desargues; Le peintre converti aux précises et universelles règles de son art.* Geneva: Minkoff Reprint.

Bourdieu, Pierre. 1984. *Distinction: A Social Critique of the Judgement of Taste.* Translated by Richard Nice. Cambridge, MA: Harvard University Press.

Bouwsma, William J. 1990. "The Two Faces of Humanism: Stoicism and Augustinianism in Renaissance Thought." In *A Usable Past: Essays in European Cultural History,* by William J. Bouwsma, pp. 19–73. Berkeley and Los Angeles: University of California Press.

Bredekamp, Horst. 2004. *Die Fenster der Monade: Gottfried Wilhelm Leibniz' Theater der Natur und Kunst.* Berlin: Akademie Verlag.

Breger, Herbert. 1986. "Leibniz' Einführung des Transzendenten." In *300 Jahre "Nova Methodus" von G. W. Leibniz,* edited by Albert Heinekamp, pp. 119–32. Studia Leibnitiana, Sonderheft, vol. 14. Stuttgart: Franz Steiner Verlag.

Brockliss, L. W. B. 1987. *French Higher Education in the Seventeenth and Eighteenth Centuries: A Cultural History.* Oxford: Clarendon Press.

Brown, Gregory. 1994. "'*Quod ostendendum suscepramus*': What Did Leibniz Undertake to Show in the *Brevis demonstratio?*" In *Gottfried Wilhem Leibniz: Critical Assessments,* vol. 3, edited by R. S. Woolhouse, 177–97. London and New York: Routledge.

——. 2004. "Leibniz's Endgame and the Ladies of the Courts." *Journal of the History of Ideas* 65:75–100.

Brown, Harcourt. 1934. *Scientific Organizations in Seventeenth Century France (1620–1680).* Baltimore: Williams & Wilkins Co.

Brown, Stuart, ed. 1999. *The Young Leibniz and His Philosophy (1646–76).* Dordrecht: Kluwer Academic Press.

Brusati, Celeste. 1995. *Artifice and Illusion: The Art and Writing of Samuel van Hoogstraten.* Chicago: University of Chicago Press.

Burke, Peter. 1993. "The Art of Conversation in Early Modern Europe." In *The Art of Conversation,* by Peter Burke, pp. 89–122. Ithaca: Cornell University Press.

Bury, Emmanuel. 1996a. *Littérature et politesse: L'invention de l'honnête homme, 1580–1750.* Paris: Presses universitaires de France.

——. 1996b. "Le sourire de Socrate ou, peut-on être à la fois philosophe et honnête homme?" In *Le loisir lettré à l'âge classique,* edited by Marc Fumaroli, Philippe-Joseph Salazar, and Emmanuel Bury, 197–212. Geneva: Librairie Droz.

Bynum, Caroline Walker. 2001. *Metamorphosis and Identity.* New York: Zone.

Camerota, Filippo. 2004. "Renaissance Descriptive Geometry: The Codification of Drawing Methods." In *Picturing Machines, 1400–1700,* edited by Wolfgang Lefèvre, pp. 175–208. Cambridge, MA: MIT Press.

Campbell, Stephen F. 1993. "Nicolas Caussin's 'Spirituality of Communication': A Meeting of Divine and Human Speech." *Renaissance Quarterly* 46:44–70.

Carr, Thomas M., Jr. 1990. *Descartes and the Resilience of Rhetoric: Varieties of Cartesian Rhetorical Theory.* Carbondale: Southern Illinois University Press.

Carraud, Vincent. 1986. "Leibniz lecteur des *Pensées* de Pascal." *XVIIe siècle,* no. 151:107–24.

——. 1992. *Pascal et la philosophie.* Paris: Presses universitaires de France.

Casaubon, Meric. 1654. *A treatise concerning enthusiasme, as it is an effect of nature: but is mistaken by many for either divine inspiration, or diabolical possession.* London.

Caton, Hiram. 1975. "Will and Reason in Descartes's Theory of Error." *Journal of Philosophy* 72:87–104.

Caussin, Nicolas. 1619. *De eloquentiae sacrae et humanae parallela, libri XVI.* Paris.

Cavaillé, Jean-Pierre. 1994. "'Le plus éloquent philosophe de derniers temps': Les stratégies d'auteur de René Descartes." *Annales HSS,* no. 2 (March–April):249–67.

Cave, Terence. 1976. "Enargeia: Erasmus and the Rhetoric of Presence in the Sixteenth Century." *L'esprit créateur* 16, no. 4:5–19.

————. 1979. *The Cornucopian Text: Problems of Writing in the French Renaissance.* Oxford: Clarendon Press.

Céard, Jean. 1996. *La nature et les prodiges: L'insolite au 16e siècle en France.* 2nd ed. Geneva: Librarie Droz.

Certeau, Michel de. 1965. "Crise sociale et réformisme spirituel au début du XVIIe siècle: Une 'Nouvelle Spiritualité' chez les Jésuites français." *Revue d'ascétique et de mystique* 41:338–86.

————. 1982. *La fable mystique, 1: XVIe–XVIIe siècle.* Paris: Gallimard.

Charron, Pierre. [1601–4] 1986. *De la sagesse.* Edited by Barbara de Negroni. Paris: Fayard.

Cherubin d'Orléans. 1671. *La Dioptrique Oculaire ou la théorique, la positive, et la méchanique de l'oculaire dioptrique, en toutes ses espèces.* Paris.

Child, J. M. 1920. *The Early Mathematical Manuscripts of Leibniz.* Chicago and London: Open Court.

Cicero. 1949. *De inventione; De optimum genere oratorum; Topica.* Translated by H. M. Hubbell. Cambridge, MA: Harvard University Press.

————. 1971. *Tusculan Disputations.* Translated by J. E. King. Cambridge, MA: Harvard University Press.

Clark, Stuart. 1997. *Thinking with Demons: The Idea of Witchcraft in Early Modern Europe.* Oxford: Clarendon Press.

Clarke, Desmond M. 1991. "Descartes's Use of 'Demonstration' and 'Deduction.'" In *René Descartes: Critical Assessments,* vol. 1, edited by George J. D. Moyal, 237–47. London and New York: Routledge.

Clavius, Christoph. [1611] 1999. *Commentaria in Euclidis Elementa geometrica.* Mainz. Reprint, with foreword by Eberhard Knobloch. Hildesheim: Olms.

Clements, Robert John. 1942. *Critical Theory and Practice of the Pléiade.* Cambridge: Harvard University Press.

Cohen, H. F. 1984. *Quantifying Music: The Science of Music at the First Stage of the Scientific Revolution, 1580–1650.* Dordrecht: D. Reidel.

Coimbra. 1984. *Commentarii collegii conimbricensis societatis Jesu in octo libros physicorum Aristotelis.* 2 vols. Hildesheim: G. Olms.

Conley, John J. 2002. *The Suspicion of Virtue: Women Philosophers in Neoclassical France.* Ithaca: Cornell University Press.

Cook, Daniel J. 1998. "Leibniz on Enthusiasm." In *Leibniz, Mysticism and Religion,* edited by Allison P. Coudert, Richard H. Popkin, and Gordon M. Weiner, pp. 107–35. Dordrecht and Boston: Kluwer.

Cooter, Roger, and Stephen Pumfrey. 1994. "Separate Spheres and Public Places: Reflections on the History of Science Popularisation and Science in Popular Culture." *History of Science* 32:237–67.

Costabel, Pierre. 1962. "Traduction française de notes de Leibniz sur les *Coniques* de Pascal." *Revue d'histoire des sciences* 15:253–68.

Cottingham, John. 1996. "Partiality and the Virtues." In *How Should One Live? Essays on the Virtues,* edited by Roger Crisp, pp. 57–76. Oxford: Clarendon Press.

————. 1998. *Philosophy and the Good Life: Reason and the Passions in Greek, Cartesian, and Psychoanalytic Ethics.* Cambridge: Cambridge University Press.

Couturat, Louis. 1901. *La logique de Leibniz d'après des documents inédits*. Paris: F. Alcan.

Couzinet, Marie-Dominique. 1996. *Histoire et méthode à la Renaissance: Une lecture de la* Methodus ad facilem historiarum cognitionem *de Jean Bodin*. Paris: J. Vrin.

Coxito, Amândio A., and Maria Luisa Couto Soares. 2001. "Pedro da Fonseca." In *História do pensamento filosófico português*, vol. 2, edited by Pedro Calafate, pp. 455–502. Lisbon: Caminho.

Crombie, Alistair. 1977. "Mathematics and Platonism in the Sixteenth-Century Italian Universities and in Jesuit Educational Policy." In *Prismata: Naturwissenschaftsgeschichtliche Studien*, edited by Y. Maeyama, pp. 63–94. Wiesbaden: W. G. Saltzer.

Dainville, François de. 1940. *La naissance de l'humanisme moderne*. Paris: Beauchesne et ses fils.

Dalibray, Charles Vion. 1653. *Les œuvres poetiques du Sr. Dalibray*. Paris.

Dascal, Marcelo. 1978. *La sémiologie de Leibniz*. Paris: Aubier-Montaigne.

———. 1987a. *Leibniz, Language, Signs, and Thought: A Collection of Essays*. Amsterdam and Philadelphia: J. Benjamins Publishing Co.

———. 1987b. "Leibniz's Early Views on Definition." In *Leibniz, Language, Signs, and Thought: A Collection of Essays*, pp. 61–79. Amsterdam and Philadelphia: J. Benjamins Publishing Co.

———. 1987c. "Signs and Thought in Leibniz's 'Paris Notes.'" In *Leibniz, Language, Signs, and Thought: A Collection of Essays*, 47–59. Amsterdam and Philadelphia: J. Benjamins Publishing Co.

———. 1993. "One Adam and Many Cultures: The Role of Political Pluralism in the Best of Possible Worlds." In *Leibniz and Adam*, edited by Marcelo Dascal and Elhanan Yakira, pp. 387–409. Tel Aviv: University Publishing Projects.

———. 1995. "Strategies of Dispute and Ethics: *Du tort* and *La place d'autruy*." In *Leibniz und Europa: Vorträge, VI. Internationaler Leibniz-Kongress*, vol. 2, pp. 108–16. Hanover: Gottfried-Wilhelm-Leibniz-Gesellschaft e.V.

———. 2000. "Leibniz and Epistemological Diversity." In *Unitá e molteplicitá nel pensiero filosofico e scientifico di Leibniz*, edited by Antonio Lamarra and Roberto Palaia, pp. 12–37. Rome: Leo S. Olschki.

———. 2001. "Nihil sine ratione [to] Blandior ratio." In vol. 1 of *Nihil sine ratione: Mensch, Natur und Technik im Wirken von G. W. Leibniz*, edited by Hans Poser, pp. 276–80. Proceedings of the VII. Internationaler Leibniz-Kongress. Berlin: Gottfried-Wilhelm-Leibniz Gesellschaft e.V.

———. 2002. "Leibniz y las technologías cognitivas." In *Ciencia, tecnologia, y el bien común: La actualidad de Leibniz*, edited by Agustín Andreu, Javier Echeverría, and Concha Roldán, pp. 359–88. Valencia: Universidad politécnica de Valencia.

———. 2003. "*Ex pluribus unum?* Patterns in 522+ Texts of Leibniz's *Sämtliche Schriften und Briefe* VI.4." *Leibniz Review* 13:104–54.

———. 2004. "*Alter et etiam*: Rejoinder to Schepers." *Leibniz Review* 14:127–51.

Daston, Lorraine. 1988. *Classical Probability in the Enlightenment*. Princeton: Princeton University Press.

———. 1994. "Historical Epistemology." In *Questions of Evidence: Proof, Practice, and Persuasion across the Disciplines*, edited by James Chandler, Arnold I. Davidson, and Harry Harootunian, pp. 282–89. Chicago: University of Chicago Press.

———. 1995. "The Moral Economy of Science." *Osiris* 10:3–24.

Daston, Lorraine, and Katharine Park. 1998. *Wonders and the Order of Nature, 1150–1750.* New York: Zone Books.

Davidson, Arnold I. 1994. "Ethics as Ascetics: Foucault, the History of Ethics, and Ancient Thought." In *Foucault and the Writing of History*, edited by Jan Goldstein, pp. 63–80. Oxford: Blackwell.

———. 2001. "Styles of Reasoning: From the History of Art to the Epistemology of Science." In *The Emergence of Sexuality: Historical Epistemology and the Formation of Concepts*, pp. 125–41. Cambridge, MA: Harvard University Press.

Davillé, Louis. [1909] 1986. *Leibniz historien: Essai sur l'activité et la méthode historiques de Leibniz.* Paris: Félix Alcan. Reprint, Aalen, Germany: Scientia Verlag.

Dear, Peter. 1995a. *Discipline and Experience: The Mathematical Way in the Scientific Revolution.* Chicago: University of Chicago Press.

———. 1995b. "Mersenne's Suggestion: Cartesian Meditation and the Mathematical Model of Knowledge in the Seventeenth Century." In *Descartes and His Contemporaries*, edited by Roger Ariew and Marjorie Grene, pp. 44–62. Chicago: University of Chicago Press.

———. 1998. "A Mechanical Microcosm: Bodily Passions, Good Manners, and Cartesian Mechanism." In *Science Incarnate: Historical Embodiments of Natural Knowledge*, edited by Christopher Lawrence and Steven Shapin, pp. 51–82. Chicago: University of Chicago Press.

Demoustier, Adrien, Dominique Julia, and Marie-Madelein Compère, eds. 1997. *Ratio studiorum: Plan raisonné et institution des études dans la Compagnie de Jésus.* Paris: Belin.

Desargues, Girard. 1639. *Brouillon proiect d'une atteinte aux evenemens des rencontres du Cone avec un Plan.* Paris.

Descartes, René. 1657. *Lettres qui traitent de plusieurs belles questions concernant la morale, la physique, la médecine et les mathématiques.* Vol. 1. Edited by Claude Clerselier. Paris.

———. 1659. *Geometria, à Renato des Cartes anno 1637 Gallicè edita: postea autem una cum notis Florimondi de Beaune Gallicè conscriptis in Latinam linguam versa & commentariis illustrata, opera atque studio Francisci à Schooten.* 2 vols. Amsterdam.

———. 1977. *Règles utiles et claires pour la direction de l'esprit en la recherche de la vérité.* Translated by Jean-Luc Marion (with the help of Pierre Costabel). The Hague: Martinus Nijhoff.

———. 1984–91. *The Philosophical Writings of Descartes.* Translated by John Cottingham, Robert Stoothoff, Dugald Murdoch, and (vol. 3 only) Anthony Kenny. 3 vols. Cambridge: Cambridge University Press. (Abbreviated as CSM.)

———. 1990. *Abrégé de musique.* Translated by Pascal Dumont. Paris: Méridiens Klincksieck.

———. 1996. *Œuvres de Descartes.* Edited by Charles Ernest Adam and Paul Tannery. 2nd ed., corrected. 11 vols. Reprint, Paris: J. Vrin. (Abbreviated as AT.)

———. 1998. *Regulae ad directionem ingenii = Rules for the Direction of the Natural Intelligence: A Bilingual Edition of the Cartesian Treatise on Method.* Translated by George Heffernan. Amsterdam and Atlanta: Rodopi.

Descartes, René, and Martin Schoock. 1988. *La querelle d'Utrecht.* Translated by Theo Verbeek. Paris: Les impressions nouvelles.

Des Chene, Dennis. 1996. *Physiologia: Natural Philosophy in Late Aristotelian and Cartesian Thought.* Ithaca: Cornell University Press.

———. 2001. *Spirits and Clocks: Machine and Organism in Descartes.* Ithaca: Cornell University Press.

Descotes, Dominique. 1990. "Espaces infinis égaux au fini." In *Le Grand et Le Petit,* edited by A. Montandon, pp. 41–67. Clermont-Ferrand: Centre de recherches . . . Université Blaise Pascal.

———. 1993. *L'argumentation chez Pascal.* Paris: Presses universitaires de France.

———. 2001. *Blaise Pascal, littérature et géométrie.* Clermont-Ferrand: Presses universitaires Blaise Pascal.

———. 2003. *Pascal: Le calcul et la théologie.* Vol. 16 of *Pour la science: Les génies de la science.* Paris: Pour la science.

Dijksterhuis, Fokko Jan. 2004. *Lenses and Waves: Christiaan Huygens and the Mathematical Science of Optics in the Seventeenth Century.* Dordrecht: Kluwer.

Donaldson, Peter Samuel. 1988. *Machiavelli and Mystery of State.* New York: Cambridge University Press.

Duchêne, Roger. 1985. *L'imposture littéraire dans les* Provinciales *de Pascal.* 2nd ed. Aix-en-Provence: Université de Provence.

Duchesneau, François. 1993. *Leibniz et la méthode de la science.* Paris: Presses universitaires de France.

———. 2001. "Leibniz, la Bible et l'ordre des vérités." *Revue de théologie et philosophie,* no. 133:267–86.

Dumont, Stephen D. 1989. "Theology as a Science and Duns Scotus's Distinction between Intuitive and Abstractive Cognition." *Speculum* 64:579–99.

du Moulin, Pierre. 1638. *La philosophie françoise de Pierre du Moulin.* 2 vols. Paris: H. Le Gras.

Dupleix, Scipion. 1984. *La logique, ou, Art de discourir et raisonner.* Paris: Fayard.

———. 1994. *L'ethique ou philosophie morale.* Edited by Roger Ariew. Paris: Fayard.

Du Vair, Guillaume. 1641. *Les œuvres de Messire Guillaume du Vair.* Final ed. Paris.

Eamon, William. 1994. *Science and the Secrets of Nature: Books of Secrets in Medieval and Early Modern Culture.* Princeton: Princeton University Press.

Echevarría, Javier. 1983. "Recherches inconnues de Leibniz sur la géométrie perspective." In *Leibniz et la Renaissance,* edited by Albert Heinekamp, pp. 191–201. Wiesbaden: Franz Steiner.

———. 1994. "Leibniz, intreprète de Desargues." In *Desargues en son temps,* edited by Jean G. Dhombres and J. Sakarovitch, pp. 283–93. Paris: Librairie scientifique A. Blanchard.

Eckert, Horst. 1971. *Gottfried Wilhelm Leibniz' Scriptores rerum Brunsvicensium: Entstehung und historiographische Bedeutung.* Frankfurt am Main: Vittorio Klostermann.

Eden, Kathy. 1986. *Poetic and Legal Fiction in the Aristotelian Tradition.* Princeton: Princeton University Press.

———. 1997. *Hermeneutics and the Rhetorical Tradition: Chapters in the Ancient Legacy and Its Humanist Reception.* New Haven: Yale University Press.

Edwards, A. W. F. 2002. *Pascal's Arithmetical Triangle.* 2nd ed. Baltimore: Johns Hopkins University Press.

————. 2003. "Pascal's Work on Probability." In *The Cambridge Companion to Pascal,* edited by Nicholas Hammond, pp. 40–52. Cambridge: Cambridge University Press.

Elkins, James. 1994. *The Poetics of Perspective.* Ithaca: Cornell University Press.

Elster, Jon. 1975. *Leibniz et la formation de l'esprit capitaliste.* Paris: Aubier-Montaigne.

Evans, R. J. W. 1979. *The Making of the Habsburg Monarchy, 1550–1700: An Interpretation.* Oxford: Clarendon.

Fabri, Honoré. 1667. *Synopsis optica: In qua illa omnia quae ad opticam, dioptricam, catoptri-cam pertinent, id est, ad triplicem radium visualem directum, refractum, reflexum, breviter quidem, accurate tamen demonstrantur.* Lyon.

Feldman, Karen S. 2004. *"Per canales Troporum:* On Tropes and Performativity in Leibniz's Preface to Nizolius." *Journal of the History of Ideas* 65:39–51.

Fenves, Peter. 2003. "Of Philosophical Style—from Leibniz to Benjamin." *boundary 2* 30:67–87.

Ferraro, Giovanni. 2000. "True and Fictional Quantities in Leibniz's Theory of Series." *Studia Leibnitiana* 32:43–67.

Ferreyrolles, Gérard. 1984. *Blaise Pascal, Les provinciales.* Paris: Presses universitaires de France.

————. 1995. *Les reines du monde: L'imagination et la coutume chez Pascal.* Paris: Champion.

Fichant, Michel. 1998a. "Mécanisme et métaphysique: Le rétablissement des formes substantielles (1679)." In *Science et métaphysique dans Descartes et Leibniz,* by Michel Fichant, pp. 163–204. Paris: Presses universitaires de France.

————. 1998b. "'Pour la beauté et pour l'harmonie': Le meilleur de la dynamique." In *Science et métaphysique dans Descartes et Leibniz,* pp. 267–86. Paris: Presses universitaires de France.

Ficino, Marsilio. 2001. *Platonic Theology.* Vol. 1. Translated by Michael J. B. Allen and John Warden. Edited by James Hankins and William R. Bowen. Cambridge, MA: Harvard University Press.

Field, Judith Veronica. 1997. *The Invention of Infinity: Mathematics and Art in the Renaissance.* Oxford: Oxford University Press.

Field, Judith Veronica, and Jeremy Gray. 1987. *The Geometrical Work of Girard Desargues.* New York: Springer-Verlag.

Filère, Joseph. 1636. *Le miroir sans tache, enrichi des merveilles de la nature dans les miroirs, rapportées aux effets de la Grace, pour voir Dieu en toutes choses et toutes choses en Dieu, et s'avancer par les degrès de la vertu jusqu'à la perfection.* Lyon.

Findlen, Paula. 1994. *Possessing Nature: Museums, Collecting, and Scientific Culture in Early Modern Italy.* Berkeley and Los Angeles: University of California Press.

————. 2003. "Scientific Spectacle in Baroque Rome: Athanasius Kircher and the Roman College Museums." In *Jesuit Science and the Republic of Letters,* edited by Mordechai Feingold, pp. 225–84. Cambridge, MA: MIT Press.

Fleck, Ludwik. 1979. *Genesis and Development of a Scientific Fact.* Edited by Thaddeus J. Trenn and Robert K. Merton. Translated by Fred Bradley and Thaddeus J. Trenn. Chicago: University of Chicago Press.

Flynn, Laurence J. 1957. "Sources and Influences of Soarez' *De arte rhetorica."* *Quarterly Journal of Speech* 43:257–65.

Fonseca, Pedro da. 1609. *Institutionum dialecticarum libri octo.* La Flèche: Iacobum Rezé.

————. 1964a. *Commentariorum in Metaphysicorum Aristotelis Stagiritae libros.* Hildesheim: G. Olms.

————. 1964b. *Instituições dialécticas: Institutionum dialecticarum libri octo.* Edited by Joaquim Ferreira Gomes. 2 vols. Coimbra: Universidade de Coimbra.

Fontenelle, Bernard Le Bovier de. 1989–97. *Œuvres complètes.* Edited by Alain Niderst. 9 vols. Paris: Fayard.

Force, James E., and Richard Henry Popkin, eds. 1990. *Essays on the Context, Nature, and Influence of Isaac Newton's Theology.* Dordrecht: Kluwer Academic Publishers.

Force, Pierre. 1989. *Le problème herméneutique chez Pascal.* Paris: J. Vrin.

————. 2005. "Innovation as Spiritual Exercise: Montaigne and Pascal." *Journal of the History of Ideas* 66:17–35.

Foucault, Michel. 1971. *The Order of Things: An Archaeology of the Human Sciences.* New York: Pantheon Books.

————. 1984. *Le souci de soi.* Vol. 3 of *Histoire de la sexualité.* Paris: Gallimard.

————. 1998. "My Body, This Paper, This Fire." In *Aesthetics, Method and Epistemology*, vol. 2 of *Essential Works of Foucault, 1954–1984*, edited by James D. Faubion, pp. 393–418. New York: New Press.

————. 2001. *L'herméneutique du sujet.* Paris: Gallimard and Seuil.

Frasca-Spada, Marina. 1999. "The Science and Conversation of Human Nature." In *The Sciences in Enlightened Europe*, edited by William Clark, Jan Golinski, and Simon Schaffer, pp. 218–45. Chicago: University of Chicago Press.

Frémont, Christiane. 2003. *Singularités: Individus et relations dans le système de Leibniz.* Paris: J. Vrin.

Fumaroli, Marc. 1979. "Pascal et la tradition rhétorique gallicane." In *Méthodes chez Pascal: Actes du colloque tenu à Clermont-Ferrand, 10–13 juin 1976*, edited by Jean Mesnard, pp. 359–72. Paris: Presses universitaires de France.

————. 1980a. "Crépuscule de l'enthousiasme au XVIIe siècle." In *Acta Conventus Neo-Latini Turonensis*, edited by Jean-Claude Margolin, pp. 1279–1305. Paris: J. Vrin.

————. 1980b. "Définition et description: Scholastique et rhétorique chez les jésuites des XVIe et XVIIe siècles." *Travaux de linguistique et de littérature* 18:37–48.

————. 1994a. *L'âge de l'éloquence: Rhétorique et "res literaria," de la Renaissance au seuil de l'époque classique.* Reprint, Paris: A. Michel.

————. 1994b. "L'art de la conversation, ou le Forum du royaume." In *La diplomatie de l'esprit: De Montaigne à La Fontaine*, by Marc Fumaroli, pp. 283–320. Paris: Hermann.

————. 1994c. "La conversation savante." In *Commercium litterarium, 1600–1750: La communication dans la République des lettres: Conférences des colloques tenus à Paris 1992 et à Nimègue 1993*, edited by Hans Bots and Françoise Waquet, pp. 67–80. Amsterdam: APA-Holland University Press.

————. 1994d. "La diplomatie au service de la méthode." In *La diplomatie de l'esprit: De Montaigne à La Fontaine*, by Marc Fumaroli, pp. 377–402. Paris: Hermann.

————. 1994e. "La prose d'état: Charles Paschal, théoricien du style royal." In *La diplomatie de l'esprit: De Montaigne à La Fontaine*, by Marc Fumaroli, pp. 59–124. Paris: Hermann.

————, ed. 1999. *Histoire de la rhétorique dans l'Europe moderne, 1450–1950.* Paris: Presses universitaires de France.

Fumaroli, Marc, Philippe-Joseph Salazar, and Emmanuel Bury, eds. 1996. *Le loisir lettré à l'âge classique*. Geneva: Librairie Droz.

Funkenstein, Amos. 1986. *Theology and the Scientific Imagination from the Middle Ages to the Seventeenth Century*. Princeton: Princeton University Press.

Gabbey, Alan. 1985. "The Mechanical Philosophy and Its Problems: Mechanical Explanations, Impenetrability, and Perpetual Motion." In *Change and Progress in Modern Science*, edited by Joseph C. Pitt, pp. 9–84. Dodrecht: D. Reidel.

Galand-Hallyn, P. 1991. "Enargeia maniériste, enargeia visionnaire des prophéties du tibe au songe d'océan." *Bibliothèque d'humanisme et Renaissance* 53:305–28.

Ganss, George E., ed. 1991. *Ignatius of Loyola: The Spiritual Exercises and Selected Works*. New York: Paulist Press.

Garber, Daniel. 1985. "Leibniz and the Foundations of Physics: The Middle Years." In *The Natural Philosophy of Leibniz*, edited by Kathleen Okruhlik and James Robert Brown, pp. 27–130. Dordrecht: D. Reidel.

———. 1988. "Descartes, the Aristotelians, and the Revolution That Did Not Happen in 1637." *Monist* 71, no. 4:471–86.

———. 1992. *Descartes' Metaphysical Physics*. Chicago: University of Chicago Press.

———. 1995. "Leibniz: Physics and Philosophy." In *The Cambridge Companion to Leibniz*, edited by Nicholas Jolley, pp. 270–352. Cambridge: Cambridge University Press.

———. 2001. "Descartes, or the Cultivation of the Intellect." In *Descartes Embodied: Reading Cartesian Philosophy through Cartesian Science*, by Daniel Garber, pp. 277–95. Cambridge: Cambridge University Press.

García-Hernández, Benjamín. 1997. "El magisterio de Séneca, reconocido por Descartes: Filosofía y poesía." In *Séneca, dos mil años después*, edited by Miguel Rodriguez-Pantoja, pp. 675–88. Córdoba: Publicaciones de la Universidad de Córdoba y Obra Social y Cultural CajaSur.

Gardies, Jean-Louis. 1982. "L'interprétation d'Euclide chez Pascal et Arnauld." *Les études philosophiques* 37:129–48.

———. 1984. *Pascal entre Eudoxe et Cantor*. Paris: J. Vrin.

———. 1995. "Arnauld et la reconstruction de la géométrie euclidienne." In *Antoine Arnauld: Philosophie du langage et de la connaissance*, edited by Jean-Claude Pariente, pp. 13–31. Paris: J. Vrin.

Gassendi, Pierre. [1658] 1964. *Opera omnia*. 6 vols. Lyon. Reprint, Hildesheim: G. Olms.

———. 2004. *Pierre Gassendi (1592–1655), Lettres latines*. Edited and translated by Sylvie Taussig. 2 vols. Turnhout: Brepols.

Gaudin, Claude. 1993. "Correspondance et responsabilité dans la philosophie des signes: Analyse critique du *Dialogus* de 1677." *Philosophie*, no. 39:83–107.

Gaukroger, Stephen. 1989. *Cartesian Logic, an Essay on Descartes's Conception of Inference*. Oxford: Oxford University Press.

———. 1992. "The Nature of Abstract Reasoning: Philosophical Aspects of Descartes' Work in Algebra." In *The Cambridge Companion to Descartes*, edited by John Cottingham, pp. 91–114. Cambridge: Cambridge University Press.

———. 1995. *Descartes: An Intellectual Biography*. Oxford: Oxford University Press.

———. 1997. "Descartes' Early Doctrine of Clear and Distinct Ideas." In *The Genealogy of Knowledge: Analytical Essays in the History of Philosophy and Science*, by Stephen Gaukroger, pp. 131–52. Aldershot, UK: Ashgate.

_____. 2001. *Francis Bacon and the Transformation of Early-Modern Philosophy.* Cambridge: Cambridge University Press.

Geertz, Clifford. 1973. "'Internal Conversion' in Contemporary Bali." In *The Interpretation of Cultures*, by Clifford Geertz, pp. 170–89. New York: Basic.

Gensini, Stefano. 1996. "The Leibnitian Concept of 'Significatio.'" In *Im Spiegel der Verstandes: Studien zu Leibniz*, edited by Stefano Gensini and Klaus D. Dutz, 69–98. Münster: Nodus-Publikationen. (Reprinted in Gensini 2000.)

_____. 2000. *"De linguis in universum": On Leibniz's Ideas on Languages.* Münster: Nodus-Publikationen.

Gil, Fernando. 1984. "Leibniz, la place d'autrui, le principe du pire, et la politique de la monadologie." *Passé présent*, no. 3:147–64.

Gilson, Etienne. 1930. *Etudes sur le rôle de la pensée médiévale dans la formation du système cartésien.* Paris: J. Vrin.

_____. 1963. *Index scolastico-cartésien.* New York: Burt Franklin.

Ginzburg, Carlo. 1989. "Montrer et citer: La vérité de l'histoire." *Le débat*, no. 56 (September–October):43–54.

_____. 1999. *History, Rhetoric, and Proof.* Hanover, NH: University Press of New England.

_____. 2001. "Distance and Perspective: Two Metaphors." In *Wooden Eyes: Nine Reflections on Distance*, by Carlo Ginzburg, pp. 139–56. Translated by Martin Ryle and Kate Soper. New York: Columbia University Press.

Giolito, Christophe. 1996. "L'éradication d'une argumentation cartésienne: Leibniz à la lecture des *Passions de l'âme*." In *Descartes et l'argumentation philosophique*, edited by Frédéric Cossutta, pp. 187–216. Paris: Presses universitaires de France.

Goclenius, Rudolph. 1964. *Lexicon philosophicum, quo tanquam clave philosophicae fores aperiuntur.* Hildesheim: G. Olms.

Goldsmith, Elizabeth C. 1988. *"Exclusive Conversations": The Art of Interaction in Seventeenth-Century France.* Philadelphia: University of Pennsylvania Press.

Goldstein, Catherine. 2001. "L'expérience des nombres de Bernard Frenicle de Bessy." *Revue de synthèse*, 4th ser., nos. 2–4:425–54.

Goodman, Dena. 1994. *The Republic of Letters: A Cultural History of the French Enlightenment.* Ithaca: Cornell University Press.

Gordin, Michael D. 2004. *A Well-Ordered Thing: Dmitrii Mendeleev and the Shadow of the Periodic Table.* New York: Basic Books.

Gorman, Michael John. 1999. "From 'The Eyes of All' to 'Useful Quarries in Philosophy and Good Literature': Consuming Jesuit Science, 1600–1650." In *The Jesuits: Cultures, Sciences and Arts, 1540–1773*, edited by John W. O'Malley, Gauvin Alexander Bailey, Steven J. Harris, and T. Frank Kenney, pp. 170–89. Toronto: University of Toronto Press.

_____. 2000. "L'académie invisible de Francesco Lana Terzi: Les Jésuites, l'expérimentation et la sociabilité scientifique au dix-septième siècle." In *Académies et sociétés savantes en Europe: 1650–1800*, edited by Daniel-Odon Hurel and Gérard Laudin, pp. 409–32. Paris: Honoré Champion.

_____. 2001. "Between the Demonic and the Miraculous: Athanasius Kircher and the Baroque Culture of Machines." In *The Great Art of Knowing: The Baroque Encyclopedia of Athanasius Kircher*, edited by Daniel Stolzenberg, pp. 59–70. Stanford, CA: Stanford University Libraries.

———. 2002. "Mathematics and Modesty in the Society of Jesus: The Problems of Christoph Grienberger." In *The New Science and Jesuit Science: Seventeenth Century Perspectives,* edited by Mordechai Feingold, pp. 1–120. Dordrecht: Kluwer.

———. 2003. "Art, Optics and History: New Light on the Hockney Thesis." *Leonardo* 36:295–301.

Gouhier, Henri. 1958. *Les premières pensées de Descartes: Contribution à l'histoire de l'anti-renaissance.* Paris: J. Vrin.

———. 1987. *L'anti-humanisme au XVIIe siècle.* Paris: J. Vrin.

Goyet, Francis. 1996. *Le sublime du "lieu commun": L'invention rhétorique dans l'Antiquité et à la Renaissance.* Paris and Geneva: Honoré Champion.

Grafton, Anthony. 1991a. *Defenders of the Text: The Traditions of Scholarship in an Age of Science, 1450–1800.* Cambridge, MA: Harvard University Press.

———. 1991b. "Renaissance Readers and Ancient Texts." In *Defenders of the Text: The Traditions of Scholarship in an Age of Science, 1450–1800,* pp. 23–46. Cambridge, MA: Harvard University Press.

———. 1996. "The New Science and the Traditions of Humanism." In *The Cambridge Companion to Renaissance Humanism,* edited by Jill Kraye, pp. 203–23. Cambridge: Cambridge University Press.

———. 2000. *Leon Battista Alberti: Master Builder of the Italian Renaissance.* New York: Hill & Wang.

———. 2001. *Bring Out Your Dead: The Past as Revelation.* Cambridge, MA: Harvard University Press.

Grafton, Anthony, and Lisa Jardine. 1986. *From Humanism to the Humanities: Education and the Liberal Arts in Fifteenth- and Sixteenth-Century Europe.* London: Duckworth.

Granger, G. G. 1981. "Philosophie et mathématique leibniziennes." *Revue de métaphysique et morale* 86:1–37.

Grosholz, Emily. 1991. *Cartesian Method and the Problem of Reduction.* Oxford: Clarendon Press.

Grotius, Hugo. 1632. *True Religion Explained and Defended Against Archenemies Thereof in These Times.* London.

Grubbs, Henry A. 1932. *Damien Mitton (1618–1690): Bourgeois honnête homme.* Princeton: Princeton University Press; Paris: Presses universitaires de France.

Guazzo, Stefano. 1581. *The ciuile conuersation of M. Steeuen Guazzo.* Translated by George Pettie. London: Richard Watkins.

Gueroult, Martial. 1953. *Descartes, selon l'ordre des raisons.* 2 vols. Paris: Editions Montaigne.

Gueydan, Edouard. 1986. *Loyola: Texte autographe des* Exercices Spirituels *et documents contemporains (1526–1615).* Paris: Desclée.

Guez de Balzac, Jean-Louis. 1995a. "Réponse à deux questions, ou Du caractère et de l'instruction de la comédie." In *Œuvres diverses (1644),* edited by Roger Zuber, pp. 117–32. Paris: Honoré Champion.

———. 1995b. "Suite d'un entretien de vive voix, ou de la conversation des Romains." In *Œuvres diverses (1644),* edited by Roger Zuber, pp. 73–96. Paris: Honoré Champion.

Guhrauer, Gottschalk E. 1966. *Gottfried Wilhelm Freiherr von Leibniz.* 2 vols. Hildesheim: Georg Olms.

Guicciardini, Niccolò. 1999. *Reading the* Principia: *The Debate on Newton's Mathematical Methods for Natural Philosophy from 1687 to 1736.* Cambridge: Cambridge University Press.

Guillén, Claudio. 1971. "On the Concept and Metaphor of Perspective." In *Literature as System: Essays toward the Theory of Literary History,* by Claudio Guillén, pp. 283–371. Princeton: Princeton University Press.

Hacking, Ian. 1975. *The Emergence of Probability: A Philosophical Study of Early Ideas about Probability, Induction and Statistical Inference.* London and New York: Cambridge University Press.

———. 1983. *Representing and Intervening: Introductory Topics in the Philosophy of Natural Science.* Cambridge: Cambridge University Press.

———. 1990. *The Taming of Chance.* Cambridge: Cambridge University Press.

———. 2002a. "Language, Truth and Reason." In *Historical Ontology,* by Ian Hacking, pp. 159–77. Cambridge, MA: Harvard University Press.

———. 2002b. "'Style' for Historians and Philosophers." In *Historical Ontology,* by Ian Hacking, pp. 178–99. Cambridge, MA: Harvard University Press.

Hadot, Pierre. 1981. *Exercices spirituels et philosophie antique.* Paris: Etudes augustiniennes.

———. 1995. *Philosophy as a Way of Life: Spiritual Exercises from Socrates to Foucault.* Edited by Arnold I. Davidson. Oxford: Blackwell.

———. 1996. "L'expérience de la méditation." *Magazine litteraire,* no. 342:73–76.

———. 1998. "La philosophie antique: Une éthique ou une pratique." In *Etudes de philosophie ancienne,* by Pierre Hadot, pp. 207–32. Paris: Les belles lettres.

———. 2002. *What Is Ancient Philosophy?* Translated by Michael Chase. Cambridge, MA: Belknap Press.

Hadot, Pierre, Jeannie Carlier, and Arnold I. Davidson. 2001. *La philosophie comme manière de vivre.* Paris: Albin Michel.

Halbfass, W. 1972. "Evidenz." In *Historisches Wörterbuch der Philosophie,* vol. 2, edited by Joachim Ritter, pp. 830–31. Basel: Schwabe & Co.

Haliczer, Stephen. 2002. *Between Exaltation and Infamy: Female Mystics in the Golden Age of Spain.* Oxford: Oxford University Press.

Hamm, E. P. 1997. "Knowledge from Underground: Leibniz Mines the Enlightenment." *Earth Sciences History* 16:77–99.

Hampton, Timothy. 1990. *Writing from History: The Rhetoric of Exemplarity in Renaissance Literature.* Ithaca: Cornell University Press.

Harrison, Peter. 2002. "Original Sin and the Problem of Knowledge in Early Modern Europe." *Journal of the History of Ideas* 63:239–59.

Harsdörffer, Georg Philipp. 1968. *Frauenzimmer Gesprächspiele.* Edited by Irmgard Böttcher. 8 vols. Tübingen: M. Niemeyer.

Harsdörffer, Georg Philipp, and Daniel Schwenter. 1991. *Deliciae physico-mathematicae, oder Mathematische und philosophische Erquickstunden.* Edited by Jörg Jochen Berns. 3 vols. Frankfurt am Main: Keip Verlag.

Harth, Erica. 1992. *Cartesian Women: Versions and Subversions of Rational Discourse in the Old Regime.* Ithaca: Cornell University Press.

Hatfield, Gary. 1986. "The Senses and the Fleshless Eye: The *Meditationes* as Cognitive Exercises." In *Essays on Descartes' Meditations,* edited by Amelie Rorty, pp. 45–76. Berkeley and Los Angeles: University of California Press.

——. 1998. "The Cognitive Faculties." In *The Cambridge History of Seventeenth-Century Philosophy*, edited by Daniel Garber, Michael Ayers, Roger Ariew, and Alan Gabbey, pp. 953–1002. Cambridge: Cambridge University Press.

Hecht, Hartmut. 1996. "Dynamik und Optik bei Leibniz." *NTM: Zeitschrift für Geschichte der Naturwissenschaft, Technik und Medizin*, n.s., 4:83–102.

Heilbron, J. L. 1982. *Elements of Early Modern Physics*. Berkeley and Los Angeles: University of California Press.

Heinekamp, Albert. 1972. "Ars characteristica und natürliche Sprache bei Leibniz." *Tijdschrift voor filosofie* 34:446–88.

Heinich, Nathalie. 1993. *Du peintre à l'artiste: Artisans et académiciens à l'âge classique*. Paris: Minuit.

Henry, John. 1986. "Occult Qualities and the Experimental Philosophy: Active Principles in Pre-Newtonian Matter Theory." *History of Science* 24:335–81.

Hermans, Michel, and Michel Klein. 1996. "Ces 'Exercices spirituels' que Descartes aurait pratiqués." *Archives de philosophie* 59:427–40.

Hess, Heinz-Jürgen. 1986. "Zur Vorgeschichte der 'Nova Methodus' (1676–1684)." In *300 Jahre "Nova Methodus" von G. W. Leibniz*, edited by Albert Heinekamp, pp. 64–102. Studia Leibnitiana, Sonderheft, vol. 14. Stuttgart: Franz Steiner Verlag.

Hevelius, Johannes. 1647. *Selenographia; sive, Lunae descriptio*. Gdansk: Typis Hünefeldianis.

Heyd, Michael. 1990. "Descartes—an Enthusiast Malgré Lui." In *Sceptics, Millenarians and Jews*, edited by David S. Katz and Jonathan I. Israel, pp. 35–58. Leiden: Brill.

——. 1995. *Be Sober and Reasonable: The Critique of Enthusiasm in the Seventeenth and Early Eighteenth Centuries*. Leiden and New York: Brill.

Hidalgo-Serna, Emilio. 1983. "*Ingenium* and Rhetoric in the Work of Vives." *Philosophy and Rhetoric* 16:228–41.

Hobbes, Thomas. 1966. *The English Works of Thomas Hobbes of Malmesbury*. Edited by William Molesworth. 11 vols. Aalen, Germany: Scientia.

——. 1994. *The Correspondence of Thomas Hobbes*. Edited by Noel Malcolm. 2 vols. Oxford: Oxford University Press.

——. 1999. *De corpore: Elementorum philosophiae sectio prima*. Edited by Karl Schuhman. Paris: J. Vrin.

Hockney, David. 2001. *Secret Knowledge: Rediscovering the Techniques of the Lost Masters*. London: Penquin Putnam.

Hofmann, Joseph E. 1970. "Über frühe mathematische Studien von G. W. Leibniz." *Studia Leibnitiana* 2:81–114.

——. 1974. *Leibniz in Paris, 1672–1676: His Growth to Mathematical Maturity*. Cambridge: Cambridge University Press.

Homann, Frederick A. 1983. "Christopher Clavius and the Renaissance of Euclidean Geometry. "*Archivum historicum Societatis Iesu* 52:233–46.

Horace. 1970. *Satires, Epistles and Ars poetica*. Translated by H. Rushton Fairclough. Cambridge, MA: Harvard University Press; London: W. Heinemann.

Hotson, Howard. 2000. *Johann Heinrich Alsted, 1588–1638: Between Renaissance, Reformation and Universal Reform*. Oxford: Clarendon.

Huarte, Juan. 1989. *Examen de ingenios para las ciencias*. Edited by Guillermo Serés. Madrid: Cátedra.

Hull, Isabel V. 1996. *Sexuality, State, and Civil Society in Germany, 1700–1815*. Ithaca: Cornell University Press.

Hume, David. 1998. *An Enquiry concerning the Principles of Morals*. Edited by Tom Beauchamp. Oxford: Clarendon.

Hunter, Ian. 2001. *Rival Enlightenments: Civil and Metaphysical Philosophy in Early Modern Germany*. Cambridge: Cambridge University Press.

Hunter, Michael, and Edward B. Davis, eds. 1999–2000. *The Works of Robert Boyle*. 14 vols. London: Pickering & Chatto.

Huppert, George. 1984. *Public Schools in Renaissance France*. Urbana: University of Illinois Press.

Huret, Grégoire. 1670. *Optique de portraiture et peinture, en deux parties*. Paris.

Hutchison, Keith. 1991. "Dormitive Virtues, Scholastic Qualities and the New Philosophies." *History of Science* 29:245–78.

Huygens, Christiaan. 1888–1950. *Œuvres completes*. 22 vols. in 23. The Hague: M. Nijhoff.

Ignatius. 1996. *Saint Ignatius of Loyola: Personal Writings*. Translated by Joseph A. Munitiz and Philip Endean. London: Penguin.

Iparraguirre, Ignatius, ed. 1955. *Directoria exercitorum spiritualium (1540–1599)*. Monumenta historica Societatis Iesu, vol. 76. Rome: Apud "Monumenta historica Societatis Iesu."

James, Susan. 1997. *Passion and Action: The Emotions in Seventeenth-Century Philosophy*. Oxford: Clarendon.

——. 1998. "Reason, the Passions, and the Good Life." In *The Cambridge History of Seventeenth-Century Philosophy*, edited by Daniel Garber, Michael Ayers, Roger Ariew, and Alan Gabbey, pp. 1358–96. Cambridge: Cambridge University Press.

Jansen, Cornelius. [1640] 1964. *Augustinus*. Louvain. Reprint, Frankfurt a. M.: Minerva.

Jardine, Lisa, and Anthony Grafton. 1990. "'Studied for Action': How Gabriel Harvey Read His Livy." *Past and Present* 129:30–78.

Jardine, Nicholas. 1988. "Epistemology of the Sciences." In *The Cambridge History of Renaissance Philosophy*, edited by Charles B. Schmitt and Quentin Skinner, pp. 685–711. Cambridge: Cambridge University Press.

Jehasse, Jean. 1977. *Guez de Balzac et le génie romain, 1597–1654*. Saint-Etienne: Université de Saint-Etienne.

Jesseph, Douglas M. 1999. *Squaring the Circle: The War between Wallis and Hobbes*. Chicago: University of Chicago Press.

Johnson, Hubert C. 1964. "The Concept of Bureaucracy in Cameralism." *Political Science Quarterly* 79:378–402.

Johnston, Stephen. 1991. "Mathematical Practitioners and Instruments in Elizabethan England." *Annals of Science* 48:314–44.

Jones, Matthew L. 2001. "Writing and *Sentiment*: Blaise Pascal, the Vacuum, and the *Pensées*." *Studies in History and Philosophy of Science* 32:139–81.

——. 2003. "Three Errors about Indifference: Pascal on the Vacuum, Sociability and Moral Freedom." *Romance Quarterly* 50:99–120.

Jungius, Joachim. n.d. *Joachimi Jungii geometria empirica*. [Rostock/Hamburg.]

——. 1929. "Über den propädeutischen Nutzen der Mathematik für das Studium der Philosophie." In *Beiträge zur Jungius-Forschung*, edited by Joahnnes Lemcke and Adolf Meyer, pp. 94–120. Hamburg: Paul Hartnung Verlag.

Kaiser, David. 2005. *Drawing Theories Apart: The Dispersion of Feynman Diagrams in Postwar Physics*. Chicago: University of Chicago Press.

Kant, Immanuel. 1996. *Critique of Pure Reason*. Translated by Werner S. Pluhar. Indianapolis, IN: Hackett Publishing Co.

———. 2000. *Critique of the Power of Judgment*. Translated by Paul Guyer. New York: Cambridge University Press.

Karger, Elizabeth. 1999. "Ockham's Misunderstood Theory of Intuitive and Abstractive Cognition." In *The Cambridge Companion to Ockham*, edited by Paul Vincent Spade, pp. 204–26. Cambridge: Cambridge University Press.

Karpinski, L. C., and F. W. Kokomoor. 1928. "The Teaching of Elementary Geometry in the Seventeenth Century." *Isis* 10:21–32.

Kauppi, Raili. [1960] 1985. *Über die Leibnizsche Logik*. Acta philosophica Fennica, fasc. 12. Helsinki: Societas philosophica. Reprint, New York and London: Garland.

Keller, Alex. 1976. "Renaissance Mathematical Duels." *History of Science* 14:208–9.

Kemp, Martin. 1977. "From 'Mimesis' to 'Fantasia': The Quattrocento Vocabulary of Creation, Inspiration and Genius in the Visual Arts." *Viator* 8:347–98.

———. 1990. *The Science of Art: Optical Themes in Western Art from Brunelleschi to Seurat*. New Haven: Yale University Press.

Kenny, Anthony. 1972. "Descartes on the Will." In *Cartesian Studies*, edited by R. J. Butler, pp. 1–31. New York: Barnes & Noble.

Keohane, Nannerl O. 1980. *Philosophy and the State in France: The Renaissance to the Enlightenment*. Princeton: Princeton University Press.

Kircher, Athanasius. 1641. *Magnes, sive de arte magnetica opvs tripartitvm*. Rome.

———. 1654. *Magnes, sive de arte magnetica opvs tripartitvm*. 3rd ed. Rome.

Kircher, Athanasius, and Gaspar Schott. 1669. *Pantometrum kircherianum, hoc est, Instrumentum Geometricum novum, à Celeberrimo Viro P. Athanasio Kirchero Ante hac inventum, nunc decem Libris, universam paenè Practicam Geometriam complectentibus explicatum, perspicuisque demonstrationibus illustratum à R. P. Gaspare Schotto*. Würzburg.

Kline, Morris. 1972. *Mathematical Thought from Ancient to Modern Times*. New York: Oxford University Press.

Knabe, Lotte. 1956. "Leibniz' Vorschläge zum Archiv- und Registraturwesen." In *Archivar und Historiker: Festschrift H. O. Meiser*, pp. 107–20. Schriftenreihe der staatlichen Archivverwaltung, vol. 7. Berlin: Rütten & Loening.

Knobloch, Eberhard. 1973. *Die mathematischen Studien von G. W. Leibniz zur Kombinatorik, auf Grund fast ausschliesslich handschriftlicher Aufzeichnungen*. Studia Leibnitiana supplementa, vol. 11. Wiesbaden: Franz Steiner.

———. 1989a. "Leibniz et son manuscrit inédit sur la quadrature des sections coniques." In *The Leibniz Renaissance*, pp. 127–51. Florence: L. S. Olschki.

———. 1989b. "Progrès et tâches futures de la recherche leibnizienne en mathématiques." *Les études philosophiques*, no. 2:161–70.

———. 1994. "The Infinite in Leibniz's Mathematics: The Historiographical Method of Comprehension in Context." In *Trends in the Historiography of Science*, edited by Kostas Gavroglu, Jean Christianidis, and Efthymios Nicolaidis, pp. 265–78. Dordrecht: Kluwer.

———. 2001. "Déterminants et élimination chez Leibniz." *Revue d'histoire des sciences et de leurs applications* 54:143–64.

———. 2002. "Leibniz's Rigorous Foundation of Infinitesimal Geometry by Means of Riemannian Sums." *Synthese* 133:59–73.

Kokomoor, F. W. 1928. "The Teaching of Elementary Geometry in the Seventeenth Century." *Isis* 11:85–110.

Krämer, Sybille. 1991. *Berechenbare Vernunft: Kalkül und Rationalismus im 17. Jahrhundert.* Berlin and New York: W. de Gruyter.

Kraye, Jill. 1988. "Moral Philosophy." In *The Cambridge History of Renaissance Philosophy,* edited by Charles B. Schmitt, Quentin Skinner, Eckhard Kessler, and Jill Kraye, pp. 303–86. Cambridge: Cambridge University Press.

———. 1996. "Philologists and Philosophers." In *The Cambridge Companion to Renaissance Humanism,* edited by Jill Kraye, pp. 142–60. Cambridge: Cambridge University Press.

Krebs, Jean-Daniel. 1991. "Harsdörffer als Vermittler des 'honnêteté'-Ideals." In *Georg Philipp Harsdörffer: Ein deutscher Dichter und europäischer Gelehrter,* edited by Italo Michele Battafarano, pp. 287–311. Bern, Berlin, and Paris: P. Lang.

Kuhn, Heinrich C. 1997. "Non-regressive Methods (and the Emergence of Modern Science)." In *Method and Order in Renaissance Philosophy of Nature: The Aristotle Commentary Tradition,* edited by Daniel A. Di Liscia, Eckhard Kessler, and Charlotte Methuen, pp. 319–36. Aldershot, UK: Ashgate.

Kulstad, Mark. 1977. "Leibniz's Conception of Expression." *Studia Leibnitiana* 9:55–76.

———. 1999. "Leibniz's *De summa rerum:* The Origin of the Variety of Things, in Connection with the Spinoza-Tschirnhaus Correspondence." In *L'actualité de Leibniz: Les deux labyrinthes,* edited by Dominique Berlioz and Frédéric Nef, pp. 69–86. Studia Leibnitiana supplementa, vol. 34. Stuttgart: Franz Steiner Verlag.

Lachtermann, David R. 1989. *The Ethics of Geometry: A Genealogy of Modernity.* New York: Routledge.

Lagrée, Jacqueline. 1991. *La raison ardente: Religion naturelle et raison au XVIIe siècle, avec traduction en appendice du Meletius de Hugo Grotius.* Paris: J. Vrin.

———. 2004. "Constancy and Coherence." In *Stoicism: Traditions and Transformations,* edited by Steven K. Strange and Jack Zupko, pp. 148–76. Cambridge: Cambridge University Press.

Laird, W. R. 2000. *The Unfinished Mechanics of Giuseppe Moletti: An Edition and English Translation of His Dialogue on Mechanics (1576).* Toronto: University of Toronto Press.

Laloubère, Antoine de. 1651. *Quadratura circuli et hyperbolae segmentorum.* Toulouse.

Lamarra, Antonio. 1978. "The Development of the Theme of the 'Logical Inventiva' during the Stay of Leibniz in Paris." In *Leibniz à Paris (1672–1676),* vol. 2, pp. 55–71. Studia Leibnitiana supplementa, vol. 18. Wiesbaden: Franz Steiner Verlag.

———. 1982. "Sur l'origine de la théorie de l'expression dans la philosophie de Leibniz." *Recherches sur le XVIIe siècle* 5:77–83.

———. 1991. "Il concetto di rappresentazione in Leibniz: Dall'algebra alla metafisica." *Bollettino del Centro di studi Vichiani* 19:41–57.

Lamy, Bernard. 1684. *Les Elemens de Géométrie, ou, de la Mesure de Corps.* Paris.

———. 1688. *La rhétorique ou L'art de parler.* 3rd ed. Paris: A. Pralard.

———. 1966. *Entretiens sur les sciences.* Edited by François Girbal and Pierre Clair. Paris: Presses universitaires de France.

Lana Terzi, Francesco. 1670. *Prodromo; ouero, Saggio di alcune inuentioni nuoue premesso all'Arte maestra, . . . Per mostrare li piu reconditi principij della naturale filosofia, riconosciuti con accurata teorica nelle piu segnalate inuentioni, ed isperienze sin'hora ritrouate da gli scrittori di questa materia & altre nuoue dell'autore medesimo.* Brescia.

Larmore, Charles. 1984. "Descartes's Psychologistic Theory of Assent." *History of Philosophy Quarterly* 1:61–74.

Lausberg, Heinrich. 1998. *Handbook of Literary Rhetoric: A Foundation for Literary Study.* Translated by Annemiek Jansen, Matthew T. Bliss, and David E. Orton. Edited by David E. Orton and R. Dean Anderson. Leiden: Brill.

Lefèvre, Wolfgang. 2004. "The Emergence of Combined Orthographic Projections." In *Picturing Machines, 1400–1700,* edited by Wolfgang Lefèvre, pp. 209–44. Cambridge, MA: MIT Press.

Le Guern, Michel. 2003. *Pascal et Arnauld.* Geneva and Paris: Honoré Champion.

Leibniz, Gottfried Wilhelm. 1671. *Notitia opticae promotiae autore G. G. L. L.* Frankfurt.

———. 1768. *Opera omnia.* Edited by L. Dutens. 6 vols. Geneva: Fratres de Tournes.

———. [1849–63] 1971. *Leibnizens mathematische Schriften.* Edited by C. I. Gerhardt. 7 vols. in 4. Berlin: A. Asher; Halle: H. W. Schmidt. Reprint, Hildesheim: G. Olms. (Abbreviated as GM.)

———. [1875–90] 1960. *Die philosophischen Schriften.* Edited by C. I. Gerhardt. 7 vols. Berlin: Weidmann. Reprint, Hildesheim: G. Olms. (Abbreviated as GP.)

———. 1899. *Der Briefwechsel von Gottfried Wilhelm Leibniz mit Mathematikern.* Edited by C. I. Gerhardt. Berlin: Mayer & Müller.

———. [1903] 1966. *Opuscules et fragments inédits de Leibniz, extraits des manuscrits de la Bibliothèque royale de Hanovre.* Edited by Louis Couturat. Paris. Reprint, Hildesheim: G. Olms. (Abbreviated as C.)

———. [1906] 1995. *Leibnizens nachgelassene Schriften physikalischen, mechanischen und technischen Inhalts.* Edited by Ernst Gerland. Abhandlungen zur Geschichte der mathematischen Wissenschaften mit Einschluss ihrer Anwendung XXI. Hft. Physikalischer Teil. Leipzig: B. G. Teubner. Reprint, Hildesheim: G. Olms. (Abbreviated as Ger.)

———. [1914–18] 1986. *Catalogue critique des manuscrits de Leibniz.* Fasc. 2, *Mars 1672– novembre 1676* [only volume that appeared]. Poitiers: Société française d'imprimerie et de librarie. Reprint, Hildesheim: G. Olms. (Abbreviated as Cc.)

———. 1923–. *Sämtliche Schriften und Briefe.* Edited by Deutsche Akademie der Wissenschaften, etc. Darmstadt, Berlin, Munich, etc. (Abbreviated as A.)

———. 1948. *Textes inédits.* Edited by Gaston Grua. 2 vols. Paris: Presses universitaires de France.

———. 1951. *Leibniz: Selections.* Edited and translated by Philip P. Wiener. New York: Charles Scribner's Sons.

———. 1966. *Logical Papers.* Edited and translated by G. H. R. Parkinson. Oxford: Clarendon.

———. 1969. *Philosophical Papers and Letters.* 2nd ed. Translated by Leroy E. Loemker. Dordrecht: D. Reidel. (Abbreviated as L.)

———. 1970. *Confessio philosophi: La profession de foi du philosophe.* Translated by Yvon Belaval. Paris: J. Vrin.

———. 1976. *Die mathematischen Studien von G. W. Leibniz zur Kombinatorik: Textband.* Edited by Eberhard Knobloch. Studia Leibnitiana Supplementa, vol. 16. Wiesbaden: F. Steiner.

———. 1980. *Der Beginn der Determinantentheorie: Leibnizens nachgelassene Studien zum Determinantenkalkül*. Edited by Eberhard Knobloch. Arbor scientiarum, Reihe B, Texte, Band 2. Hildesheim: Gerstenberg.

———. 1988. *Leibniz: Political Writings*. 2nd ed. Edited and translated by Patrick Riley. Cambridge: Cambridge University Press.

———. 1992. *De summa rerum: Metaphysical Papers, 1675–1676*. Edited and translated by G. H. R. Parkinson. New Haven: Yale University Press. (Abbreviated as Pk.)

———. 1993. *De quadratura arithmetica circuli ellipseos et hyperbolae cujus corollarium est trigonometria sine tabulis*. Edited by Eberhard Knobloch. Abhandlungen der Akademie der Wissenschaften in Göttingen, Mathematische-physikalische Klasse, vol. 43. Göttingen: Vanderhoeck & Ruprecht. (Abbreviated as K.)

———. 1994a. *Le droit de la raison*. Edited and translated by René Sève. Paris: J. Vrin.

———. 1994b. *La réforme de la dynamique: "De corporum concursu" (1678) et autres textes inédits*. Edited and translated by Michel Fichant. Paris: J. Vrin.

———. 1995. *L'estime des apparences: 21 manuscrits de Leibniz sur les probabilités, la théorie des jeux, l'espérance de vie*. Edited and translated by Marc Parmentier. Paris: J. Vrin.

———. 1998. *Recherches générales sur l'analyse des notions et des vérités: 24 thèses métaphysiques et autres textes logiques et métaphysiques*. Edited by Jean-Baptiste Rauzy. Paris: Presses universitaires de France.

———. 2001. *The Labyrinth of the Continuum: Writings on the Continuum Problem, 1672–1686*. Edited and translated by Richard T. W. Arthur. New Haven: Yale University Press. (Abbreviated as LC.)

Lemaistre de Sacy, Isaac. 1654. *Les enluminures du fameux almanach des PP. Jésuites intitulé La déroute et confusion des Jansénistes ou Triomphe de Molina Jésuite sur S. Augustin*. N.p.

Le Moyne, Pierre. 1645. *Les peintures morales, où les passions sont représentées par tableaux, par charactères et par questions nouvelles et curieuses*. 2nd ed. Paris.

Lenoir, Timothy. 1979. "Descartes and the Geometrization of Thought: The Methodological Background of Descartes' *Géométrie*." *Historia mathematica* 6:355–79.

———, ed. 1998. *Inscribing Science: Scientific Texts and the Materiality of Communication*. Stanford, CA: Stanford University Press.

Levey, Samuel. 1998. "Leibniz on Mathematics and the Actually Infinite Division of Matter." *Philosophical Review* 107:49–96.

Levi, Anthony. 1963. "La psychologie des facultés dans les discussions théologiques du XVIIe siècle." In *L'homme devant Dieu: Mélanges offerts au père Henri de Lubac*, vol. 2, edited by Henri de Lubac, pp. 293–302. Paris: Aubier.

———. 1964. *French Moralists: The Theory of the Passions, 1585 to 1649*. Oxford: Clarendon Press.

Lewalter, Ernst. 1935. *Spanisch-Jesuitische und Deutsch-Lutherische Metaphysik des 17. Jahrhunderts: Ein Beitrag zur Geschichte der Iberisch-Deutschen Kulturbeziehungen und zur Vorgeschichte des Deutschen Idealismus*. Hamburg: Ibero-Amerikanisches Institut.

Licoppe, Christian. 1996. *La formation de la pratique scientifique: Le discours de l'expérience en France et en Angleterre, 1630–1820*. Paris: La Découverte.

Lines, David A. 2002. *Aristotle's Ethics in the Italian Renaissance (1300–1600): The Universities and the Problem of Moral Education*. Leiden and Boston: Brill.

Locke, John. 1975. *An Essay concerning Human Understanding*. Edited by Peter H. Nidditch. Oxford: Oxford University Press.

Lodolini, Elio. 1998. "Archivio e registratura (archivistica e gestione dei documenti) nel pensiero di Leibniz." *Rassegna degli archivi di Stato* 58:245–67.

Loeffel, Hans. 1987. *Blaise Pascal, 1623–1662*. Basel: Birkhäuser.

Lubac, Henri de. 1969. *Augustinianism and Modern Theology*. Translated by Lancelot Sheppard. London: G. Chapman.

Lukàcs, Ladislaus, ed. 1965–. *Monumenta paedagogica Societatis Iesu*. New series, 7 vols. to date. Rome: Institutum historicum Societatis Iesu.

Maat, Jaap. 2004. *Philosophical Languages in the Seventeenth Century: Dalgarno, Wilkins, Leibniz*. Dordrecht: Kluwer.

Maclean, Ian. 1977. *Woman Triumphant: Feminism in French Literature, 1610–1652*. Oxford: Clarendon Press.

——. 1992. *Interpretation and Meaning in the Renaissance: The Case of Law*. Cambridge: Cambridge University Press.

——. 2002. *Logic, Signs and Nature in the Renaissance: The Case of Learned Medicine*. Cambridge: Cambridge University Press.

——. 2005. "White Crows, Graying Hair, and Eyelashes: Problems for Natural Historians in the Reception of Aristotelian Logic and Biology from Pomponazzi to Bacon." In *Historia: Empiricism and Erudition in Early Modern Europe*, edited by Gianna Pomata and Nancy G. Siraisi, pp. 147–80. Cambridge, MA: MIT Press.

Maeda, Yoichi. 1964. "Le premier jet du fragment pascalian sur les deux infinis." *Etudes de langue et littéraire française* (Tokyo), no. 4:1–19.

Magendie, Maurice. 1925. *La politesse mondaine et les théories de l'honnêteté, en France au XVIIe siècle, de 1600 à 1660*. Paris: F. Alcan.

Mahnke, Dietrich. 1926. "Neue Einblicke die Entdeckungsgeschichte der höheren Analysis." *Abhandlungen der Preussischen Akademie der Wissenschaften, Physicalisch-mathematische Klasse*, Jahrgang 1925, no. 1:1–67.

Mahoney, Michael S. 1980. "The Beginnings of Algebraic Thought in the Seventeenth Century." In *Descartes: Philosophy, Mathematics and Physics*, edited by Stephen Gaukroger, pp. 141–68. Sussex, UK: Harvester Press.

——. 1993. "Algebraic vs. Geometric Techniques in Newton's Determinations of Planetary Orbits." In *Action and Reaction*, edited by Paul Theerman and Adele F. Seeff, pp. 183–205. Newark: University of Delaware Press.

——. 1994. *The Mathematical Career of Pierre de Fermat, 1601–1665*. 2nd ed. Princeton: Princeton University Press.

Malcolm, Noel. 2002. "The Title Page of *Leviathan*, Seen in a Curious Perspective." In *Aspects of Hobbes*, by Noel Malcolm, pp. 200–233. New York: Oxford University Press.

——. 2003. "Leibniz, Oldenburg, and Spinoza, in the Light of Leibniz's Letter to Oldenburg of 18/28 November 1676." *Studia Leibnitiana* 35:225–43.

Malebranche, Nicolas. 1991. *De la recherche de la vérité*. Edited by Geneviève Rodis-Lewis. 3rd ed. Vol. 1 of *Œuvres Complètes*, edited by André Robinet. Paris: J. Vrin.

Mancosu, Paolo. 1992. "Descartes's *Géométrie* and Revolutions in Mathematics." In *Revolutions in Mathematics*, edited by Donald Gillies, pp. 83–116. Oxford: Clarendon.

——. 1996. *The Philosophy of Mathematics and Mathematical Practice in the Seventeenth Century*. New York: Oxford University Press.

Manders, Ken. 1995. "Descartes et Faulhaber." *Bulletin cartésien, Archives de philosophie* 58:1–12.

Maravall, José Antonio. 1986. *Culture of the Baroque: Analysis of a Historical Structure.* Translated by Terry Cochran. Minneapolis: University of Minnesota Press.

Marion, Jean-Luc. 1975. *Sur l'ontologie grise de Descartes: Science cartésienne et savoir aristotélicien dans les Regulae.* Paris: J. Vrin.

Mariotte, Edme. 1678. *Essay de logique, contenant les principes des sciences et la manière de s'en servir pour faire de bons raisonnemens.* Paris: E. Michallet.

Marshall, John. 1998. *Descartes's Moral Theory.* Ithaca: Cornell University Press.

Martin, A. Lynn. 1988. *The Jesuit Mind: The Mentality of an Elite in Early Modern France.* Ithaca: Cornell University Press.

Mates, Benson. 1986. *The Philosophy of Leibniz: Metaphysics and Language.* Oxford: Oxford University Press.

Mazauric, Simone. 1997. *Savoirs et philosophie à Paris dans la première moitié du XVIIe siècle: Les conférences du Bureau d'adresse de Théophraste Renaudot, 1633–1642.* Paris: Publications de la Sorbonne.

Mazzantini, C. 1967. "Evidenza." In *Enciclopedia filosofica*, vol. 2, pp. 1159–66. Florence: G. C. Sansoni.

McGrade, Arthur Stephen. 1988. "Some Varieties of Skeptical Experience: Ockham's Case." In *Die Philosophie im 14. und 15. Jahrhundert*, edited by Olaf Pluta, pp. 421–38. Amsterdam: B. R. Grüner.

Méchoulan, Henry, ed. 1988. *Problématique et réception du* Discours de la méthode *et des essais.* Paris: J. Vrin.

Mehl, Edouard. 2001. *Descartes en Allemagne, 1619–1620.* Strasbourg: Presses universitaires de Strasbourg.

Mercer, Christia. 2001. *Leibniz's Metaphysics: Its Origins and Development.* Cambridge: Cambridge University Press.

———. 2004. "Leibniz, Aristotle, and Ethical Knowledge." In *The Impact of Aristotelianism on Modern Philosophy*, edited by Riccardo Pozzo, pp. 113–47. Washington, DC: Catholic University of America.

Mercer, Christia, and R. C. Sleigh. 1995. "Metaphysics: The Early Period to the *Discourse on Metaphysics*." In *The Cambridge Companion to Leibniz*, edited by Nicholas Jolley, pp. 67–123. Cambridge: Cambridge University Press.

Méré, Antoine Gombaud. 1682. *Lettres de M. le chevalier de Méré.* 2 vols. Paris.

———. 1687. *Maximes, sentences et réflexions morales et politiques.* Paris: E. du Castin.

Mersenne, Marin. [1634] 1972a. *Questions inouyes, ou récréation des sçavans.* Stuttgart and Bad Cannstatt: F. Frommann.

———. [1634] 1972b. *Questions harmoniques: Dans lesquelles sont contenües plusieurs choses remarquables pour la physique, pour la morale, et pour les autres sciences.* Stuttgart and Bad Cannstatt: F. Frommann.

———. 1933–88. *Correspondance du P. Marin Mersenne: Religieux minime.* Edited by Paul Tannery and Cornelis de Waard. 17 vols. Paris: G. Beauchesne, etc.

Mesnard, Jean. 1953. "Jansénisme et mathématiques: Autour des écrits de Pascal sur la roulette." *Annales Universitatis Saraviensis: Philosophie-lettres* 2, no. 1/2:3–30.

———. 1963. "Pascal à l'Académie Le Pailleur." *Revue d'histoire des sciences* 16:1–10.

———. 1965. *Pascal et les Roannez.* Paris: Desclée De Brouwer.

_____. 1976. *Les* Pensées *de Pascal*. Paris: Société d'édition d'enseignement supérieur.

_____. 1978. "Leibniz et les papiers de Pascal." In *Leibniz à Paris (1672–1676)*, vol. 1, pp. 45–58. Studia Leibnitiana supplementa, vol. 17. Wiesbaden: Franz Steiner Verlag.

Michon, Hélène. 1996. *L'ordre du cœur: Philosophie, théologie et mystique dans les* Pensées *de Pascal*. Paris: Honoré Champion.

Milhaud, Gaston. 1921. *Descartes savant*. Paris: F. Alcan.

Miller, Jon, and Brad Inwood, eds. *Hellenistic and Early Modern Philosophy*. Cambridge: Cambridge University Press, 2003.

Miller, Peter N. 2000. *Peiresc's Europe: Learning and Virtue in the Seventeenth Century*. New Haven: Yale University Press.

Moll, Konrad. 2002. "Der Enzyklopädiegedanke bei Comenius und Alsted, seine Übernahme und Umgestaltung bei Leibniz: Neue Perspektiven der Leibniz-forschung." *Studia Leibnitiana* 34:1–30.

Molland, A. G. 1976. "Shifting the Foundations: Descartes's Transformation of Ancient Geometry." *Historia mathematica* 3:21–49.

Montaigne, Michel de. 1652. *Les essais de Michel seigneur de Montaigne*. Paris.

_____. 1992. *Les essais*. 2nd ed. Edited by Pierre Villey. Paris: Presses universitaires de France.

Mora, José Ferrater. 1953. "Suarez and Modern Philosophy." *Journal of the History of Ideas* 14:528–47.

Moran, Bruce T. 1991a. *The Alchemical World of the German Court: Occult Philosophy and Chemical Medicine in the Circle of Moritz of Hessen, 1572–1632*. Sudhoffs Archiv Beihefte, Heft 29. Stuttgart: F. Steiner Verlag.

_____, ed. 1991b. *Patronage and Institutions: Science, Technology, and Medicine at the European Court, 1500–1750*. Rochester, NY: Boydell Press.

Moreau, Pierre-François, ed. 1999. *Le stoïcisme au XVIe et au XVIIe siècle*. Paris: A. Michel.

_____, ed. 2001. *Le scepticisme au XVIe et au XVIIe siècle*. Paris: A. Michel.

Morhof, Daniel Georg. 1708. *Danielis Georgi Morhofi Polyhistor, in tres tomos, literarium (cujus soli tres libri priores hactenus prodiere, nunc autem quatuor reliqvi. & mss. accedunt) philosophicum et practicum divisus*. Lübeck.

Moss, Jean Dietz. 2001. "Sacred Rhetoric and Appeals to the Passions: A Northern Italian View." In *Perspectives on Early Modern and Modern Intellectual History: Essays in Honor of Nancy S. Struever*, edited by Joseph Marino and Melinda W. Schlitt, pp. 375–400. Rochester: University of Rochester Press.

Mouchel, Christian. 1999. "Les rhétoriques post-tridentines (1570–1600)." In *Histoire de la rhétorique dans l'Europe moderne, 1450–1950*, edited by Marc Fumaroli, pp. 431–97. Paris: Presses universitaires de France.

Naërt, Emilienne. 1985. "Double infinité chez Pascal et Monade." *Studia Leibnitiana* 17:44–51.

Nef, Frédéric. 2000. *Leibniz et le langage*. Paris: Presses universitaires de France.

Nero, Valerio Del. 1992. "Pedagogia e psicologia nel pensiero di Vives." In *Ioannis Lodovici Vivis Valentini opera omnia*, vol. 1, edited by Antonio Mestre, pp. 179–216. Valencia: Edicions Alfons el Magnànim.

Newton, Isaac. 1959–77. *The Correspondence of Isaac Newton*. Edited by H. W. Turnbull and J. F. Scott. 7 vols. Cambridge: Cambridge University Press.

_____. 1967–81. *The Mathematical Papers of Isaac Newton*. Edited by Derek T. Whiteside. 8 vols. Cambridge: Cambridge University Press.

Niceron, Jean François. 1652. *La perspective curieuse*. Paris: Vve F. Langlois.

Nicole, Pierre. 1970. *Œuvres philosophiques et morales, comprenant un choix de ses essais et publiées avec des notes et une introduction*. Edited by Charles Jourdain. Hildesheim: G. Olms.

_____. 1996. *La vraie beauté et son fantôme, et autres textes d'esthétique*. Translated and edited by Béatrice Guion. Paris: Honoré Champion.

Nuchelmans, Gabriël. 1983. *Judgment and Proposition: From Descartes to Kant*. Amsterdam and New York: North-Holland Publishing Co.

Ockham, William of. 1967–88. *Opera theologica*. Edited by Franciscan Institute (St. Bonaventure University). 10 vols. St. Bonaventure, NY: Editiones Instituti Franciscani Universitatis S. Bonaventurae.

Ogilvie, Brian. 2003. "The Many Books of Nature: Renaissance Naturalists and Information Overload." *Journal of the History of Ideas* 64:29–40.

Olaso, Ezequiel de. 1975. "Leibniz et l'art de disputer." In *Akten des II. Internationalen Leibniz-Congresses*, vol. 4, pp. 207–28. Studia Leibnitiana supplementa, vol. 15. Wiesbaden: Franz Steiner Verlag.

Oldenburg, Henry. 1965–86. *The Correspondence of Henry Oldenburg*. Edited and translated by A. Rupert Hall and Marie Boas Hall. 13 vols. Madison: University of Wisconsin Press; London: Mansell; London: Taylor & Frances.

Oliazola, Ruth. 1999. "Les jésuites et l'utopie du 'comédien honnête' aux XVIe et XVIIe siècles." *Revue de synthèse*, 4th ser., nos. 2–3:381–407.

O'Malley, John W. 1979. *Praise and Blame in Renaissance Rome: Rhetoric, Doctrine, and Reform in the Sacred Orators of the Papal Court, c. 1450–1521*. Durham, NC: Duke University Press.

_____. 1993. *The First Jesuits*. Cambridge, MA: Harvard University Press.

O'Neill, Eileen. 1999. "Women Cartesians, 'Feminine Philosophy,' and Historical Exclusion." In *Feminist Interpretations of René Descartes*, edited by Susan Bordo, pp. 232–57. University Park: Pennsylvania State University Press.

Ore, Øystein. 1953. *Cardano, the Gambling Scholar*. Princeton: Princeton University Press.

Osler, Margaret J. 1994. *Divine Will and the Mechanical Philosophy: Gassendi and Descartes on Contingency and Necessity in the Created World*. Cambridge: Cambridge University Press.

Ott, Johannes. 1671. *Dissertatio inauguralis de Propriorum oculorum defectibus ad leges mechanices revocatis, quam . . . in . . . Rauracorum Universitate. subjicit Johannes Ott Scaphusa Helvetius*. Basel.

Panofsky, Erwin. 1991. *Perspective as Symbolic Form*. Translated by Christopher S. Wood. New York: Zone.

Pardies, Ignace-Gaston. 1710. *Elémens de géométrie, où l'on peut apprendre ce qu'il faut sçavoir d'Euclide, d'Archimède, d'Apollonius et les plus belles inventions des anciens et des nouveaux géomètres*. 5th ed. The Hague: A. Moetjens.

Parish, Richard. 1989. *Pascal's* Lettres Provinciales: *A Study in Polemic*. Oxford: Clarendon Press.

Parkinson, G. H. R. 1986. "Leibniz's *De summa rerum*: A Systematic Approach." *Studia Leibnitiana* 18:132–51.

Parmentier, Marc. 2001. "Démonstrations et infiniment petits dans la *Quadratura arithmetica* de Leibniz." *Revue d'histoire des sciences* 54:275–89.

Pascal, Blaise. 1923. *Œuvres de Blaise Pascal.* 2nd ed. Edited by Léon Brunschvicg and Pierre Boutroux. 14 vols. Paris: Hachette et cie.

———. 1963. *Pensées.* Edited by Louis Lafuma. In *Œuvres complètes.* Paris: Seuil.

———. 1964–. *Œuvres complètes.* Edited by Jean Mesnard. 4 vols. to date. Paris: Desclée de Brouwer. (Abbreviated as M.)

———. 1992. *Les provinciales.* Edited by Louis Cognet and Gérard Ferreyrolles. Paris: Bordas. (Abbreviated as CF.)

———. 1998. *Œuvres complètes.* Edited by Michel Le Guern. 2 vols. Paris: Gallimard. (Abbreviated as LG.)

———. 2000. *Pensées.* Edited by Gérard Ferreyrolles and Philippe Sellier. Paris: Livre de Poche. (Abbreviated as S.)

Pasini, Enrico. 1997. "*Arcanum Artis Inveniendi:* Leibniz and Analysis." In *Analysis and Synthesis in Mathematics,* edited by M. Otte and M. Panza, pp. 35–46. Dordrecht: Kluwer.

Paulo, Eustachius a Sancto. 1640. *Summa philosophiae quadripartita: De rebus dialecticis, ethicis, physicis, & metaphysicis.* Cambridge.

Peckhaus, Volker. 1997. *Logik, Mathesis Universalis und allgemeine Wissenschaft: Leibniz und die Wiederentdeckung der formalen Logik im 19. Jahrhundert.* Berlin: Akademie Verlag.

Pereiro, Benito. 1576. *De communibus omnium rerum naturalium principiis et affectionibus libri quindecem.* Rome.

Pérez-Gómez, Alberto, and Louise Pelletier. 1995. *Anamorphosis: An Annotated Bibliography, with Special Reference to Architectural Representation.* Montreal: McGill University Libraries.

Pérez-Ramos, Antonio. 1988. *Francis Bacon's Idea of Science and the Maker's Knowledge Tradition.* Oxford: Clarendon Press.

Perkins, Franklin. 2002. "Virtue, Reason and Cultural Exchange: Leibniz's Praise of Chinese Morality." *Journal of the History of Ideas* 63:447–64.

Pérouse, Gabriel A. 1970. *L'examen des esprits du docteur Juan Huarte de San Juan: Sa diffusion et son influence en France aux XVIe et XVIIe siècles.* Paris: Les Belles lettres.

Petey-Girard, Bruno. 1999. "Les stratégies de *Traité de la pénitence* de Guillaume du Vair." *Bibliothèque d'humanisme et Renaissance* 61:71–84.

Pintard, René. 1943. *Le libertinage érudit dans la première moitié du XVIIe siècle.* Paris: Boivin.

Piro, Francesco. 1999. "Leibniz and Ethics: The Years 1669–72." In *The Young Leibniz and His Philosophy (1646–76),* edited by Stuart Brown, pp. 147–67. Dordrecht: Kluwer.

pseudo-Cicero. 1954. *Rhetorica ad herennium.* Translated by Harry Caplan. Cambridge, MA: Harvard University Press.

Pucelle, Jean. 1963. "La 'lumière naturelle' et le Cartésianisme dans 'L'esprit géométrique' et 'L'art de persuader.'" In *Pascal: Textes du Tricentenaire,* pp. 50–61. Paris: Le Signe, Librarie Arthème Fayard.

Quintilian. 1920–22. *The* Institutio oratoria *of Quintilian.* Translated by H. E. Butler. 4 vols. Cambridge: Harvard University Press.

Rabuel, Claude. 1730. *Commentaires sur la Geometrie de M. Descartes, Par le R. P. Claude Rabuel, de la Compagnie de Jesus.* Lyon.

Raeff, Marc. 1983. *The Well-Ordered Police State: Social and Institutional Change through Law in the Germanies and Russia, 1600–1800.* New Haven: Yale University Press.

Ramati, Ayval. 1996. "Harmony at a Distance: Leibniz's Scientific Academies." *Isis* 87:430–52.

Rauzy, Jean-Baptiste. 2001. *La doctrine leibnizienne de la vérité: Aspects logiques et ontologiques.* Paris: J. Vrin.

Recker, Doren A. 1993. "Mathematical Demonstration and Deduction in Descartes's Early Methodological and Scientific Writings." *Journal of the History of Philosophy* 31:223–44.

Reese, Armin. 1967. *Die Rolle der Historie beim Aufstieg des Welfenhauses, 1680–1714.* Quellen und Darstellung zur Geschichte Niedersachsens, vol. 71. Hildesheim: August Lax Verlagsbuchhandlung.

Reiss, Timothy J. 1997. *Knowledge, Discovery, and Imagination in Early Modern Europe: The Rise of Aesthetic Rationalism.* Cambridge: Cambridge University Press.

Rescher, Nicholas. 1992. "Leibniz Finds a Niche: Settling In at the Court of Hanover, 1676–1677." *Studia Leibnitiana* 24:25–48.

Ribe, Neil M. 1997. "Cartesian Optics and the Mastery of Nature." *Isis* 88:42–61.

Riley, Patrick. 1996. *Leibniz' Universal Jurisprudence: Justice as the Charity of the Wise.* Cambridge, MA: Harvard University Press.

Risse, Wilhelm. 1964. *Die Logik der Neuzeit.* 2 vols. Stuttgart and Bad Cannstatt: F. Frommann.

Robinet, André. 1994. *G. W. Leibniz: Le meilleur des mondes par la balance de l'Europe.* Paris: Presses Universitaires de France.

————. 1996. *Aux sources de l'esprit cartesien: L'axe La Ramée–Descartes: De la* Dialectique *des 1555 aux* Regulae. Paris: J. Vrin.

Rochemonteix, Camille de. 1889. *Un collège de Jésuites aux XVIIe et XVIIIe siècles: Le Collège Henri IV de La Flèche.* 4 vols. Le Mans: Leguicheux.

Rodis-Lewis, Geneviève. 1987. "Descartes et les mathématiques au collège." In *Le Discours et sa méthode: Colloque pour le 350e anniversaire du* Discours de la méthode, edited by Nicolas Grimaldi and Jean-Luc Marion, pp. 187–211. Paris: Presses universitaires de France.

————. 1991. "Le premier registre de Descartes." *Archives de philosophie,* nos. 3–4:353–77, 639–57.

————. 1998. *Descartes: His Life and Thought.* Translated by Jane Marie Todd. Ithaca: Cornell University Press.

Romano, Antonella. 1999. *La contre-réforme mathématique: Constitution et diffusion d'une culture mathématique jésuite à la Renaissance (1540–1640).* Rome: École française de Rome.

Rose, Paul Lawrence. 1975. *The Italian Renaissance of Mathematics: Studies on Humanists and Mathematicians from Petrarch to Galileo.* Geneva: Librarie Droz.

Rossi, Paolo. 2000. *Logic and the Art of Memory: The Quest for a Universal Language.* Translated by Stephen Clucas. Chicago: University of Chicago Press.

Rousseau, Jean-Jacques. 1959. *Confessions; Autre textes autobiographiques.* Edited by Bernard Gagnebin and Marcel Raymond. Vol. 1 of *Œuvres complètes.* Paris: Gallimard.

Roy, Ranjan. 1990. "The Discovery of the Series Formula for π by Leibniz, Gregory and Nilakantha." *Mathematics Magazine* 63:291–306.

Rubidge, Bradley. 1990. "Descartes's *Meditations* and Devotional Meditations." *Journal of the History of Ideas* 51:27–49.

Rummel, Erika. 1995. *The Humanist-Scholastic Debate in the Renaissance and Reformation.* Cambridge, MA: Harvard University Press.

Rutherford, Donald. 1995. *Leibniz and the Rational Order of Nature.* Cambridge: Cambridge University Press.

———. 2003. "*Patience sans espérance:* Leibniz's Critique of Stoicism." In *Hellenistic and Early Modern Philosophy,* edited by Jon Miller and Brad Inwood, pp. 62–89. Cambridge: Cambridge University Press.

Saint-Evremond. 1965. *Œuvres en prose.* Vol. 2. Edited by René Ternois. Paris: Marcel Didier.

Salas Ortueta, Jaime. 1991. "La verdad del otro y la práctica ecuménica en Leibniz." *Theoria,* 2nd ser., 6, nos. 14–15:161–73.

Sarasohn, Lisa T. 1996. *Gassendi's Ethics: Freedom in a Mechanistic Universe.* Ithaca: Cornell University Press.

Sargent, Rose-Mary. 1995. *The Diffident Naturalist: Robert Boyle and the Philosophy of Experiment.* Chicago: University of Chicago Press.

Sasaki, Chikara. 2003. *Descartes's Mathematical Thought.* Dordrecht: Kluwer.

Scaglione, Aldo D. 1986. *The Liberal Arts and the Jesuit College System.* Amsterdam and Philadelphia: J. Benjamins Publishing Co.

Schaffer, Simon. 1983. "Natural Philosophy and Public Spectacle in the Eighteenth Century." *History of Science* 21:1–43.

———. 1987. "Godly Men and Mechanical Philosophers: Souls and Spirits in Restoration Natural Philosophy." *Science in Context* 1:55–85.

———. 1994. "Machine Philosophy: Demonstration Devices in Georgian Mechanics." *Osiris,* 2nd ser., 9:157–82.

———. 1995. "The Show That Never Ends: Perpetual Motion in the Early Eighteenth Century." *British Journal for the History of Science* 28:157–89.

Scheel, Günter. 1983. "Kurbraunschweig und die übrigen welfischen Lande." In *Deutsche Verwaltungsgeschichte,* vol. 1, edited by Kurt G. A. Jeserich, Hans Pohl, and Georg-Christoph von Unruh, pp. 741–63. Stuttgart: Deutsche Verlags-Anhalt.

Schepers, Heinrich. 2004. "Non alter, sed etiam Leibnitius: Reply to Dascal's Review *Ex plubius unum.*" *Leibniz Review* 14:117–35.

Schneider, Hans-Peter. 1999. "Leibniz und der moderne Staat." In *Leibniz und Niedersachen,* edited by Herbert Breger and Friedrich Niewöhner, pp. 23–34. Studia Leibnitiana, Sonderheft 28. Stuttgart: Franz Steiner Verlag.

Schneider, Ivo. 1970. *Der Proportionalzirkel: Ein universelles Analogrecheninstrument der Vergangenheit.* Munich: R. Oldenbourg.

———. 1997. "Wie Huren und Betrüger: Die Begegnung des Jungen Descartes mit der Welt der Pratiker der Mathematik." *Berichte zur Wissenschaftsgeschichte* 20:173–88.

Schoock, Martinus. 1643. *Admiranda methodus novae philosophiae Renati Des Cartes.* Utrecht.

Schott, Gaspar. 1664. *Technica curiosa, sive Mirabilia artis, libris XII comprehensa.* Nuremberg.

Schouls, Peter A. 2000. *Descartes and the Possibility of Science.* Ithaca: Cornell University Press.

Schüling, Hermann. 1969. *Die Geschichte der axiomatischen Methode in 16. und beginnenden 17. Jahrhundert.* Hildesheim: G. Olms.

Schuster, John. 1977. "Descartes and the Scientific Revolution, 1618–1634: An Interpretation." Ph.D. diss., Princeton University.

———. 1984. "Methodologies as Mythic Structures: A Preface to the Future Historiography of Method." *Metascience* 1/2:15–36.

———. 1993. "Whatever Should We Do with Cartesian Method? Reclaiming Descartes for the History of Science." In *Essays on the Philosophy and Science of René Descartes,* edited by Stephen Voss, pp. 195–224. Oxford: Oxford University Press.

Schuster, John A., and Alan B. H. Taylor. 1997. "Blind Trust: The Gentlemanly Origins of Experimental Science." *Social Studies of Science* 27:504–36.

Schuurman, Paul. 2001. "Locke's Logic of Ideas in Context: Content and Structure." *British Journal for the History of Philosophy* 9:439–65.

Schyrleus de Rheita, Antonius Maria. 1645. *Oculus Enoch et Eliae, sive Radius sidereomysticus.* 2 vols. Antwerp: H. Verdussii.

Seckendorff, Veit Ludwig von. 1660. *Teutscher Fürsten-Stat, oder gründliche und kurtze Beschreibung. . . .* Frankfurt.

Secord, James A. 2000. *Victorian Sensation: The Extraordinary Publication, Reception, and Secret Authorship of* Vestiges of the Natural History of Creation. Chicago: University of Chicago Press.

Seigel, Jerrold E. 2005. *The Idea of the Self: Thought and Experience in Western Europe since the Seventeenth Century.* Cambridge: Cambridge University Press.

Sellier, Philippe. 1999. "Rhétorique et apologie: 'Dieu parle bien de Dieu.'" In *Port-Royal et la littérature,* vol. 1, *Pascal,* by Philippe Sellier, 117–26. Paris: H. Champion.

Seneca, Lucius Annaeus. 1965. *Ad Lucilium epistulae morales.* Edited by L. D. Reynolds. 2 vols. Oxford: Clarendon.

———. 1969. *Letters from a Stoic.* Translated by Robin Campbell. Baltimore, MD: Penguin.

Sepper, Dennis L. 1996. *Descartes's Imagination: Proportion, Images, and the Activity of Thinking.* Berkeley and Los Angeles: University of California Press.

———. 2000. "The Texture of Thought: Why Descartes' *Meditiationes* Is Meditational, and Why It Matters." In *Descartes' Natural Philosophy,* edited by Stephen Gaukroger, John A. Schuster, and John Sutton, pp. 736–50. London and New York: Routledge.

Serfati, Michel. 1993. "Les compas cartésiens." *Archives de philosophie* 56:197–230.

Serres, Michel. 1982. *Le système de Leibniz et ses modèles mathématiques.* 2nd ed. Paris: Presses universitaires de France.

Shapin, Steven. 1988. "Robert Boyle and Mathematics: Reality, Representation, and Experimental Practice." *Science in Context* 2:23–58.

———. 1991. "'The Mind in Its Own Place': Science and Solitude in Seventeenth-Century England." *Science in Context* 4:191–218.

———. 1994. *A Social History of Truth: Civility and Science in Seventeenth-Century England.* Chicago: University of Chicago Press.

———. 2000. "Descartes the Doctor: Rationalism and Its Therapies." *British Journal for the History of Science* 33:131–54.

Shapin, Steven, and Simon Schaffer. 1985. *Leviathan and the Air-Pump: Hobbes, Boyle, and the Experimental Life.* Princeton: Princeton University Press.

Shapiro, Barbara J. 2000. *A Culture of Fact: England, 1550–1720*. Ithaca: Cornell University Press.

Shapiro, Lisa. 1999. "Cartesian Generosity." In *Norms and Modes of Thinking in Descartes*, edited by Tuomo Aho and Mikko Yrjönsuuri, pp. 249–75. Acta philosophica Fennica, vol. 64. Helsinki: Societas philosophica.

Shea, William R. 2003. *Designing Experiments and Games of Chance: The Unconventional Science of Blaise Pascal*. Canton, MA: Science History Publications.

Shuger, Debora K. 1988. *Sacred Rhetoric: The Christian Grand Style in the English Renaissance*. Princeton: Princeton University Press.

Siegert, Bernhard. 2003. *Passage des Digitalen: Zeichenpraktiken der neuzeitlichen Wissenschaften, 1500–1900*. Berlin: Brinkmann & Bose.

Sirven, J. 1928. *Les années d'apprentissage de Descartes (1596–1628)*. Albi: Imprimerie coöperative du sud-ouest.

Skinner, Quentin. 1996. *Reason and Rhetoric in the Philosophy of Hobbes*. Cambridge: Cambridge University Press.

Sleigh, R. C. 1990. *Leibniz and Arnauld: A Commentary on Their Correspondence*. New Haven: Yale University Press.

Small, Albion Woodbury. 1909. *The Cameralists, the Pioneers of German Social Polity*. Chicago: University of Chicago Press.

Smith, Pamela H. 1994. *The Business of Alchemy: Science and Culture in the Holy Roman Empire*. Princeton: Princeton University Press.

———. 2004. *The Body of the Artisan: Art and Experience in the Scientific Revolution*. Chicago: University of Chicago Press.

Soarez, Cipriano. 1569. *De arte rhetorica libri tres: ex Aristotele, Cicerone, at Quintiliano praecipue deprompti*. Seville: Ex officina A. Escruani.

Solomon, Julie Robin. 1998. *Objectivity in the Making: Francis Bacon and the Politics of Inquiry*. Baltimore, MD: Johns Hopkins University Press.

Sorel, Charles. 1671. *De la connoissance des bons livres, ou Examen de plusieurs autheurs*. Paris.

Spink, John Stephenson. 1960. *French Free-Thought from Gassendi to Voltaire*. London: University of London, Athlone Press.

Stedall, Jacqueline A. 2002. *A Discourse concerning Algebra: English Algebra to 1685*. Oxford: Oxford University Press.

Stewart, Larry. 1992. *The Rise of Public Science: Rhetoric, Technology, and Natural Philosophy in Newtonian Britain, 1660–1750*. Cambridge: Cambridge University Press.

Stolleis, Michael. 1988. *Geschichte des öffentlichen Rechts in Deutschland: Erster Band: Reichspublizistik und Policeywissenschaft, 1600–1800*. Munich: Beck, 1988.

———. 1987. "Veit Ludwig von Seckendorf." In *Staatsdenker im 17. und 18. Jahrhundert*, edited by Michael Stolleis, 148–171. Frankfurt a. M.: Alfred Metzner Verlag, 1987.

Stone, M. W. F., and T. Van Houdt. 1999. "Probabilism and Its Methods: Leonardus Lessius and His Contribution to the Development of Jesuit Casuistry." *Ephemerides theologicae Lovanienses* 75:359–94.

Strowski, Fortunat Joseph. 1921. *Pascal et son temps*. 3 vols. Paris: Plon-Nourrit et cie.

Struever, Nancy S. 1993. "Rhetoric and Medicine in Descartes' *Passions de l'âme*: The Issue of Intervention." In *Renaissance-Rhetorik*, edited by Heinrich F. Plett, pp. 196–212. Berlin and New York: De Gruyter.

Suárez, Francisco. 1740. *De divina gratia*. Vols. 6–8 of *Opera omnia hactenus edita*. Venice.

Sutton, Geoffrey V. 1995. *Science for a Polite Society: Gender, Culture, and the Demonstration of Enlightenment*. Boulder, CO: Westview Press.

Swoyer, Chris. 1995. "Leibnizian Expression." *Journal of the History of Philosophy* 33:65–99.

Tacquet, Andreas. 1651. *Andreae Tacquet e Societate Iesu Cylindricorum et annularium libri iv item de circulorum volutione per planum dissertatio Physiomathica ad Sereniss. Principem Fredericvm Ducem sleswici Holsatiae, etc.* Antwerp.

Tallemant des Réaux, Gédéon, III. 1961. *Histoirettes*. Edited by Antoine Adam. 2 vols. Paris: Gallimard.

Taton, René. 1978. "L'initiation de Leibniz à la géométrie (1672–1676)." In *Leibniz à Paris (1672–1676)*, vol. 1, pp. 103–30. Studia Leibnitiana supplementa, vol. 17. Wiesbaden: Franz Steiner Verlag.

Taussig, Sylvie. 2003. *Pierre Gassendi (1592–1655), introduction à la vie savante*. Turnhout: Brepols.

Taylor, Charles. 1989. *Sources of the Self: The Making of Modern Identity*. Cambridge, MA: Harvard University Press.

Terrall, Mary. 1995. "Gendered Spaces, Gendered Audiences: Inside and Outside the Paris Academy of Sciences." *Configurations* 3:207–32.

———. 2002. *The Man Who Flattened the Earth: Maupertuis and the Sciences in the Enlightenment*. Chicago: University of Chicago Press.

Thewes, Alfons. 1983. *Oculus Enoch: Ein Beitrag zur Entdeckungsgeschichte des Fernrohrs*. Oldenburg: Verlag Isensee.

Thirouin, Laurent. 1991. *Le hasard et les règles: Le modèle du jeu dans la pensée de Pascal*. Paris: J. Vrin.

Thom, Johan Carl. 1995. *The Pythagorean Golden Verses, with Introduction and Commentary*. Leiden and New York: E. J. Brill.

Toledo, Francisco de. 1985a. *Commentaria in universam Aristotelis logicam*. Hildesheim: G. Olms.

———. 1985b. *Introductio in universam Aristotelis logicam*. Hildesheim: G. Olms.

Tomlinson, Gary. 1993. *Music in Renaissance Magic: Toward a Historiography of Others*. Chicago: University of Chicago Press.

Tourneur, Zacharie. 1933. *"Beauté poëtique": Histoire critique d'une "pensée" de Pascal et des ses annexes*. Melun: R. Rozelle.

Trimpi, Wesley. 1983. *Muses of One Mind: The Literary Analysis of Experience and Its Continuity*. Princeton: Princeton University Press.

Trinkaus, Charles. 1970. *In Our Image and Likeness: Humanity and Divinity in Italian Humanist Thought*. 2 vols. Chicago: University of Chicago Press.

Tuck, Richard. 1993. *Philosophy and Government, 1572–1651*. Cambridge: Cambridge University Press.

Tully, James. 1993. "Governing Conduct: Locke on the Reform of Thought and Behavior." In *An Approach to Political Philosophy: Locke in Contexts*, pp. 179–241. Cambridge: Cambridge University Press.

Van Damme, Stéphane. 2002. *Descartes: Essai d'histoire culturelle d'une grandeur philosophique*. Paris: Presses de sciences Po.

Van De Pitte, Frederick. 1988a. "Intuition and Judgment in Descartes' Theory of Truth." *Journal of the History of Philosophy* 26:453–70.

————. 1988b. "Some of Descartes' Debts to Eustachius a Sancto Paulo." *Monist* 71:487–97.

————. 1991. "The Dating of Rule IV-B in Descartes's *Regulae ad directionem ingenii*." *Journal of the History of Philosophy* 29:375–95.

Varani, Giovanna. 1995. "Ramistische Spuren in Leibniz' Gestaltung der Begriffe 'dialectica,' 'topica' und 'ars inveniendi.'" *Studia Leibnitiana* 27:135–56.

————. 2003. "Il 'textus' fra 'actor' e 'lector': Elementi di ermeneutica nel pensiero del giovane Leibniz." *Rivista di storia della filosofia*, no. 4:637–54.

Vasoli, Cesare. 1999. "Le rapport entre les 'Olympica' et la culture de la renaissance." In *Descartes et la renaissance*, edited by Emmanuel Faye, pp. 187–208. Paris: Honoré Champion.

Verbeek, Theo. 1992. *Descartes and the Dutch: Early Reactions to Cartesian Philosophy, 1637–1650*. Carbondale and Edwardsville: Southern Illinois University Press.

Viala, Alain. 1985. *Naissance de l'écrivain: Sociologie de la littérature à l'âge classique*. Paris: Editions de Minuit.

Vickers, Brian. 1988. *In Defence of Rhetoric*. Oxford: Clarendon Press.

Vinci, Thomas C. 1998. *Cartesian Truth*. New York: Oxford University Press.

Vives, Juan Luis. [1782] 1964. *Opera omnia*. 8 vols. Reprint, London: Gregg Press.

Voelke, André-Jean. 1993. *La philosophie comme thérapie de l'âme: Etudes de philosophie hellénistique*. Paris: Cerf.

Wakefield, Andre. 2005. "Books, Bureaus, and the Historiography of Cameralism." *European Journal of Law and Economics* 19:311–20.

Walker, D. P. 1985. *Music, Spirit and Language in the Renaissance*. Edited by Penelope Gouk. London: Variorum Reprints.

Walters, Alice N. 1997. "Conversation Pieces: Science and Politeness in Eighteenth-Century England." *History of Science* 35:121–54.

Waquet, Françoise. 2003. *Parler comme un livre: L'oralité et le savoir, XVIe–XXe siècle*. Paris: Albin Michel.

Warwick, Andrew. 2003. *Masters of Theory: Cambridge and the Rise of Mathematical Physics*. Chicago: University of Chicago Press.

Westerhoff, Jan C. 1999. "'Poeta Calculans': Harsdörffer, Leibniz, and the 'Mathesis Universalis.'" *Journal of the History of Ideas* 60:449–67.

Westman, Robert S. 1980. "The Astronomer's Role in the Sixteenth Century: A Preliminary Study." *History of Science* 18:105–47.

————. 1990. "Proof, Poetics and Patronage: Copernicus's Preface to *De revolutionibus*." In *Reappraisals of the Scientific Revolution*, edited by David C. Lindberg and Robert S. Westman, pp. 167–206. Cambridge: Cambridge University Press.

Wetsel, David. 1981. *L'écriture et le reste: The* Pensées *of Pascal in the Exegetical Tradition of Port-Royal*. Columbus: Ohio State University Press.

————. 1994. *Pascal and Disbelief: Catechesis and Conversion in the* Pensées. Washington, DC: Catholic University of America Press.

Wiedeburg, Paul Herrmann Arthur. 1962. *Der junge Leibniz, das Reich und Europa*. 6 vols. Wiesbaden: F. Steiner.

Williston, Byron. 2003. "The Cartesian Sage and the Problem of Evil." In *Passion and Virtue in Descartes*, edited by Byron Williston and André Gombay, pp. 301–31. Amherst, NY: Humanity Books.

Wilson, Catherine. 1995. *The Invisible World: Early Modern Philosophy and the Invention of the Microscope*. Princeton: Princeton University Press.

Wilson, R. N. 2004. *Reflecting Telescope Optics*. Vol. 1, *Basic Design Theory and Its Historical Development*. 2nd ed. Berlin: Springer.

Wintroub, Michael. 1997. "The Looking Glass of Facts: Collecting, Rhetoric and Citing the Self in the Experimental Natural Philosophy of Robert Boyle." *History of Science* 35:189–217.

Witt, Ronald G. 2000. *In the Footsteps of the Ancients: The Origins of Humanism from Lovato to Bruni*. Leiden: Brill.

Wojcik, Jan W. 1997. *Robert Boyle and the Limits of Reason*. Cambridge: Cambridge University Press.

Yates, Frances A. 1947. *The French Academies of the Sixteenth Century*. London: Warburg Institute.

——. 1964. *Giordano Bruno and the Hermetic Tradition*. Chicago: University of Chicago Press.

Zacher, Hans J. 1973. *Die Hauptschriften zur Dyadik von G. W. Leibniz: Eine Beitrag zur Geschichte des binären Zahlensystems*. Frankfurt am Main: Vittorio Klostermann.

Zammito, John H. 2002. *Kant, Herder, and the Birth of Anthropology*. Chicago: University of Chicago Press.

Zedelmaier, Helmut. 2001. "Lesetechniken: Die Pratiken der Lektüre in der Neuzeit." In *Die Praktiken der Gelehrsamkeit in der frühen Neuzeit*, edited by Helmut Zedelmaier and Martin Mulsow, pp. 11–30. Tübingen: M. Niemeyer.

Zuber, Roger. 1968. *Les "Belles infidèles" et la formation du goût classique*. Paris: A. Colin.

——. 1981. "Guez de Balzac et les deux antiquités." *XVIIe siècle*, no. 131:135–48.

——. 1993. "Die Theorie der Honnêteté." In *Die Philosophie des 17. Jahrhunderts*, vol. 2, *Frankreich und Niederlände*, edited by Jean-Pierre Schöbinger, pt. 1, pp. 156–66, 194–95. Basel: Schwabe.

——. 1997. "L'urbanité française." In *Les émerveillements de la raison*, pp. 151–61. Paris: Klincksieck.

Index